Engineering Experimentation
Ideas, Techniques and Presentation

Engineering Experimentation

Ideas, Techniques and Presentation

Martyn Spencer Ray

Lecturer in Chemical Engineering
Curtin University of Technology
Perth, Western Australia

McGRAW-HILL BOOK COMPANY

London · New York · St Louis · San Francisco · Auckland
Bogotá · Caracas · Hamburg · Lisbon · Madrid · Mexico
Milan · Montreal · New Delhi · Panama · Paris · San Juan
São Paulo · Singapore · Sydney · Tokyo · Toronto

Published by
McGRAW-HILL Book Company Europe
Shoppenhangers Road · Maidenhead Berkshire SL6 2QL · England
Tel: 0628 23432; Fax: 0628 770224

British Library Cataloguing in Publication Data
Ray, Martyn S.
  Engineering experimentation: ideas,
techniques and presentation.
  1. Engineering design
  I. Title
  620′.00425   TA174
  ISBN 0-07-084184-5

Library of Congress Cataloging-in-Publication Data
Ray, Martyn S., 1949–
  Engineering experimentation: ideas, techniques, and presentation
/ Martyn Spencer Ray.
      p.   cm.
  Includes bibliographies and index.
  ISBN 0-07-084184-5
  1. Engineering—Experiments. I. Title.
TA152.R29   1988
620′.00724—dc 19   87-31793

234 CUP 92

Typeset by Eta Services (Typesetters) Ltd, Beccles, Suffolk
Printed and bound in Great Britain at the University Press, Cambridge.

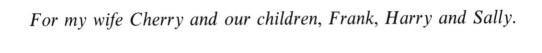

For my wife Cherry and our children, Frank, Harry and Sally.

CONTENTS

PREFACE

This book has been written for use in two parts of an engineering or science course. First, as a textbook for a formal taught unit (lecture and tutorial classes), typically titled 'Engineering Experimentation', 'Information Presentation' or similar. Such a unit is usually included in the first year of a degree course. It provides an introduction to the many and diverse topics with which students need to be familiar in order to perform their experimental work correctly, and to make sense of the data they obtain.

Second, it has been written specifically for students who must perform laboratory work as part of their course. It is intended to provide background information and ideas to help these students get the maximum benefit from the practical classes.

Laboratory classes are often seen as an entity in themselves, intended to supplement lecture material and to develop the students' appreciation of the skills necessary to perform reliable experimental studies. However, too often students are left to gain this appreciation with little useful guidance from staff (except for operating instructions of particular items of equipment), and no suitable book for general reference.

This book is intended to fill this void and provide useful information and ideas for the students. Even if a formal unit is not taught, the book is written in a straightforward style to facilitate self-study.

The book does not cover all the equipment that will be found in a laboratory, nor does it describe in detail all the techniques of data analysis. It provides a broad introductory coverage of several topics including common instruments (Chapters 3 to 5), aspects of measurements (Chapter 2), report writing (Chapter 10), etc. The presentation is deliberately general and non-specialized, making it suitable for many disciplines. Except for the discussion of statistical techniques in Chapter 7, no advanced mathematics are required beyond that taught in sixth forms. Chapter 11 provides some useful ideas and discussion related to the design and planning of experimental work. This is intended to assist students faced with laboratory work for the first time.

The book is suitable for engineers and scientists at both diploma and degree level. It would also be a useful book in schools for science sixth-form students, providing an introduction to the practical aspects of engineering and the subjects covered in an engineering course.

Finally, remember that experimentation is very much like design or project studies: you can always improve upon what you have done, and there are always new techniques to be used—if only you can find them!

January 1988   Martyn Ray

ACKNOWLEDGEMENTS

The material presented in Chapter 10 was originally published in a shorter form in the book *Elements of Engineering Design: An Integrated Approach* by M. S. Ray, published by Prentice-Hall International (UK) Ltd (1985), and reproduced with the permission of the publisher.

Permission for the reproduction of the material as shown, is acknowledged from the following bodies:

Prentice-Hall International (UK) Ltd, for Secs 7.2.1, 7.2.2, 7.2.8, 7.2.9, 7.2.10, Fig. 7.1, Fig. 7.3, Fig. 7.4, Table 11.1; all reproduced from the book quoted above.

The British Standards Institution, 2 Park Street, London W1A 2BS, from whom complete copies of the documents can be obtained:

Figs 9.22, 9.23, 9.24, from BS 5070 (1974);
Figs 9.26, 9.27, 9.28, from BS 308: Part 1 (1984).

I would like to thank Dr Jeffrey Claflin for his advice and ideas regarding the presentation of engineering reports (Section 10.3).

Finally, I would like to thank my wife, Cherry, for her understanding, encouragement and patience during the preparation of this book.

ABBREVIATIONS USED FOR STANDARDS ORGANIZATIONS

AASHTO	American Association of State Highway and Transportation Officials
AGA	American Gas Association
AGMA	American Gear Manufacturers Association
AIA/NAS	Aerospace Industries Association of America, Inc.
API	American Petroleum Institute
ASAE	American Society of Agricultural Engineers
ASHRAE	American Society of Heating, Refrigerating and Air-Conditioning Engineers
ASME	American Society of Mechanical Engineers
ASTM	American Society for Testing and Materials
AWWA	American Water Works Association
BSI	British Standards Institution
CGA	Compressed Gas Association, Inc.
EIA	Electronic Industries Association
IEEE	Institute of Electrical and Electronic Engineers
IPC	Institute for Interconnecting and Packaging Electronic Circuits
ISA	Instrument Society of America
ISO	International Organization for Standardization
NFP(A)	National Fluid Power Association
PD	Published Document (British Standards Institution)
SAE	Society of Automotive Engineers
TAPPI	Technical Association of the Pulp and Paper Industry
UL	Underwriters Laboratories, Inc.

BASIC PRINCIPLES
OF ENGINEERING EXPERIMENTATION

SCOPE

Part One comprises the first two chapters and aims to describe the International System of Units (SI) and the terminology and standards associated with instruments. The characteristics and performance of instrument systems are also discussed.

Chapter 1 contains useful reference material including definitions of the basic units; tables of conversion factors, important physical constants, etc., are also presented. The difference between units and dimensions is discussed. Some conversion factors are derived from first principles, i.e. from conversion factors associated with the basic units.

Chapter 2 provides a general introduction to instruments and experimentation. The static and dynamic characteristics of instruments are discussed and mathematical models are used to predict the dynamic performance of instrument systems. Instrument specification and the selection of instruments are also considered. The elements comprising an instrument system are discussed, as well as the general considerations which influence the techniques to be used and the design of such a system.

STANDARDS, UNITS AND CONVERSION FACTORS

CHAPTER OBJECTIVES

1. To provide information regarding the basic units of mass, length, time, temperature and electrical quantities.
2. To discuss the SI system of units.
3. To provide tables of data, e.g. conversion factors, for reference purposes.
4. To derive some conversion factors from first principles, i.e. from conversion factors for the basic units, as examples of the method to be employed.

QUESTIONS

- What is the difference between a dimension and a unit?
- What units are used when describing measurements?
- How are these units defined?
- What are the basic units?
- How can a measured value be checked or compared against a standard?
- What notation/abbreviations should be used when writing units?
- What are the commonly used derived units?
- What are the systems of units normally encountered?
- Which conversion factors do you frequently use? How are they obtained?

1.1 STANDARDS

Certain standard units of mass, length, time, temperature and electrical quantities have been established; these enable measurements made in different locations to be compared on a consistent basis. Standards are maintained by the National Physical Laboratory (UK) and the National Bureau of Standards (USA).

The *kilogram* is the standard for the unit of mass. It is defined by a cylinder of platinum–iridium alloy that is kept at the International Bureau of Weights and Measures in Paris. Duplicates are maintained in the UK and USA. The kilogram is the only base unit that is still defined by an artefact. The pound mass (lb_m) is exactly defined as

$$1 \text{ pound mass} = 453.592\,427\,7 \text{ grams}$$

The *metre* is the standard for the unit of length. It was defined by the General Conference on Weights and Measures (1960) as 1 650 763.73 wavelengths in vacuum of the orange–red line of the spectrum of a krypton-86 lamp. The standard measurement system uses an interferometer to define the length of a standard metre on a metal bar. The inch is exactly defined as

$$1 \text{ inch} = 25.4 \text{ millimetres}$$

The *second* is the fundamental unit of time. It has previously been defined as (1/86 400) of a mean solar day; the mean solar year is 365 days 5 hours 48 minutes 48 seconds. Although this definition is exact, it is inconvenient because it depends upon astronomical observations. Standard units of time can be established in terms of known frequencies of oscillation of certain devices, e.g. a pendulum, a tuning fork, a torsional–vibrational system. The second is now defined as the duration of 9 192 631 770 cycles of the radiation associated with a specific transition of the caesium atom. In the 'atomic clock' an accuracy of 1 part in 10^{11}, or 1 second in 3000 years, is obtained.

The absolute temperature scale was proposed by Lord Kelvin (1854) and has its origin or zero point at absolute zero. It has a fixed point at 273.16 kelvins or 0.01° celsius (this is approximately 32.02° on the Fahrenheit scale). The Kelvin temperature scale forms the basis for thermodynamic calculations (for further details see Sec. 5.2). A *triple-point cell* is used to define a known, fixed temperature and for the calibration of thermometers. The cell is cooled until a layer of ice is formed around an inner well; the temperature at the interface of solid, liquid and vapour is 0.01°C.

Note: The triple point of water (0.01°C) is *not* the same as the ice point or freezing point (0°C).

The International Practical Temperature Scale of 1968[1] provides a close approximation to the absolute thermodynamic temperature scale, and is a useful experimental scale. There are 11 primary temperature points ranging from the triple point of equilibrium hydrogen at −259.34°C to the normal freezing point of gold at 1064.43°C. Secondary fixed points are also established ranging from the triple point of normal hydrogen at −259.194°C to the freezing point of tungsten at 3387°C. Precise procedures have been established for interpolation between these points.

The Celsius (°C) and Fahrenheit (°F) temperature scales are widely used; it is important to be able to work in either and to make quick conversions. The Rankine (°R) scale is the absolute Fahrenheit scale and the Kelvin (K, *not* °K) scale is designated as the absolute Celsius. Their relationships are given by

$$K = {}^{\circ}C + 273.15$$

$$^{\circ}R = {}^{\circ}F + 459.67$$

The conversion between the Celsius and Fahrenheit scales is given (approximately) by the relationships

$$^{\circ}C = [({}^{\circ}F - 32) \div 9] \times 5$$

$$^{\circ}F = (1.8 \times {}^{\circ}C) + 32$$

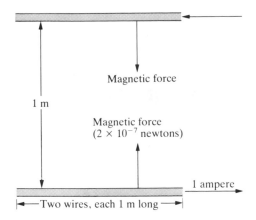

Figure 1.1 Definition of the ampere.

When converting temperatures between different scales, the following (approximate) equivalents are useful:

$$0°C \approx 32°F$$

$$100°C \approx 212°F$$

$$-40°C \approx -40°F$$

For an order-of-magnitude conversion (mostly for quick engineering estimates), the °C temperature is approximately half of the °F temperature.

The *ampere* is the fundamental unit of electric current. It is defined as the magnitude of the current which, when flowing through each of two long parallel wires separated by a distance of 1 metre in free space, results in a force between the two wires (caused by their magnetic fields) of 2×10^{-7} newton for each metre of length. The definition is illustrated in Fig. 1.1.

Other standard absolute units of electrical quantities can be derived from the basic mechanical units of mass, length and time. An international system of electrical units has been accepted (1948) which enables a standard cell to be established. Conversions between the international and absolute systems are:

$$1 \text{ international ohm} = 1.000\,49 \text{ absolute ohm}$$

$$1 \text{ international volt} = 1.000\,330 \text{ absolute volt}$$

$$1 \text{ international ampere} = 0.998\,35 \text{ absolute ampere}$$

The *candela* is the fundamental unit of luminous intensity. The candela is defined as the luminous intensity of (1/600 000) of a square metre of a radiating cavity at the temperature of freezing platinum (2042 K). A thorium oxide tube (ThO_2) is used as a black body that absorbs all radiation falling upon it, becoming incandescent with the constant radiation during the long solidification process.

These are the six basic units used by most scientists and engineers. In addition, an amount of substance known as the *mole* is used, particularly by chemists and chemical engineers. Two supplementary base units for measurement of a plane angle and a solid angle have also been defined. These units are discussed in the next section.

1.2 INTERNATIONAL SYSTEM OF UNITS

The International System of Units (Système International d'Unités) is officially abbreviated as SI and is a modernized version of the metric system. This system was established by the Eleventh General Conference of Weights and Measures (1960). It was declared to be UK government policy in 1965, and it is explained in detail in British Standard *BS 3763: The International System of Units (SI)*.

Other useful British Standards include:

BS 350: *Conversion factors and tables*
Part 1: Basis of tables. Conversion factors
(Part 2 has now been withdrawn)
Supplement No. 1 (PD 6203) to BS 350: Part 2, Additional tables for SI conversions
BS 2856: Precise conversion of inch and metric sizes on engineering drawings
BS 5775: Specification for quantities, units and symbols (Parts 0 to 13)

Details of US standards are included in the Bibliography.

1.2.1 Basic SI units

The International Organization for Standardization (ISO) defines seven basic units from which a wide range of derived units can be expressed in the form of products and/or quotients of these basic units. The seven basic units are given in Table 1.1. These units are defined in the ISO publication *ISO R.31*, 1965 (the mole was included in the list in 1971) as follows:

Metre (m) The metre is the length equal to 1 650 763.73 wavelengths in vacuum of the radiation corresponding to the transition between levels $2p_{10}$ and $5d_5$ of the krypton-86 atom.

Table 1.1 SI base units

Quantity	Name of unit	Recommended unit symbol
mass	kilogram	kg
length	metre	m
time	second	s
thermodynamic temperature	kelvin*	K†
electric current	ampere*	A
luminous intensity	candela	cd
amount of substance	mole	mol‡

* According to the ISO, if the unit is named after a person it should be written in small letters, and a capital letter should be used for the symbol.
† Note that it is *not* written as °K.
‡ The units mol and kmol should be used, *not* gmol or kg mol.

Kilogram (kg)   The kilogram is the unit of mass: it is equal to the mass of the international prototype of the kilogram.

Second (s)   The second is the duration of 9 192 631 770 periods of the radiation corresponding to the transition between the two hyperfine levels of the ground state of the caesium-133 atom.

Ampere (A)   The ampere is that constant current which, if maintained in two straight parallel conductors of infinite length, of negligible circular cross-section, and placed 1 metre apart in vacuum, would produce between these conductors a force equal to 2×10^{-7} newton per metre of length.

Kelvin (K)   The kelvin, unit of thermodynamic temperature, is the fraction (1/273.15) of the thermodynamic temperature of the triple point of water.

Candela (cd)   The candela is the luminous intensity, in the perpendicular direction, of a surface of (1/600 000) square metre of a black body at the temperature of freezing platinum under a pressure of 101 325 newtons per square metre.

Mole (mol)   The mole is defined as an amount of substance of a system which contains as many elementary units as there are carbon atoms in 0.012 kg (exactly) of the pure nucleide carbon-12.

The supplementary base units radian and steradian are also defined:

Plane angle (radian)   The angle subtended at the centre of a circle of radius 1 metre by an arc of length 1 metre along the circumference.

Solid angle (steradian)   The solid angle subtended at the centre of a sphere of radius 1 metre by an area of 1 square metre on the surface.

1.2.2 Magnitude of SI units

The *SI system* is a decimal system in which calculations are performed using the numeral 10, multiplied or divided by itself. It will be useful to remember the following simple mathematical rules for such calculations:

$$10^m = 10 \times 10 \times 10 \ldots \text{to } m \text{ factors}$$

$$10^m \times 10^n = 10^{(m+n)}$$

$$10^m \times 10^n \times 10^p = 10^{(m+n+p)}$$

$$\frac{10^m}{10^n} = 10^{(m-n)} \text{ if } m > n$$

$$= 10^{-(n-m)} \text{ if } n > m$$

$$(10^m)^n = 10^{(mn)}$$

$$10^0 = 1$$

$$10^{-n} = \frac{1}{10^n}$$

$$(10a)^m = 10^m a^m$$

$$\left(\frac{10^x}{10^y}\right)^m = \frac{10^{xm}}{10^{ym}}$$

$$(10)^{m/n} = \sqrt[n]{(10^m)}$$

Table 1.2 Standard multiples and sub-multiples

Multiplication factor		Prefix	Symbol
One million million million	10^{18}	exa	E
One thousand million million	10^{15}	peta	P
One million million	10^{12}	tera	T
One thousand million	10^{9}	giga	G
One million	10^{6}	mega	M
One thousand	10^{3}	kilo	k
One hundred	10^{2}	hecto	h*
Ten	10^{1}	deca	da*
Unity	10^{0}	—	—
One-tenth	10^{-1}	deci	d*
One-hundredth	10^{-2}	centi	c*
One-thousandth	10^{-3}	milli	m
One-millionth	10^{-6}	micro	μ
One-thousand millionth	10^{-9}	nano	n
One-million millionth	10^{-12}	pico	p
One-thousand million millionth	10^{-15}	femto	f
One-million million millionth	10^{-18}	atto	a

* It is suggested that all SI units be expressed in 'preferred standard form' in which the multiplier is 10^{3n} where n is a positive or negative whole number. Consequently the use of hecto, deca, deci and centi is to be avoided wherever possible.

To obtain multiples and sub-multiples of the units, standard prefixes have been recommended by the ISO as shown in Table 1.2.

1.2.3 Rules for multiples and sub-multiples

The following rules have been recommended for the use of SI units:

(a) The basic SI units are to be preferred but it is impracticable to limit the usage to these alone; therefore, their decimal multiples and sub-multiples are also required.

(b) In order to avoid errors in calculations, it is preferable to use coherent units. It is, therefore, strongly recommended that in calculations only SI units are used and not their multiples and sub-multiples. (For example: use $N/m^2 \times 10^6$ rather than MN/m^2 or N/mm^2.)

(c) The use of prefixes representing 10 raised to the power '$3n$', where 'n' is a positive or negative integer, is especially recommended. (For example: μF, mm, kg are preferred while hm, cm and dag are to be avoided.)

(d) When expressing a quantity by a numerical value of a unit, it is advisable to use quantities that result in numerical values between 0 and 1000. (For example: 15 kN or 15×10^3 N rather than 15 000 N; 2.75 mm or 2.75×10^{-3} m rather than 0.002 75 m.)

(e) Compound prefixes are not used. (For example: write nm (nanometre) rather than mμm (millimicrometre).)

(f) Multiplying prefixes are printed immediately adjacent to the SI symbol with which they are associated. The multiplication of symbols is usually indicated by leaving a small gap between them. (Example: mN means millinewton whereas m N would indicate metre newton.)

(g) The numbers should be grouped in threes on either side of the decimal marker and the use of commas should be avoided. (For example: 8 475.2 rather than 8,475.2; 0.002 52 rather than 0.00252).

(h) A prefix applied to a unit becomes part of that unit and is subject to any applied power. (For example: $1 \text{ mm}^3 = 1 \text{ (mm)}^3 = 10^{-9} \text{ m}^3$ rather than $1 \text{ mm}^3 = 1 \text{ m(m)}^3 = 10^{-3} \text{ m}^3$.)

1.2.4 Derived SI units

All *derived units* can be expressed in terms of the basic units. Some derived units are named units such as the newton, joule, lumen, etc. Some of these derived units are given in Table 1.3. However, other systems of units are used throughout the world, the most common being the British (Imperial) or US system based upon the pound, foot, second, °F units. This system is widely used in the USA, often in 'parallel' with the SI system. Units in common use in the USA are given in Table 1.4. The older system will continue to be used for many years, but changeover is likely to be quicker in the sciences than in some fields of engineering. Instruments are not replaced because a

Table 1.3 Derived SI units in science and engineering

Quantity	SI unit	Symbol	Definition
(a) *Applied mechanics units*			
Force	newton	N	kg m/s^2
Work, energy, quantity of heat	joule	J	N m
Power, heat flow rate	watt	W	J/s
Moment of force	newton metre	—	N m
Pressure, stress	pascal	Pa	N/m^2
Temperature (basic)	kelvin	K	—
Temperature (common use)	celsius	°C	$0°C = 273.15 \text{ K}$
Surface tension	newtons per metre	—	N/m
Thermal coefficient of linear expansion	reciprocal kelvin	—	K^{-1}
Heat flux density, irradiance	watt per square metre	—	W/m^2
Thermal conductivity	watt per metre kelvin	—	W/m K
Coefficient of heat transfer	watt per square metre kelvin	—	$\text{W/m}^2 \text{ K}$
Heat capacity	joules per kelvin	—	J/K
Specific heat capacity	joules per kilogram kelvin	—	J/kg K
Entropy	joules per kelvin	—	J/K
Specific entropy	joules per kilogram kelvin	—	J/kg K
Specific energy: specific latent heat	joules per kilogram	—	J/kg
Viscosity (kinematic)	metre squared per second	—	m^2/s
Viscosity (dynamic)	pascal second	—	Pa s
(b) *Electrical units*			
Electric resistance	ohm	Ω	V/A
Electric charge	coulomb	C	A s
Electric potential difference or voltage or e.m.f.	volt	V	W/A
Electric conductance	siemens	S	A/V
Electric capacitance	farad	F	A s/V
Luminance	candela per square metre (nit)	—	cd/m^2
Illumination	lux	lx	lm/m^2
Luminous flux	lumen	lm	cd sr
Frequency	hertz	Hz	s^{-1}
Electric field strength	volts per metre	—	V/m
Electric flux density	coulombs per square metre	—	C/m^2
Magnetic flux	weber	Wb	V s
Magnetic flux density	tesla	T	Wb/m^2
Inductance	henry	H	Wb/A

Table 1.4 American National Standard Letter Symbols for units used in science and technology

Unit	Symbol	Notes
ampere	A	SI unit of electric current
ampere per metre	A/m	SI unit of magnetic field strength
ångstrom	Å	$1 \text{ Å} = 10^{-10} \text{ m}$
atmosphere, standard	atm	$1 \text{ atm} = 101\,325 \text{ N/m}^2$
atto	a	SI prefix for 10^{-18}
barrel	bbl	$1 \text{ bbl} = 9702 \text{ in.}^3 = 0.158\,99 \text{ m}^3$
British thermal unit	Btu	
calorie	cal	$1 \text{ cal} = 4.1868 \text{ J}$
candela	cd	SI unit of luminous intensity
centi	c	SI prefix for 10^{-2}
centimetre	cm	
coulomb	C	SI unit of electric charge
cubic centimetre	cm^3	
cubic foot	ft^3	
cubic inch	in.^{3}	
cubic metre	m^3	
curie	Ci	$1 \text{ Ci} = 3.7 \times 10^{10}$ disintegrations per second. Unit of activity in the field of radiation dosimetry
cycle per second	Hz, c/s	*See* hertz. The name 'hertz' is internationally accepted for this unit; the symbol Hz is preferred to c/s
day	d	
decibel	dB	
degree (plane angle)	...$^\circ$	
degree (temperature)		
degree Celsius	$^\circ$C	Note that there is no space between the symbol $^\circ$ and the letter. The use
degree Fahrenheit	$^\circ$F	of the word 'centigrade' for the Celsius temperature scale was abandoned by the Conférence Générale des Poids et Mesures (CGPM) in 1948
degree Kelvin	K	*See* kelvin
degree Rankine	$^\circ$R	
dyne	dyn	
electronvolt	eV	
erg	erg	
farad	F	SI unit of capacitance
foot	ft	
foot per second	ft/s	
foot pound-force	ft lb$_f$	
gallon	gal	The gallon, quart and pint differ in the USA and the UK, and their use in science and technology is deprecated
gauss	G	The gauss is the electromagnetic c.g.s. unit of magnetic flux density. Use of the SI unit, the tesla, is preferred
gram	g	
henry	H	SI unit of inductance
hertz	Hz	SI unit of frequency
horsepower	hp	The horsepower is an anachronism in science and technology. Use of the SI unit of power (the watt) is preferred
hour	h	
joule	J	SI unit of energy
kelvin	K	In 1967 the CGPM gave the name kelvin to the SI unit of temperature which had formerly been called degree Kelvin, and it was assigned the symbol K (without the symbol $^\circ$)
kilo	k	SI prefix for 10^3
kilogram	kg	SI unit of mass
kilogram-force	kg$_f$	In some countries the name kilopond (kp) has been adopted for this unit

Unit	Symbol	Notes
lambert	L	$1 \text{ L} = (1/\pi) \text{ cd/cm}^2$. A c.g.s. unit of luminance. One lumen per square centimetre leaves a surface whose luminance is 1 lambert in all directions within a hemisphere. Use of the SI unit of luminance, the candela per square metre, is preferred.
litre	l	$1 \text{ l} = 10^{-3} \text{ m}^3$
lumen	lm	SI unit of luminous flux
mega	M	SI prefix for 10^6
megahertz	MHz	
metre	m	SI unit of length
mho	mho	CIPM accepted the name siemens (S) for this unit and it was approved by the 14th CGPM (1971)
micro	μ	SI prefix for 10^{-6}
micron	μm	The name micron was abrogated by the CGPM (1967)
mile (statute)	mi	$1 \text{ mi} = 5280 \text{ ft}$
mile per hour	mi/h	Although use of m.p.h. as an abbreviation is common, it should not be used as a symbol
milli	m	SI prefix for 10^{-3}
minute (time)	min	Time may also be designated by means of superscripts as in the following example: $9^\text{h}46^\text{m}30^\text{s}$
mole	mol	SI unit of amount of substance
nano	n	SI prefix for 10^{-9}
newton	N	SI unit of force
newton per square metre	N/m^2	SI unit pressure or stress; *see* pascal
oersted	Oe	The oersted is the electromagnetic c.g.s. unit of magnetic field strength. Use of the SI unit, the ampere per metre, is preferred
ohm	Ω	SI unit of resistance
pascal	Pa	$\text{Pa} = \text{N/m}^2$ SI unit of pressure or stress. This name was accepted by the CIPM in 1969 and it was approved by the 14th CGPM (1971)
pico	p	SI prefix for 10^{-12}
poise	P	SI unit of absolute viscosity
pound	lb	
pound-force	lb$_f$	The symbol lb, without a subscript, may be used for pound-force where no confusion is foreseen
pound-force per square inch	lb/in.^{2}	Although use of the abbreviation p.s.i. is common, it should not be used as a symbol. Refer to the note on pound-force regarding subscript to the symbol
radian	rad	SI unit of plane angle
revolution per minute	r/min	Although use of the abbreviation r.p.m. is common, it should not be used as a symbol
revolution per second	r/s	
roentgen	R	Unit of exposure in the field of radiation dosimetry
second (time)	s	SI unit of time
siemens	S	$S = \Omega^{-1}$
slug	slug	$1 \text{ slug} = 14.5939 \text{ kg}$
steradian	sr	SI unit of solid angle
stokes	St	SI unit of dynamic viscosity
tesla	T	$T = \text{N}/(\text{A m}) = \text{Wb/m}^2$; SI unit of magnetic flux density (magnetic induction)
ton	ton	$1 \text{ ton} = 2000 \text{ lb}$
volt	V	SI unit of voltage
voltampere	VA	IEC name and symbol for the SI unit of apparent power
watt	W	SI unit of power
watt-hour	W h	
weber	Wb	$\text{Wb} = \text{V s}$; SI unit of magnetic flux
yard	yd	

new system of units is introduced; they continue to be used until they *need* replacing for more efficient or accurate measurement. This means that engineers and scientists must continue to be familiar with the units used in different systems, and the conversions of quantities between the systems. The use and understanding of published work also requires familiarity with a wide range of units.

Factors for the conversion of non-SI units into the appropriate SI units are given in Table 1.5. These factors are arranged according to the particular physical quantity, i.e. in subject categories. Some of the conversion factors are exact in value and these are marked by an asterisk (*).

Table 1.5 Conversion factors to SI units

Length

1 ångstrom (Å)	$= *10^{-10}$ m
1 microinch (1 μin.)	$= *0.0254$ μm
1 thou; milli-inch or mil	$= *25.4$ μm
1 inch (in.)	$= *25.4$ mm
1 foot (ft)	$= *0.3048$ m
1 yard (yd)	$= *0.9144$ m
1 fathom (6 ft)	$= *1.8288$ m
1 furlong (220 yd)	$= *0.201\,168$ km
1 mile	$= *1.609\,344$ km

Area

1 circular mil $[(\pi/4) \times 10^{-6}$ in.$^2]$	$= 506.707$ μm^2
1 square inch (in.^{2})	$= *645.16$ mm^2
1 square foot (ft^2)	$= 0.092\,903$ m^2
1 square yard (yd^2)	$= 0.836\,127$ m^2
1 acre (4840 yd^2)	$= 4046.86$ m^2
1 square mile (mile^2)	$= 2.589\,99$ km^2

Capacity and volume

1 cubic inch (in.^{3})	$= 16.3871$ cm^3
1 cubic foot (ft^3)	$= 0.028\,316\,8$ m^3
1 cubic yard (yd^3)	$= 0.764\,555$ m^3
1 fl oz (1/20 pint)	$= 28.413$ cm^3
1 gill (1/4 pint)	$= 0.142\,065$ dm^3
1 pint (pt)	$= 0.568\,261$ dm^3
1 quart (qt)	$= 1.136\,52$ dm^3
1 imperial gallon	$= 4.546\,09$ dm^3
1 US gallon (231 in.^{3})	$= 3.785\,41$ dm^3

Second moment of area

1 ft^4	$= 0.008\,630\,97$ m^4
1 in.^{4}	$= 0.416\,231 \times 10^{-6}$ m^4
1 in.^{3} (section modulus)	$= 16.3871 \times 10^{-6}$ m^3

Velocity and acceleration

1 in./min	$= 0.423\,333$ mm/s
1 in./s	$= *25.4$ mm/s
1 ft/min	$= *0.005\,08$ m/s
1 ft/s	$= *0.3048$ m/s
1 mile/h (m.p.h.)	$= 1.609\,34$ km/h
1 mile/h (m.p.h.)	$= 0.447\,040$ m/s
1 ft/s^2	$= *0.3048$ m/s^2
1 in./s^2	$= *25.4$ m/s^2

(continued)

Mass
1 grain (gr) $= 64.7989$ mg
1 dram (dr) $= 1.771\,85$ g
1 ounce (oz) $= 28.3495$ g
1 pound (lb) $= *453.592\,427$ g
1 stone (14 lb) $= 6.350\,29$ kg
1 quarter (qr) (28 lb) $= 12.7006$ kg
1 slug $= 14.5939$ kg
1 cental (ctl) (100 lb) $= 45.3592$ kg
1 hundredweight (cwt) (112 lb) $= 50.8023$ kg
1 ton (2240 lb) $= 1016.05$ kg
1 US ton (2000 lb) $= 907.18$ kg

Mass per unit length
1 oz/in. $= 1.116\,12$ kg/m
1 lb/in. $= 17.8580$ kg/m
1 lb/ft $= 1.488\,16$ kg/m
1 lb/yd $= 0.496\,055$ kg/m
1 ton/1000 yd $= 1.111\,16$ kg/m
1 ton/mile $= 0.631\,342$ t/km (Mg/km)

Mass per unit area
1 oz/ft^2 $= 305.152$ g/m^2
1 oz/yd^2 $= 33.9057$ g/m^2
1 lb/in.^{2} $= 703.070$ kg/m^2
1 lb/ft^2 $= 4.882\,43$ kg/m^2

Specific volume
1 ft^3/ton $= 0.027\,869\,6$ dm^3/kg
1 ft^3/lb $= 62.428$ dm^3/kg
1 in.^{3}/lb $= 36.1273$ cm^3/kg
1 gal/lb $= 10.0224$ dm^3/kg

Mass rate of flow
1 ton/h $= 1.016\,05$ t/h (Mg/h)
1 lb/h $= 0.453\,592$ kg/h
1 lb/h $= 0.125\,998$ g/s
1 lb/min $= 7.559\,87$ g/s
1 lb/s $= 0.453\,592$ kg/s

Volume rate of flow
1 ft^3/s (cusec) $= 0.028\,316\,8$ m^3/s
1 gal/h $= 4.546\,09$ dm^3/h
1 gal/min $= 0.075\,768\,2$ dm^3/s
1 gal/s $= 4.546\,09$ dm^3/s

Fuel consumption
1 gal/mile $= 2.824\,81$ dm^3/km $= 282.481$ litre/100 km
1 US gal/mile $= 2.352\,15$ dm^3/km $= 235.215$ litre/100 km
1 mile/gal $= 0.354\,006$ km/dm^3
1 mile/US gal $= 0.425\,144$ km/dm^3

Density
1 lb/in.^{3} $= 27.6799$ g/cm^3
1 lb/in.^{3} $= 27.6799$ Mg/m^3
1 lb/ft^3 $= 16.0185$ kg/m^3
1 slug/ft^3 $= 515.379$ kg/m^3
1 lb/gal $= 99.7763$ kg/m^3
1 ton/yd^3 $= 1328.94$ kg/m^3

(continued)

Table 1.5 *continued*

Force

1 poundal (pdl)	= 0.138 255 N
1 ozf	= 0.278 014 N
1 lbf	= 4.448 22 N
1 tonf	= 9.964 02 kN

Force per unit length

1 lbf/in.	= 175.127 N/m
1 lbf/ft	= 14.5939 N/m
1 tonf/ft	= 32.6903 kN/m

Moment of force (torque)

1 lbf in.	= 0.112 985 N m
1 pdl ft	= 0.042 140 N m
1 lbf ft	= 1.355 82 N m
1 tonf ft	= 3.037 03 kN m

Moment of inertia

1 slug ft^2	= 1.355 82 kg m^2
1 oz in.^{2}	= 18.290 kg mm^2
1 lb in.^{2}	= 292.640 kg mm^2
1 lb ft^2	= 0.042 140 1 kg m^2

Pressure and stress

1 pdl/ft^2	= 1.488 16 Pa
1 lbf/in.^{2} (p.s.i.)	= 6.894 76 kPa
1 lbf/ft^2	= 47.880 3 Pa
1 tonf/in.^{2} (t.s.i.)	= 15.444 3 MPa
1 tonf/ft^2	= 107.252 kPa
1 ft water	= 2.989 07 kPa
1 in. water	= 249.089 Pa
1 in. mercury	= 3.386 39 kPa
1 bar	= *10^5 Pa = 10^5 N/m^2
1 p.s.i.a.	= 1 lbf/in.^{2}
1 p.s.i.a.	= 2.0360 in. Hg at 0°C
1 p.s.i.a.	= 2.311 ft H$_2$O at 70°F
1 p.s.i.a.	= 51.715 mm Hg at 0°C
	(ρ_{Hg} = 13.5955 g/cm^3)
1 atm	= 14.696 p.s.i.a.
	= 1.013 25 × 10^5 N/m^2
	= 1.013 25 bar
1 atm	= 760 mm Hg at 0°C
	= 1.013 25 × 10^5 Pa
1 atm	= 29.921 in. Hg at 0°C
1 atm	= 33.90 ft H$_2$O at 4°C
1 p.s.i.a.	= 6.894 76 × 10^4 g/cm s^2
1 p.s.i.a.	= 6.894 76 × 10^4 dyn/cm^2
1 dyn/cm^2	= 2.0886 × 10^{-3} lbf/ft^2
1 p.s.i.a.	= 6.894 76 × 10^3 N/m^2
1 lbf/ft^2	= 4.7880 × 10^2 dyn/cm^2
	= 47.880 N/m^2
1 mm Hg (0°C)	= 1.333 224 × 10^2 N/m^2
	= 0.133 322 4 kPa

(*continued*)

Dynamic viscosity

1 lbf h/ft^2	= 0.172 369 MPa s
1 lbf s/ft^2	= 47.8803 Pa s
1 pdl s/ft^2	= 1.488 16 Pa s
1 pdl s/ft^2	= 1488.16 cP (centipoise)
1 lb/ft s	= 1.488 16 kg/m s
1 lb/ft s	= 1488.16 cP
1 slug/ft s	= 47.8803 km/s

Kinematic viscosity

1 ft^2/h	= 0.092 903 m^2/h
1 ft^2/h	= *25.8064 cSt (centistokes)
1 ft^2/s	= 0.092 903 m^2/s
1 in.^{2}/h	= 0.179 211 × 10^{-6} m^2/s
1 in.^{2}/h	= *6.4516 cm^2/h
1 in.^{2}/s	= *645.16 mm^2/s
1 in.^{2}/s	= *645.16 cSt
1 in.^{2}/s	= 10.7527 mm^2/s
1 in.^{2}/min	= 10.7527 cSt

Heat, energy and work

1 J	= 1 N m = 1 kg m^2/s^2
1 kg m^2/s^2 (joule)	= 10^7 g cm^2/s^2 (erg)
1 Btu	= 1055.06 J
	= 1.055 06 kJ
1 Btu	= 252.16 cal (thermochemical)
1 kcal (thermochemical)	= 1000 cal
	= 4.1840 kJ
1 cal (thermochemical)	= 4.1840 J
1 cal (IT)	= 4.1868 J
1 Btu	= 251.996 cal (IT)
1 Btu	= 778.17 ft lbf
1 hp h	= 0.7457 kW h
1 hp h	= 2544.5 Btu
1 ft lbf	= 1.355 82 J
1 ft lbf/lb$_m$	= 2.9890 J/kg
1 ft pdl	= 0.042 140 J
1 ft lbf	= 1.355 82 J
1 kW h	= *3.6 MJ
1 horsepower-hour (hp-h)	= 2.684 52 MJ
1 therm	= 105.506 MJ

Power

1 horsepower (hp)	= 745.700 W
1 hp	= 550 ft lbf/s
1 hp	= 0.7068 Btu/s
1 W	= 14.340 cal/min
1 Btu/h	= 0.293 07 W
1 J/s	= 1 W
1 ft lbf/s	= 1.355 82 W

Heat-flow rate and heat flux

1 Btu/h	= 0.293 071 W
1 ton of refrigeration (288 000 Btu/day)	= 3.516 85 kW
1 Btu/h ft^2	= 3.1546 W/m^2

(continued)

Table 1.5 *continued*

Thermal conductivity

1 Btu ft/ft^2 h °F	= 1.730 73 W/m °C
1 Btu in./ft^2 h °F	= 0.144 228 W/m °C
1 Btu in./ft^2 s °F	= 519.220 W/m °C
1 Btu/h ft^2 °F	= 4.1365 × 10^{-3} cal/s cm °C

Temperature

1 deg F (Fahrenheit)	= $\frac{5}{9}$ deg C
T K (Kelvin)	= T°C + 273.15
T°R (Rankine)	= T°F + 459.67
T°C (Celcius)	= (T°F − 32) × $\frac{5}{9}$

Heat transfer coefficient

1 Btu/h ft^2 °F	= 1.3571 × 10^{-4} cal/s cm^2 °C
1 Btu/h ft^2 °F	= 5.6783 × 10^{-4} W/cm^2 °C
1 Btu/h ft^2 °F	= 5.6783 W/m^2 K
1 kcal/h m^2 °F	= 0.2048 Btu/h ft^2 °F

Viscosity

1 cp	= 10^{-2} g/cm s (poise)
1 cp	= 2.4191 lb$_m$/ft h
1 cp	= 6.7197 × 10^{-4} lb$_m$/ft s
1 cp	= 10^{-3} Pa s
	= 10^{-3} kg/m s
	= 10^{-3} N s/m^2
1 cp	= 2.0886 × 10^{-5} lbf s/ft^2
1 Pa s	= 1 N s/m^2
	= 1 kg/m s
	= 1000 cp

Diffusivity

1 cm^2/s	= 3.875 ft^2/h
1 cm^2/s	= 10^{-4} m^2/s
1 m^2/h	= 10.764 ft^2/h
1 m^2/s	= 3.875 × 10^4 ft^2/h
1 centistoke	= 10^{-2} cm^2/s
1 Pa s	= 1 N s/m^2
	= 1 kg/m s
	= 1000 cp

Mass flux and molar flux

1 g/s cm^2	= 7.3734 × 10^3 lb$_m$/h ft^2
1 mol/s cm^2	= 7.3734 × 10^3 lb mol/h ft^2
1 mol/s cm^2	= 10 kmol/s m^2
	= 1 × 10^4 mol/s m^2
1 lb mol/h ft^2	= 1.3562 × 10^{-3} kmol/s m^2

Heat capacity and enthalpy

1 Btu/lb$_m$ °F	= 4.1868 kJ/kg K
1 Btu/lb$_m$ °F	= 1.000 cal/g °C
1 Btu/lb$_m$	= 2326.0 J/kg
1 ft lbf/lb$_m$	= 2.9890 J/kg
1 cal (IT)/g °C	= 4.1868 kJ/kg K

(continued)

Mass transfer coefficient

$1\ k_c$ cm/s	$= 10^{-2}$ m/s
$1\ k_c$ ft/h	$= 8.4668 \times 10^{-5}$ m/s
$1\ k_x$ mol/s cm^2 mol frac.	$= 10$ kmol/s m^2 mol frac.
$1\ k_x$ mol/s m^2 mol frac.	$= 1 \times 10^4$ mol/s m^2 mol frac.
$1\ k_x$ lb mol/h ft^2 mol frac.	$= 1.3562 \times 10^{-3}$ kmol/s m^2 mol frac.
$1\ k_x a$ lb mol/h ft^3 mol frac.	$= 4.449 \times 10^{-3}$ kmol/s m^3 mol frac.
$1\ k_G$ kmol/s m^2 atm	$= 0.98692 \times 10^{-5}$ kmol/s m^2 Pa
$1\ k_G a$ kmol/s m^3 atm	$= 0.98692 \times 10^{-5}$ kmol/s m^3 Pa

Illumination

1 lumen/ft^2 (ft-candle)	$= 10.7639$ lx
1 cd/ft^2	$= 10.7639$ cd/m^2
1 cd/in.^{2}	$= 1550.00$ cd/m^2
1 ft lambert	$= 3.42626$ cd/m^2

Electric units

(c.g.s. units to SI units; c is velocity of light in free space and numerically equal to 2.997925×10^8 m/s)

Current	1 e.m.u. = *10 A
	1 e.s.u. = *1/10c A
e.m.f. (potential difference)	1 e.m.u. = *10^{-8} V
	1 e.s.u. = *$10^{-6}c$ V
Field strength	1 e.m.u. = *10^{-6} V/m
	1 e.s.u. = $10^{-4}c$ V/m
Charge (quantity of electricity)	1 e.m.u. = *10 C
	1 e.s.u. = *1/10c C
Resistance	1 e.m.u. = *10^{-9} Ω
	1 e.s.u. = *$10^{-5}c^2$ Ω
Capacitance	1 e.m.u. = *10^9 F
	1 e.s.u. = *$10^5/c^2$ F
Inductance	1 e.m.u. = *10^{-9} H
Magnetic flux	1 e.m.u. = *10^{-8} Wb (maxwell)
Magnetic flux density	1 e.m.u. = *10^{-4} T (gauss)
Permeability	1 e.m.u. = *$4\pi \times 10^{-7}$ H/m
Magnetic field strength	1 e.m.u. = *$10^3/4\pi$ A/m (oersted)
Electric flux density	1 e.m.u. = *10^5 C/m^2
	1 e.s.u. = *$10^3/c$ C/m^2
Permittivity	1 e.s.u. = *$10^7/4\pi c^2$ F/m

* These conversion factors are exact in value.

Other useful information concerning SI units is presented here in the following tables:

Table 1.6 SI values of important physical constants

Description	Symbol*	Numerical value in SI units
Avogadro's number	N	6.023×10^{26} kmol
Bohr magneton	β	9.27×10^{-24} A m^2
Boltzmann's constant	k	1.380×10^{-23} J/K
Characteristic impedance of free space	Z_0	$(\mu_0/E_0)^{1/2} = 120\ \pi\Omega$
Electron volt	eV	1.602×10^{-19} J
Electric charge	e	1.602×10^{-19} C
Electronic charge-to-mass ratio	e/m_e	1.759×10^{11} C/kg
Electronic rest mass	m_e	9.109×10^{-31} kg
Energy for ground state atom (Rydberg energy)	H	13.60 eV
Energy for $T = 290$ K	kT	4×10^{-21} J
Faraday constant	F	9.65×10^7 C/kmol
Permeability of vacuum or inductance	μ_0	$4\pi \times 10^{-7}$ H/m
Permittivity of vacuum or dielectric constant	E_0	$\dfrac{1}{36\pi} \times 10^{-9}$ F/m $= 8.854 \times 10^{-12}$ F/m
Planck's constant	h	6.626×10^{-34} J s
Proton mass	m_p	1.672×10^{-27} kg
Proton-to-electron mass ratio	m_p/m_e	1836.1
Radius of first H orbit (Bohr atom)		$0.529 \times 10^{-10} = 0.529$ Å
Standard gravitational acceleration (free fall)	g	9.807 m/s^2
Stefan–Boltzmann constant	σ	5.67×10^{-8} J/m^2 s K^4
Universal constant of gravitation	G	6.67×10^{-11} N m^2/kg^2
Universal gas constant	R	8.314 kJ/kmol K
Velocity of light *in vacuo*	c	2.9979×10^8 m/s
Volume of 1 kmol of ideal gas at STP†		22.42 m^3

* Symbol often used, not unit symbols.
† At STP, approximately 273.15 K and 101.325 kPa.

1.3 DIMENSIONS

It is important to realize that there is a difference between dimensions and units. In most situations the difference does not affect us as we talk about an object of mass 10 kg or a rod of length 5 m. However, it is pertinent to mention the difference here.

A *dimension* is a physical variable that is used to specify some characteristic of a particular system, such as the two examples mentioned above, or the temperature of a gas (which is one of its thermodynamic dimensions). In the above examples, we have stated the units in which the dimension is measured. Therefore, a dimension is the type of quantity being measured, and this is expressed in units.

The use of dimensional analysis as an experimental technique is discussed in detail in Chapter 8. The basic SI units of mass, length, time, temperature and electric current are each denoted by a dimension, the symbols M, L, t, T and A, respectively, being normally used. Derived units can then be considered in terms of these dimensions. For example, the derived unit of velocity has the dimensions of length divided by time; this can be written as

$$[v] = [\text{length/time}] = L\,t^{-1}$$

where [ ] represents 'the dimensions of'.

Similarly

$$\text{density } [\rho] = [\text{mass/volume}] = M\,L^{-3}$$

Table 1.7 General data (SI units)

Standard gravitational acceleration $= 9.81$ m/s^2

International standard atmosphere (ISA): Pressure $= 1.013$ bar $= 101.3$ kPa
Temperature $= 15°C = 288$ K

Universal gas constant $(\bar{R}) = 8.314$ kJ/kmol K

Molal volume $(\bar{V}) = 22.41$ m^3/kmol at ISA pressure and 0°C

Composition of air	Volumetric analysis	Gravimetric analysis
Nitrogen (N$_2$; 28.013 kg/kmol)	0.7809	0.7553
Oxygen (O$_2$; 31.999 kg/kmol)	0.2095	0.2314
Argon (Ar; 39.948 kg/kmol)	0.0093	0.0128
Carbon dioxide (CO$_2$; 44.010 kg/kmol)	0.0003	0.0005

Properties of air
Molecular mass $(\bar{M}) = 29$
Specific gas constant $(R) = 0.287$ kJ/kg K (0.0685 Btu/lb °R; 53.3 ft lbf/lb °R)
Specific heat at constant pressure $(C_p) = 1.005$ kJ/kg K (0.240 Btu/lb °R)
Specific heat at constant volume $(C_v) = 0.718$ kJ/kg K (0.1715 Btu/lb °R)

Ratio of specific heats $(\gamma) = \dfrac{C_p}{C_v} = 1.40$

Thermal conductivity $(k) = 0.0253$ W/m K at ISA conditions
Dynamic viscosity $(\mu) = 17.9$ μPa s
Density at ISA conditions $(\rho) = 1.225$ kg/m^3
Sonic velocity at ISA conditions $(a) = 340$ m/s

Properties of water
Molecular mass $(\bar{M}) = 18$
Specific heat at constant pressure at 15°C $(C_p) = 4.186$ kJ/kg K
Thermal conductivity $(k) = 595$ kW/m K at 15°C
Dynamic viscosity $(\mu) = 1.14$ mPa s at 15°C
Density at 4°C $(\rho) = 1000$ kg/m^3

Table 1.8 Values of the Universal Gas Law Constant (R)

Numerical value	Units
1.9872	g cal/mol K
1.9872	Btu/lb mol °R
82.057	cm^3 atm/mol K
8314.34	J/kmol K
82.057×10^{-3}	m^3 atm/kmol K
8314.34	kg m^2/s^2 kmol K
10.731	ft^3 lbf/in.^{2} lb mol °R
0.7302	ft^3 atm/lb mol °R
1545.3	ft lbf/lb mol °R
8314.34	m^3 Pa/kmol K

Many quantities have dimensions that are obtained from an established physical law, e.g. the dimensions of force are obtained from Newton's second law. This may be written in the form

$$F \propto ma \qquad (1.1)$$

Table 1.9 Common non-SI units

Quantity	SI unit	Associated non-SI units*
Angle	rad	degree °, minute ', second "
Time	s	hour (h); minute (min)
Mass	kg	tonne (t) = 10^3 kg = 1 Mg
Area	m^2	hectare (ha) = 10^4 m^2
Volume	m^3	litre (l) = 10^{-3} m^3
Pressure	Pa	bar (bar) = 10^5 N/m^2 = 10^5 Pa
Energy	J	kWh = 3.6 MJ
Stress	Pa; N/m^2	hectobar (hbar) = 10^7 N/m^2
		= 10^7 Pa

* These non-SI units are in common use and will probably remain so for a very long time.

Table 1.10 Mathematical constants

e	= 2.7183	$\log \pi^2$	= 0.9943
$\dfrac{1}{e}$	= 0.3679	$\sqrt{\pi}$	= 1.7725
$\log_e 10$	= 2.3026	$\dfrac{\pi}{4}$	= 0.7854
$\log_{10} x$	= 0.4343 ln x	$\sqrt{2}$	= 1.414
π	= 3.1416	$\sqrt{3}$	= 1.732
π^2	= 9.8696	$\sqrt{5}$	= 2.236
$\dfrac{1}{\pi}$	= 0.3183	$\sqrt{10}$	= 3.162
		1 rad	= 57°39'
ln x	= 2.3026 $\log_{10} x$	1°	= 0.017 45 rad
$\log \pi$	= 0.4971		

Hence

$$\text{force } [F] = [\text{mass} \times \text{acceleration}] = M\,L\,t^{-2}$$

Other quantities can then be expressed in terms of the quantity of 'force'; for example

$$\text{work } [W] = [\text{force} \times \text{distance}] = M\,L^2\,t^{-2}$$

In some situations, the heat flow is a convenient quantity to use, and heat has the same dimensions as work. Therefore

$$\text{heat flow } [\dot{Q}] = [\text{heat/time}] = M\,L^2\,t^{-3}$$

Similarly

$$\text{thermal conductivity } [k] = [\text{heat per unit area per unit time per unit}$$
$$\text{temperature gradient}]$$

$$= M\,L\,t^{-3}\,T^{-1}$$

From Eq. (1.1), if the mass is constant, Newton's law can be written as

$$F = k\,m\,a \tag{1.2}$$

or

$$F = \frac{1}{g_c}\,m\,a \tag{1.3}$$

Table 1.11 The Greek alphabet

Capital letter	Small letter	Name
A	α	alpha
B	β	beta
Γ	γ	gamma
Δ	δ	delta
E	ε	epsilon
Z	ζ	zeta
H	η	eta
Θ	θ	theta
I	ι	iota
K	κ	kappa
Λ	λ	lambda
M	μ	mu
N	ν	nu
Ξ	ξ	xi
O	o	omicron
Π	π	pi
P	ρ	rho
Σ	σ	sigma
T	τ	tau
Y	υ	upsilon
Φ	ϕ	phi
X	χ	chi
Ψ	ψ	psi
Ω	ω	omega

Equation (1.3) is used to define different systems of units for mass, force, length and time. Some common systems of units are:

Force	Mass	Acceleration	Constant (g_c)
1 N	1 kg	1 m s^{-2}	1 kg m/N s^{-2}
1 kg$_f$	1 kg$_m$	9.806 65 m s^{-2}	9.806 65 kg$_m$ m/kg$_f$ s^2
1 lb$_f$	1 lb$_m$	32.174 ft s^{-2}	32.174 lb$_m$ ft/lb$_f$ s^2
1 lb$_f$	1 slug$_m$	1 ft s^{-2}	1 slug$_m$ ft/lb$_f$ s^2
1 dyne$_f$	1 g$_m$	1 cm s^{-2}	1 g$_m$ cm/dyn s^2

In the SI system, the concept of 'g_c' is not normally used and the newton is expressed as

$$1 \text{ newton} = 1 \text{ kg m s}^{-2}$$

However, the relationship given by Newton's second law and by Eqs (1.2) and (1.3) should be remembered.

The weight (W) of an object (of mass m) is defined as the force exerted on it as a result of the acceleration due to gravity (g); therefore

$$W = \frac{g}{g_c} m$$

The weight of a body has the dimensions of a force and, in the systems previously described, 1 pound mass (lb$_m$) will weigh 1 pound force (lb$_f$) at sea level, and also 1 kg$_m$ will weigh 1 kg$_f$. The suffixes (m and f) are usually omitted when confusion is unlikely to occur. It does not really matter which system is used, provided that the definitions are consistent.

1.4 CONVERSION FACTORS FROM FIRST PRINCIPLES

Usually a conversion factor is obtained from a table of values such as Table 1.5; sometimes a table may not be available or the particular factor may not be included. In these cases it is possible to derive the value of the conversion factor from first principles. It is necessary to know the values of the conversion factors for some of the basic units of mass, length, time, etc. The following examples will illustrate the procedure to be used.

Example 1.1 To obtain the conversion factor for cubic feet to cubic metres:

$$1 \text{ ft}^3 = 12 \text{ in.} \times 12 \text{ in.} \times 12 \text{ in.} = 12^3 \text{ in.}^3 = 1728 \text{ in.}^3$$

To convert ft^3 to in.^3, multiply (ft^3) by 1728.

To convert in.^3 to ft^3, divide (in.^3) by 1728
or
multiply by $1/1728 = 0.000\,579$ (to 3 significant figures).
Note that conversion factors are usually expressed as the value to be multiplied rather than divided.

To convert ft^3 to cm^3:

$$1 \text{ ft}^3 = 12 \text{ in.} \times 12 \text{ in.} \times 12 \text{ in.}$$

$$= (12 \times 2.54) \text{ cm} \times (12 \times 2.54) \text{ cm} \times (12 \times 2.54 \text{ cm})$$

$$[\text{Since } 1 \text{ in.} = 2.54 \text{ cm}]$$

$$= 12^3 \times 2.54^3 \text{ cm}^3$$

$$= 28\,317 \text{ cm}^3 \text{ (to 5 significant figures)}$$

Since $100 \text{ cm} = 1 \text{ m}$ (or $1 \text{ cm} = 0.01 \text{ m}$):

$$1 \text{ ft}^3 = 12^3 \times \left(\frac{2.54}{100}\right)^3 \text{ m}^3$$

$$= 28\,317 \times 10^{-6} \text{ m}^3 \ (1 \text{ cm}^3 \times (0.01)^3 = 10^{-6} \text{ m}^3)$$

Therefore to convert ft^3 to m^3, multiply by $0.028\,317$ (to 5 significant figures; compare Table 1.5).

Example 1.2 Convert a density of 62.4 lb/ft^3 to kg/m^3
From tables, look up

$$1 \text{ lb} = 453.59 \text{ g} = 0.453\,59 \text{ kg}$$

$$1 \text{ ft} = 0.3048 \text{ m}$$

Therefore

$$1 \text{ ft}^3 = 0.3048^3 \text{ m}^3 = 0.028\,317 \text{ m}^3$$

hence

$$62.4 \text{ lb/ft}^3 \text{ is equivalent to } \frac{62.4 \times 0.453\,59}{0.028\,317} \frac{\text{kg}}{\text{m}^3}$$

$$\text{i.e., } 62.4 \text{ lb/ft}^3 = 62.4 \times 16.0183 \text{ kg/m}^3$$

$$= 999.54 \text{ kg/m}^3 \text{ (to 5 significant figures)}$$

The conversion factor for lb/ft^3 to kg/m^3 is 16.018 (to 5 significant figures).

Alternatively

$$62.4 \frac{\text{lb}}{\text{ft}^3} = 62.4 \text{ lb} \times \left(\frac{\text{kg}}{\text{lb}}\right) \times \left(\frac{1}{\text{ft}^3}\right) \times \left(\frac{\text{ft}^3}{\text{m}^3}\right)$$

$$= 62.4 \frac{\text{lb}}{\text{ft}^3} \times 0.453\,59 \frac{\text{kg}}{\text{lb}} \times 35.314 \frac{\text{ft}^3}{\text{m}^3}$$

$$= 999.54 \text{ kg/m}^3$$

Note: Density is dependent upon the temperature and pressure at which measurements are taken; these values should always be stated. For example, $\rho_{\text{Hg}} = 13.5955 \text{ g/cm}^3$ at $0°\text{C}$ and 10^5 Pa.

Example 1.3 Convert values of specific heat from kJ/kg °C to Btu/lb °F.
From Table 1.5:

$$1 \text{ Btu} = 1.055\,06 \text{ kJ (i.e., } 1.055\,06 \text{ kJ per Btu)}$$

$$1 \text{ lb} = 453.59 \text{ g} = 0.453\,59 \text{ kg}$$

$$9°\text{F} = 5°\text{C}$$

Note: This is the conversion for a temperature difference, i.e. a rise or fall of temperature. The difference between the ice point ($0°\text{C}$ and $32°\text{F}$) on the two temperature scales has to be considered only for the conversion of actual temperature values, it has no effect on the conversion factors for quantities such as specific heat, thermal conductivity, etc.

Therefore

$$1 \text{ kg} = (1/0.453\,59) \text{ lb} = 2.2046 \text{ lb (i.e. approximately 2.2 lb/kg)}$$

$$1°\text{C} = 1.8°\text{F (i.e. } 1.8°\text{F/°C)}$$

Since

$$\frac{\text{kJ}}{\text{kg °C}} = \frac{\text{kJ}}{\text{Btu}} \times \frac{\text{lb}}{\text{kg}} \times \frac{°\text{F}}{°\text{C}} \frac{\text{Btu}}{\text{lb °F}}$$

$$1 \frac{\text{kJ}}{\text{kg °C}} = 0.947\,813 \times 2.2046 \times 1.8 \frac{\text{Btu}}{\text{lb °F}}$$

$$1 \frac{\text{kJ}}{\text{kg °C}} = 4.1868 \frac{\text{Btu}}{\text{lb °F}} \text{ (to 5 significant figures)}$$

Compare this value with the conversion factor given in Table 1.5:

$$\text{As derived: } 1 \text{ kJ/kg }°C = 4.1868 \text{ Btu/lb }°F$$

$$\text{From Table 1.5: } 1 \text{ Btu/lb }°F = 4.1868 \text{ kJ/kg K!}$$

Why?

Assuming that the conversion tables are correct (check in another book), then there must be a fault in the logic of our derivation.

The mistake is shown as follows:

$$1 \frac{\text{kJ}}{\text{kg }°C} = p \frac{\text{kJ}}{\text{Btu}} \times q \frac{\text{lb}}{\text{kg}} \times r \frac{°F}{°C} \times s \frac{\text{Btu}}{\text{lb }°F}$$

$$= 4.1868 \frac{\text{kJ/kg }°C}{\text{Btu/lb }°F} \times s \frac{\text{Btu}}{\text{lb }°F}$$

$$(p \times q \times r)$$

$$= 4.1868 \frac{\text{kJ/kg }°C}{\text{Btu/lb }°C} \times \frac{1}{4.1868} \frac{\text{Btu}}{\text{lb }°F}$$

$$= 1 \text{ kJ/kg }°C$$

Therefore the conversion factor is

$$1 \text{ Btu/lb }°F = 4.1868 \text{ kJ/kg }°C \text{ (as given in Table 1.5).}$$

Note 1. The purpose of including the first error-ridden example was to demonstrate the need for careful reasoning when deriving conversion factors—although you are probably already aware of these pitfalls!

Note 2. Since a temperature *difference* of one degree on both the Celsius and Kelvin scales represents the same temperature change, the specific heat can be written as either 1 kJ/kg °C or 1 kJ/kg K (although K is the SI unit).

An alternative approach will be shown in the next example.

Example 1.4 Convert values of thermal conductivity from W/m K to Btu/ft h °F. [Note that the definition of thermal conductivity is heat per unit area per unit time per unit temperature gradient, i.e. Btu/ft^2 h (°F/ft).]

Since values in Table 1.5 are for British-to-metric conversions, it will be easier to perform the conversion in that manner. From Table 1.5:

$$1 \text{ Btu} = 1.055\,06 \text{ kJ}$$

$$1 \text{ ft} = 0.3048 \text{ m}$$

$$1 \text{ h} = 3600 \text{ s}$$

$$1°F = (5/9)°C = (5/9) \text{ K}$$

Therefore

$$\frac{1 \text{ Btu}}{1 \text{ ft} \times 1 \text{ h} \times 1°F} = \frac{1.055\,06 \text{ kJ}}{0.3048 \text{ m} \times 3600 \text{ s} \times (5/9) \text{ K}}$$

$$= \frac{1.055\,06}{0.3048 \times 3600 \times (5/9)} \frac{kJ}{m\,s\,K}$$

$$= 0.001\,731 \frac{kW}{m\,K}$$

$$1\frac{Btu}{ft\,h\,°F} = 1.731 \frac{W}{m\,K} \text{ (to 4 significant figures)}$$

This is in agreement with the value given in Table 1.5. Therefore

$$1\,W/m\,K = 0.5778\,Btu/ft\,h\,°F$$

Example 1.5 Convert the value of the universal gas constant (R) from $10.731\,ft^3\,lbf/in.^2$ lbmol °R to kJ/kmol K.
The universal gas law

$$Pv = nRT$$

defines the units of R:

$$R = \frac{Pv}{nT}$$

$$= \frac{(lbf/in.^2) \times (ft^3)}{lbmol \times °R}$$

From Table 1.5:

$$1\,ft\,lbf = 1.355\,82\,J$$

$$1\,ft^2 = 0.092\,903\,m^2$$

$$1\,in.^2 = 645.16 \times 10^{-6}\,m^2$$

$$1\,lbmol = 0.453\,592\,kmol$$

$$1°R = (5/9)\,K$$

Therefore

$$10.731\frac{ft^3\,lbf}{in.^2\,lbmol\,°R} = \frac{10.731 \times 0.092\,903 \times 1.355\,82}{645.16 \times 10^{-6} \times 0.453\,592 \times (5/9)} \frac{m^2\,J}{m^2\,kmol\,K}$$

$$= 0.008\,314 \times 10^{-6} \frac{J}{kmol\,K}$$

$$= 8.314\,kJ/kmol\,K \text{ (compare with value in Table 1.8)}$$

Note: Conversion factor from lbmol to kmol is the same as from lb to kg; conversion factor from °R to K is the same as from °F to °C.

EXERCISES

1.1 What are the basic units of measurement?
1.2 List some of the important derived units for your engineering discipline.

1.3 What is a mole of a substance?
What is a lbmol of zinc?
What is a kmol of zinc?
What is the difference between a kg and a kmol of tin?

1.4 What is the difference between a dimension and a unit?

1.5 Derive some conversion factors from first principles, i.e. from the basic units, and check the answers against the values given in Table 1.5. Choose conversion factors that are commonly encountered in your branch of engineering, e.g. heat-transfer coefficients, energy and power, velocity and acceleration, etc.

1.6 Convert some of the important physical constants (Table 1.6) and the values of the universal gas constant (Table 1.8) into other units.

REFERENCE

1. 'International Practical Temperature Scale of 1968', *Metrologia*, vol. 5, no. 2, April, 1969.

BIBLIOGRAPHY

Standards

The following standards relate to the use of the SI system of units. Abbreviations used to designate the standards organizations are listed at the beginning of the book (page xiii).

British Standards
BS 5555: *Specification for SI units and recommendations for the use of their multiples and of certain other units.*
PD 5686: *The use of SI units.* (Incorporates information from *ISO R.1000.*)

US Standards
AASHTO R1–77: Standard Metric Practice Guide (ASTM E380–76).
AGMA 600.01–79: Metric Usage.
AIA/NAS NAS 10000–77: Documents Preparation and Maintenance in SI (Metric) Units (Rev. 1).
API Manual of Petroleum Measurement Standards, Chapter 15—Guidelines for the Use of the International System of Units (SI) in the Petroleum and Allied Industries (1980).
ASAE EP285.6–85: Use of SI (Metric) Units.
ASHRAE Ch37–85: Units and Conversions (ASHRAE Handbook, 1985: Fundamentals, SI edn).
ASME SI–9–81: Guide for Metrication of Codes and Standards SI (Metric) Units.
ASTM E380–84: Standard for Metric Practice.
IEEE 268–82: Metric Practice.
IEEE 945–84: Recommended Practice for Preferred Metric Units for Use in Electrical and Electronics Science and Technology.
IPC 5.4.2–73 to 5.4.6–73, and 5.5–73: Conversion Factors (Test Methods Manual).

Other sources

Cheremisinoff, N. P. and P. N. Cheremisinoff, *Unit Conversions and Formulas Manual*, Butterworth, London 1980.
Chiswell, B. and E. M. C. Grigg, *SI Units*, Wiley, Brisbane, 1971.
Galyer, J. and C. Shotbolt, *Metrology for Engineers*, 4th edn, Cassell, London, 1980.
Green, M. H., *International and Metric Units of Measurement*, Chemical Publishing Co., New York, 1973.
Lewis, R., *Engineering Quantities and Systems of Units*, Elsevier Scientific Publishing Co., New York, 1972.
Wandmacher, C., *Metric Units in Engineering—Going SI*, Industrial Press, New York, 1978.

QUALITIES OF MEASUREMENTS

CHAPTER OBJECTIVES

To present information concerning the following aspects of instruments and instrument systems:
1. the functional elements;
2. associated terminology;
3. characteristics and performance;
4. dynamic response;
5. specification and selection;
6. calibration;
7. different types of transducers, their characteristics and performance;
8. devices and techniques for the acquisition, transmission and presentation of data.

QUESTIONS

- What is an instrument (or instrument system)?
- What task does an instrument perform?
- How does an instrument work? (Select one and explain.)
- How would you choose an instrument for a particular application?
- What determines the accuracy, precision, discrimination, etc., of an instrument?
- How is an instrument calibrated?
- What are instrument characteristics?
- How is the response of an instrument defined?
- How is the performance of an instrument defined?
- What measuring instruments are commonly encountered in the laboratory?
- How do these instruments operate?
- What is a transducer?
- How is an instrument reading transferred from the initial measuring instrument to the final display/recording equipment?

2.1 ELEMENTS OF AN INSTRUMENT SYSTEM

All industrial processes and many 'everyday' situations, e.g. cooking or driving a car, require the use of instruments to make measurements. Instruments may be categorized as one of the following types depending upon the function they perform:

(a) Indicating instruments, e.g. pressure gauge, car speedometer.
(b) Controlling instruments, e.g. thermostat, pressure-relief valve.
(c) Recording instruments, e.g. pen and chart recorder.

Therefore, an instrument (or an instrument system) senses a variable quantity and then indicates or records its value, or controls its value. The signal from the sensing element may be changed to another form, e.g. the Bourdon gauge where a pressure change is converted to a scale movement (see Sec. 3.5), or it may be amplified. It may be necessary to transmit the signal some distance from the sensing element to the recording instrument.

The elements of an instrument system are shown in Fig. 2.1. The general term *transducer* is used to describe a device that converts the measured value (sensing element input) into a convenient form of *signal* (the output). The output is a different physical quantity from the input. Sometimes the term 'transducer' is used to describe a sensing element having an electrical output, in which case other terms such as 'primary element' or 'measuring element' are used for non-electrical devices. A *primary transducer* reacts directly to changes in the measured value, whereas a *secondary transducer* is included in the system between the initial sensing point and the indicating or recording elements. An example of such a system would be the mercury-in-steel thermometer (see Sec. 5.4.2) where the thermometer bulb is the primary transducer and the Bourdon tube is the secondary transducer. Transducers are considered in more detail in Sec. 2.8, and transmission and display equipment in Sec. 2.9.

The term *transmission* is used here in the broad sense to include any modification or conditioning of the transducer signal, as well as the transfer of the signal between the sensing and indicating elements. The three main functional elements shown in Fig. 2.1 are not necessarily physically separate units. For example, an amplifier may be built into the transducer unit, and the whole unit is then usually referred to as the transducer. An amplifying element is used when the signal from the sensing element is very small. The signal is amplified to enable it to be transmitted over some distance, or to make it large enough to operate an indicating element. The operation of an amplifying element may be based upon mechanical, optical, fluid, or electrical principles, or on a combination of these.

The increased power available from an amplified output signal may be used to drive an indicator or recorder. The display element should be capable of accurately following the signal from the amplifier; it should also respond quickly to any variations of the signal. Any pens and pointers should be as light as possible to reduce inertia; this effect can be eliminated by using laser

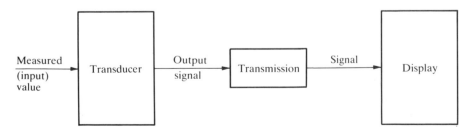

Figure 2.1 Elements of an instrument system.

beams or electron beams. A signal can be shown on an oscilloscope or recorded using either an ink pen or a heated stylus and heat-sensitive paper. Indicators may be classified broadly into two types according to the following brief definitions.

An *analogue* display instrument uses the magnitude of a pointer displacement to represent (i.e. to be analogous to) the measured value. A *digital* display instrument indicates the measurement by a series of digits (figures), which either appear on a screen or are printed on paper. The measurement is obtained directly as a number and this display has the advantage of avoiding interpolation between scale divisions.

A signal may be transmitted within an instrument (over short distances) by the movement of a fluid column or a lever, or by changes in various electrical quantities. In industrial applications, remote indication of a measurement is often required and either a pneumatic or an electrical transmission system is employed. A pneumatic system is operated by compressed air (known as 'instrument air lines'), whereas electrical transmitters normally use changes in resistance and induction values to vary the current in the transmission circuit. Some systems use the making or breaking of a circuit to indicate a quantity, or the time that a circuit is open or closed, or the number of pulses in a given time.

2.2 INSTRUMENT SELECTION

When designing a laboratory experiment, or in a practical industrial situation, a particular instrument is specified primarily because it can measure the variable over its intended ranges with appropriate accuracy. In some cases it may be necessary to use several instruments, each one making measurements over a range of values, e.g. when an instrument has poor accuracy in the first and fourth quartiles of its scale.

There are very many factors that need to be considered when selecting an instrument for a particular task. Those factors that are related directly to the performance of an instrument, e.g. accuracy, sensitivity, precision, etc., are discussed in Sec. 2.3. More general factors are mentioned briefly here. Assuming that the instrument(s) being evaluated can measure the required variable with appropriate accuracy, then the major consideration is usually the cost. This includes not only the purchase price (capital cost) of the instrument but also maintenance costs, servicing and repair costs, costs of replacement parts, etc., all of which will be incurred over the expected life of the instrument. These operating or running costs can form a significant portion of the annual budget for a project. It is important to establish the 'quality' (i.e. accuracy, etc.) of the measurement that is *actually* required. Since the cost of an instrument usually increases as the accuracy improves, an instrument is required that is capable of making measurements that are both adequate and suitable. Also, when the readings from several instruments are to be combined, the accuracy of the final value is less than that obtained from the least accurate instrument. Therefore, all instruments should have similar accuracy.

All instruments (if operated as intended) should possess a high degree of reliability, however it is the consequences of failure that need to be considered. The reliability also affects the cost and in certain cases, such as nuclear installations, it may be necessary to duplicate certain instruments to ensure continuous operation in the event of a failure.

Instruments are usually designed and manufactured to operate under particular conditions. If an instrument must be modified for use in unusual environmental conditions, these changes will incur additional costs. The most common situations requiring special modifications are the presence of corrosive media, and use of an instrument at high or low temperatures. Other less-common conditions are vibration, high acceleration, nuclear radiation and explosive atmospheres. Adverse environmental conditions can affect not only the instrument itself but also the measurements that are made.

If a visible output is required from an instrument, it will be necessary to specify an appropriate type of output, e.g. a visible instantaneous display or a recorded signal. The safe working limits of the measured value may also need to be included on the display, and sometimes additional audible alarms are required.

The possible sources of errors in instruments and measurement systems should be considered when instruments are being selected. However, errors are more often evaluated with hindsight, i.e. after a system has been established. Errors may occur for the following reasons:

(a) Manufacturing imperfections or an inappropriate design may cause errors in a measurement system. Examples are incorrect graduation of scales, hysteresis effects (i.e. different measurements during unloading from those obtained on loading), friction, and backlash in gears.

(b) Operating errors occur for a variety of reasons, e.g. parallax error with scale reading, deformation due to contact between a measured component and the measuring instrument, or misalignment of the instrument.

(c) Environmental errors occur because of variations in local values of temperature, pressure, acceleration, etc. If the conditions cannot be standardized, then compensation in the measuring instrument is required.

(d) Application errors occur when the introduction of an instrument alters the variable it is measuring. A temperature-measuring device inserted into a hot fluid causes heat flow to the instrument, and the indicated fluid temperature is not then the true value. Similarly, an ammeter connected into an electric circuit causes a slight increase in the resistance, and hence a small decrease in the current flowing. The current indicated by the ammeter is not the true value in the original circuit, even if the instrument itself is perfectly accurate.

The selection of a particular instrument, or the decision to use a particular measurement technique, is based upon considerations of the 'quality' of the measurements obtained (i.e. accuracy, etc.), speed of measurement response, cost, environmental conditions, reliability, physical size, etc. The techniques to be used for measuring similar variables may depend upon the particular application. However, the basic principles of measurement are common to all branches of engineering. *Metrology* is the field of measurement associated with the physical dimensions of an object. It is of particular importance in mechanical and production engineering where high degrees of accuracy are usually required and speed of measurement is of secondary importance.

The measuring equipment used on chemical process plant and other continuous industrial processes is known as *process instrumentation*. These measurements are used to control the operation of the process; this may be achieved by manual adjustment (by an operator) or by automatic process control systems. The most important feature of any instrument or a control system for a chemical process is repeatability of measurements; this is even more important than absolute accuracy. Accuracy is very desirable, but product quality can usually be accurately assessed by manual sampling and subsequent correction. The main process requirement is the ability to produce a consistent product continuously for 24 hours. The speed of response will be less important if the measured quantities themselves cannot change quickly. The measuring equipment needs to be robust for use in unfavourable environmental conditions, reliable for continuous operation over long periods, and satisfactory in terms of safety.

The speed of response of an instrument is of primary importance in situations where the measured quantity is either transient or cyclical. The equipment must be capable of measuring the variable as changes occur. In research and development situations, versatility is often a requirement for a measuring instrument, e.g. several ranges of measurement. The aim is to avoid the need for additional instruments if the project specification changes, and instruments that are sufficiently versatile may be used on several different projects.

Electronic equipment is now commonplace in most measurement fields because of its very rapid response, and the ability to amplify and transmit electrical signals over large distances. The measured value is often converted into an electrical signal at an early stage of the measurement process, rather than transmitting signals of pressure, displacement, etc. The advent of microelectronics technology has enabled data acquisition systems and control systems to become more compact, as well as increasing the complexity and capability of electronic signal processing devices.

2.3 TERMINOLOGY AND STANDARDS

For the system shown in Fig. 2.1, the input is the quantity being measured (known as the *measurand*), and the output is the observed response indicated by the display. The output of an ideal measuring system would always have a known relationship to the input, and it would also respond instantaneously to any change in the input. In practice this ideal performance cannot be achieved and there will always be some *error* or inaccuracy in the relationship.

Standards exist for all the physical units and these have been discussed in Chapter 1. There are also standards that relate to practices and specifications, and these standards also possess agreed nomenclature. Unfortunately measurement science is not considered as a separate discipline and, although measurement is practised in all branches of science and engineering, there has been little coordination between the disciplines. However, some standard terminology that covers the general terms and definitions is now emerging.

The definitions of some common terms are presented in this section; some of these are taken from British Standards, some are not universally accepted throughout industry, and others are subject to change in usage. The definitions of some less-common terms are included at the end of this chapter. The performance of an instrument is discussed in relation to its static characteristics and dynamic characteristics in Secs 2.4 and 2.5 respectively.

The Organisation Internationale de Métrologie Légale (OIML) issued a second edition of the document entitled *Vocabulary of Legal Metrology, Fundamental Terms* in 1978. ('Metrology' is the term often used to describe measurement science.) The British Standards Institution issued a dual language version (same title) in 1980 designated as PD 6461.[1] The official definitions are those expressed in the French language, although the English language listings are used in many countries. Additional useful standards are BS 2643[2] and BS 5233[3]. Other countries have produced their own standards related to metrology and measurement science, most of which are based upon the OIML document.

Accuracy is the quality that characterizes the ability of a measuring instrument to give indications approximating to the true value of the measured quantity. It is a qualitative term describing the deviation (within declared probability limits) of the reading from the known input. It is specified as an inaccuracy (error or uncertainty) and is the sum of errors contributed by several factors, e.g. hysteresis, temperature, non-linearity, drift and vibration. The lack of accuracy arises from both the instrument and the imperfect standard of the unit. Accuracy is usually expressed as a percentage of the full-scale reading, e.g. a 100 p.s.i. pressure gauge having an accuracy of 1 per cent would be accurate within ± 1 p.s.i. over the entire range of the gauge.

This is the accuracy based upon the *full-scale deflection* (*f.s.d.*) and may be expressed as:

$$\text{error} = \{(\text{measured value} - \text{true value})/\text{maximum scale value}\} \times 100\%$$

The accuracy of an instrument can be specified in several different ways. The *point accuracy* is stated for one or more points in its range, e.g. temperature-measuring devices at the fixed points (see Sec. 5.2). The accuracy can also be expressed as a *percentage of the true value* where

$$\text{error} = \{(\text{measured value} - \text{true value})/\text{true value}\} \times 100\%$$

The value of this error is the maximum for any point in the range.

A *complete accuracy statement* specifies the accuracy at a large number of points, either in tabular or graphical form, e.g., for pyrometers. Instruments having the same percentage accuracy but based upon the f.s.d. or the true value are more accurate in the latter case.

An *error* is the algebraic difference between the value indicated by the transducer (or the instrument output) and the true value of the quantity presented to the input. No instrument is absolutely accurate when indicating the value of a measured quantity. It is impossible to obtain the absolute value. An instrument that provides a primary standard of a unit always introduces an error when it is used to determine a measured quantity.

Repeatability is defined in the OIML document as the ability of a measuring instrument to give the same value of the measured quantity, not taking into consideration the systematic errors associated with variations of the indications. An alternative definition is the ability to reproduce a certain reading with a given accuracy. *Precision* is also widely used to describe this feature, although it is not defined in the OIML standard. Repeated application of a known input pressure of 1 bar would have a repeatability of ± 1 per cent if all the input readings were within the limits of 0.99–1.01 bar.

Discrimination is the ability of the measuring instrument to react to small changes of the measured quantity. This is commonly called the *resolution*, although it is not referred to in PD 6461. It is sometimes incorrectly called the 'sensitivity'.

Sensitivity is the ratio of the change in transducer output to the corresponding change in the measured quantity.

The *range* of an instrument is the scale values on the display between the lowest and highest points of continuous calibration. For a thermometer calibrated from 200°C to 350°C, these readings represent the range.

The *span* is the difference between the highest and lowest points of continuous calibration; in the above example the span is $(350 - 200)$ degrees, i.e. 150 degrees.

Traceability is the process that relates accuracies from the measuring instrument to the primary standard.

Linearity expresses how values lie on a linear, proportional scale. It is often confused with the term 'accuracy', however a very linear scale that is biased in slope (or offset from the assigned value) is not accurate.

Calibration is the comparison of the output from a measuring instrument against a common standard of the unit (see Secs 1.1, 1.2 and 2.7).

Drift is the slow but persistent variation in transducer output which is not caused by any change of input. It may be caused by poor design or internal temperature changes; it is normally considered separately from ambient temperature effects which can often be calculated.

Threshold of an instrument is the minimum input necessary to cause a detectable change from zero output or indication; it is a particular case of discrimination. It may be caused by backlash or internal noise. In a digital system, the threshold value is the signal necessary to cause the least significant digit of the output reading to change.

Backlash is the minimum distance or angle through which any part of a mechanical system may be moved in one direction, without applying appreciable force or motion to the next part in a mechanical sequence. It is present in all instruments and systems (analogue and digital) and may appear as a 'dead zone' in which small input changes produce no change in the display reading.

Zero stability is a measure of the ability of an instrument to restore a zero reading in the absence of input stimulus after there has been a change in that stimulus, and after removal of variations in external factors.

Live zero is a transducer output signal corresponding to zero input signal. For example, a 4–20 mA transducer output produces a 4 mA output current when the input is zero, and a 20 mA output for the full-scale input value.

Suppressed zero is a means of obtaining higher discrimination of the reading by using a scale representing only the necessary part of the range; for example, a mains-voltage scale reading from 220 to 260 V, where the initial 0–220 V is 'backed-off' so that the meter scale begins with the reading of 220 V.

Hysteresis is the algebraic difference between the average errors at corresponding points of measurement when approached from opposite directions, i.e. rising as opposed to falling values of the input. It may be caused by backlash, friction, or mechanical effects.

Static friction is the force or torque necessary to begin motion from rest (sometimes called 'stiction'). It is often expressed as a fraction of full scale.

Dynamic friction is the frictional torque that opposes motion of the output (sometimes called 'coulomb friction'). It is independent of velocity and the value is usually less than the corresponding value of static friction.

Viscous friction is the frictional component that varies as a function of the velocity of a transducer mechanism. It produces damping, and affects the response of the output because it introduces a lag in the motion.

Additional definitions of selected terms are included in the Glossary at the end of this chapter.

2.4 STATIC CHARACTERISTICS

The static characteristics of an instrument must be considered when the measured quantity does not vary with time. These characteristics have been defined in Sec. 2.3 and are discussed here in more detail. Desirable static characteristics are accuracy, repeatability and discrimination; the opposite (and undesirable) qualities are static error, drift and hysteresis (or dead zone) respectively.

A measurement can be exact only if it is a count of a number of separate items, e.g. number of components or number of electrical impulses, since these are discrete measurements. In other situations (i.e. continuous measurement values) there will be a measurement error. The *static error* of an instrument is the difference between the true value of a quantity (that is not changing with time) and the indicated value. The static error can be positive or negative; an instrument reads high for positive static errors and vice versa. *Repeatability* is the ability of an instrument to reproduce a reading when it receives the same input signal *repeatedly over a short period*. *Stability* is this same ability obtained *repeatedly* over a long period, and *constancy* is this same ability obtained *continuously*. The accuracy and repeatability (or precision) of an instrument are often confused. An example will help to clarify the difference between these terms. Suppose that measurement of a known temperature of 100°C resulted in five measurements of 105°C, 103°C, 105°C, 104°C and 103°C. The accuracy of the measuring instrument would not be better than 5 per cent (5°C). However, the repeatability of ± 1 per cent is achieved because the maximum deviation from the mean reading of 104°C is only 1°C. The instrument could be calibrated and used to measure temperatures reliably within ± 1°C, therefore the accuracy can be improved by calibration only up to the value of the repeatability of the instrument. Instrument drift, i.e. change of accuracy, may occur after some time owing to wear, friction, etc., although the repeatability may be maintained.

The discrimination of an instrument means that a *dead zone* exists. This is the largest range of values of a measured variable to which the instrument does not respond; it is sometimes called 'dead spot' or 'hysteresis'. A dead zone may occur in an indicating or recording instrument because of friction. The characteristic hysteresis loop occurs when an instrument is calibrated in the forward and then the reverse directions, and is caused by friction and backlash. The hysteresis effect as a dead zone is important when a measured variable does not change significantly over a long period.

The sensitivity of a system can be expressed in a wide variety of units, e.g. mm/°C for a thermometer output representing the displacement of a fluid column (mm) corresponding to an input of temperature change (°C). If the input and output values are of the same form, then the sensitivity is usually referred to as the *magnification* or *gain*. The former term applies to mechanical devices and the latter term to electronic and mechanical devices. An amplifier producing a 100 mV output for a 20 mV input has a voltage gain of 5. The sensitivity is represented by the slope of the graph of output against input. The instrument or system is said to be *linear* (having a linear response or linear characteristic) if the slope of this graph is constant; then the sensitivity is also constant. An instrument should have a range close to the values to be measured to ensure that the sensitivity is high.

2.5 DYNAMIC CHARACTERISTICS

An instrument having a good dynamic response produces an output signal that closely follows a rapidly fluctuating input signal. Instruments that must possess dynamic characteristics are specially designed and the cost is usually higher than if static values only are to be measured. Instruments designed specifically for either static or dynamic measurements are often unsuitable for use in the alternative situation. When dynamic measurements are made, it is necessary to distinguish between *static* and *dynamic accuracy*. An instrument used to measure a steady value of a quantity over a long period of time is subject to static error. However, if the same instrument is transferred rapidly to another system, it will not immediately record the true measurement. The difference between the actual and instantaneous true readings is the *dynamic error* (obtained after allowing for any static errors). The magnitude of a dynamic error depends upon the way in which the input signal changes. One method of specifying this change is to define the error in terms of the response to a sinusoidal input of known frequency. This input produces a sinusoidal output having an amplitude error, and the magnitude of this error at certain frequencies provides an indication of the dynamic performance of the system. This is known as 'frequency-response testing', details of which are given in more advanced instrument handbooks.

To determine the dynamic behaviour of an instrument, the primary element is subjected to a known change in the measured quantity. Three common variations are:

(a) Step change, i.e. an instantaneous and finite change in the measured variable.
(b) Linear change, i.e. the measured variable changes linearly with time.
(c) Sinusoidal change, i.e. the change in the measured variable is a sinusoidal function of constant amplitude.

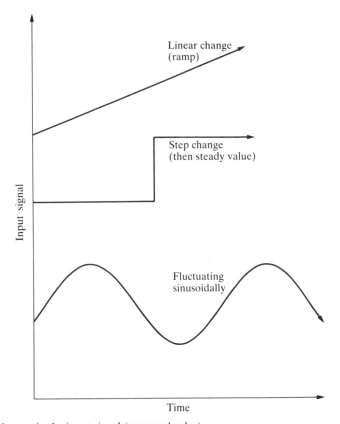

Figure 2.2 Types of changes in the input signal (measured value).

These three types of changes are illustrated in Fig. 2.2.

The desirable dynamic characteristics of an instrument are the speed of response to a change in the measured variable and a minimum dynamic error in the output reading. The *dynamic error* is the difference between the true value of the quantity changing with time and the value indicated by the instrument if no static error is assumed. An undesirable dynamic characteristic is the *measuring lag*; this is a retardation or delay in response to changes in the measured quantity.

The measuring lag may occur immediately with a change in the measured variable (i.e. retardation), or there may be a time delay as shown in Fig. 2.3. The time delay is known as *dead time* and its effect is to shift the instrument response along the time scale, causing a corresponding dynamic error. The dead time is usually very small, typically a fraction of 1 second.

2.6 DYNAMIC PERFORMANCE

A *mathematical model* is an equation or set of equations that describe a particular situation. We are interested in the model that describes the response of an instrument, and how this model can be used (solved) for various inputs. Our requirements for the model would probably be: (a) an equation that describes the performance exactly; and (b) a simple equation, so that closed (analytical) solutions can be easily obtained. Unfortunately, these two requirements are usually not compatible and a compromise is necessary in order to obtain useful (but not exact) results from closed solutions.

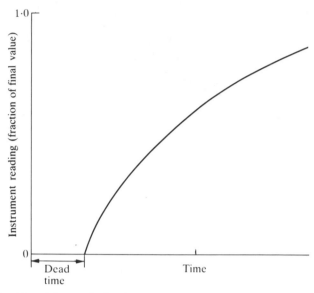

Figure 2.3 Measuring lag (time-delay type) of an instrument.

A general model that describes the performance of most instruments to a satisfactory degree is the linear differential equation with constant coefficients, that is

$$a_n \frac{d^n y}{dt^n} + a_{n-1} \frac{d^{n-1} y}{dt^{n-1}} + \cdots + a_1 \frac{dy}{dt} + a_0 y = b_0 x$$

where x and y are the instrument input and output respectively; t is the time; $a_n, \ldots, a_0$, and b_0 are the physical parameters of a particular instrument. For a fluid-filled thermometer, the parameters would be a function of the thermometer and the system characteristics. These include the mass and specific heat of the fluid in the bulb, surface area, volume, coefficient of expansion, etc. An instrument is classified according to the order of the differential equation required to describe its performance adequately, e.g. zero-order, first-order, etc., instruments. Certain measures of performance can be defined for each order of instrument.

2.6.1 Zero-order model

A zero-order instrument is described by the equation

$$a_0 y = b_0 x$$

or

$$y = \frac{b_0}{a_0} x$$

Therefore, the output is proportional to the input. The constant of proportionality (designated K, equal to b_0/a_0) is the *static sensitivity*. Instruments behave as zero order when subjected to a static input. The static sensitivity is constant if the coefficients (a_0, b_0) are constant. Some instruments behave as zero order even when dynamic inputs are applied, others approximate to zero-order behaviour over a wide range of operating conditions.

2.6.2 First-order model

The performance of a first-order instrument is described by the equation

$$a_1 \frac{dy}{dt} + a_0 y = b_0 x$$

This equation may be written as

$$\frac{a_1}{a_0} \frac{dy}{dt} + y = \frac{b_0}{a_0} x$$

The ratio (a_1/a_0) has units of time and is known as the *time constant* (symbol τ). Therefore, the first-order equation can be written as

$$\tau \frac{dy}{dt} + y = Kx$$

and K is known as the *process steady-state gain* or *static gain* of the system.

It can be shown (this would be a useful exercise for the student) that the mercury-in-glass thermometer is a first-order model. The time constant is (mC_p/hA) and the static sensitivity is $(\alpha V/A_c)$ where m, C_p and α are the mass, specific heat and coefficient of expansion of the fluid in the bulb respectively; h is the film heat-transfer coefficient; A is the surface area of the bulb; V is the fluid volume; A_c is the cross-sectional area of the fluid column.

Most other temperature sensors can also be modelled as first order. The static sensitivity of a thermocouple is not constant but, because the non-linearity (i.e. output (mV) against temperature input) is small, the error caused by using a simple first-order model is usually acceptable. The parameters comprising the time constant and the static sensitivity depend upon the particular device. Some pneumatic devices may also be considered as first order, even though their static sensitivity is variable.

2.6.3 Second-order model

The second-order model can be written as

$$\frac{a_2}{a_0} \frac{d^2 y}{dt^2} + \frac{a_1}{a_0} \frac{dy}{dt} + y = \frac{b_0}{a_0} x$$

Two new terms are defined; these are:

(a) The *natural frequency* or *natural period of oscillation* of the system, $\omega_n = \sqrt{(a_0/a_2)}$; the process time constant $(\tau = \omega_n^{-1})$ is also used in some books.
(b) The *damping ratio* or *damping coefficient* or *damping factor*, $\zeta = a_1/\{2\sqrt{(a_0 a_2)}\}$.

The second-order model becomes

$$\frac{1}{\omega_n^2} \frac{d^2 y}{dt^2} + \frac{2\zeta}{\omega_n} \frac{dy}{dt} + y = Kx$$

Many instruments can be modelled as second order, including dial indicators, pressure transducers and most load cells.

2.6.4 Dynamic response of first-order instruments

The dynamic response depends upon the type of input that is applied. The step function, the pulse and the terminated ramp are illustrated in Fig. 2.4, and their mathematical expressions are also given. These inputs are often referred to as *forcing functions*. A sinusoidal input is also used; this has the form: $x = X \sin \omega t$ where X is the amplitude of the input and ω is the frequency (radian/second); the response to this type of forcing function is known as the *frequency response* of the system.

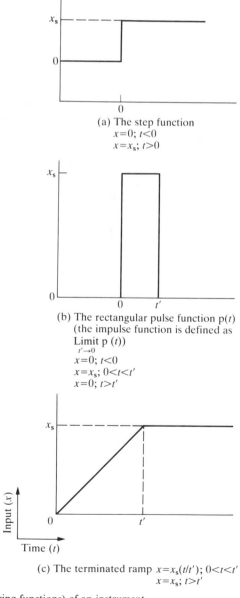

(a) The step function
$$x=0; \ t<0$$
$$x=x_s; \ t>0$$

(b) The rectangular pulse function p(t)
(the impulse function is defined as
Limit p (t))
$$_{t'\to 0}$$
$$x=0; \ t<0$$
$$x=x_s; \ 0<t<t'$$
$$x=0; \ t>t'$$

(c) The terminated ramp $x=x_s(t/t')$; $0<t<t'$
$$x=x_s; \ t>t'$$

Figure 2.4 Types of inputs (forcing functions) of an instrument.

For a first-order instrument receiving a step input (x_s in Fig. 2.4a), the mathematical model is

$$\tau \frac{dy}{dt} + y = Kx$$

$$x = 0; \quad t < 0$$

$$x = x_s; \quad t > 0$$

The solution to this model is

$$y = C \exp(-t/\tau) + Kx_s$$

Using an appropriate initial condition such as $y = 0$ at $t = 0$, then $C = -Kx_s$ and the solution becomes

$$y = K[1 - \exp(-t/\tau)]x_s$$

This solution is shown graphically in Fig. 2.5(a) for various values of the time constant, and it is plotted with dimensionless axes in Fig. 2.5(b). A first-order instrument responds immediately it

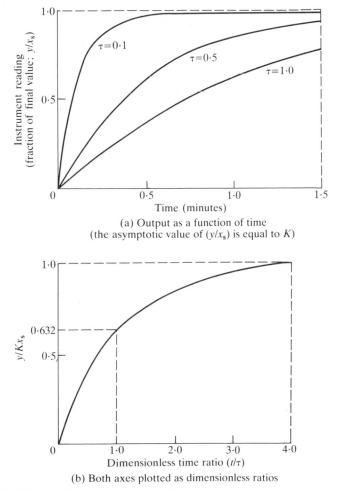

(a) Output as a function of time
(the asymptotic value of (y/x_s) is equal to K)

(b) Both axes plotted as dimensionless ratios

Figure 2.5 Response of a first-order instrument to a step change.

receives an input signal, however a considerable time may elapse before the output reaches the final or steady-state value. Figure 2.5(a) shows that some time is required for the output to reach its final value, and this time depends upon the value of the time constant. The time constant is a dynamic characteristic that indicates how quickly an instrument will respond. The smaller the time constant, the more rapidly an instrument responds. The term $[\exp(-t/\tau)]$ in the first-order model above becomes negligibly small after a time equivalent to a few time constants. The output signal will then be approximately equal to the input.

It should be apparent that the time constant is the measuring lag discussed in Sec. 2.5. Figure 2.5(a) shows that as the time constant becomes larger, the response (while maintaining the same shape) becomes proportionately slower. The time constant is the time required to indicate 63.2 per cent of the complete change, as shown in Fig. 2.5(b). Sometimes the measuring lag is specified as the time required to attain 90, 95, or 99 per cent of the full change. The time constant can be thought of as the product of a capacitance and a resistance. For the mercury-in-glass thermometer discussed earlier, the thermal capacity (capacitance) is equal to $(m\,C_\mathrm{p}/A)$ and the thermal resistance is $(1/h)$.

The dynamic response of a first-order instrument to a linear change (ramp function) for given initial conditions is shown in Fig. 2.6. The mathematical solution is left as an exercise for the student. It is apparent from Fig. 2.6 that after the transient has disappeared, the instrument lags behind by a constant time and a constant value. The transient disappears after a time approximately equal to three time constants; by then 95 per cent of the lag is attained. The measuring lag of the instrument is given directly by the time constant. The dynamic error increases directly with the rate of change of the measured variable, and it is equal to the product of the time constant and the rate of change of the true value of the measured variable. If a small dynamic error is specified, this means that either the instrument must have a small lag or the measured quantity must change slowly.

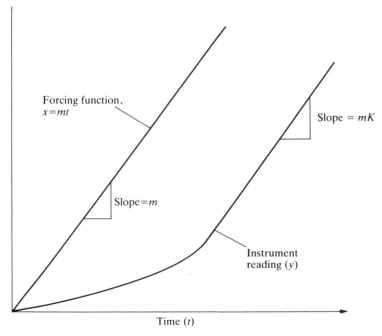

Figure 2.6 Dynamic response of a first-order instrument to a linear change (ramp function).

2.6.5 Second-order instruments

The dynamic response of many instruments cannot be predicted by the first-order model because they have an additional lag. This may be due to the existence of a mass that must be accelerated, or the presence of more than one element of fluid capacity. Examples are the use of a protective sheath around a thermometer (providing an additional thermal capacitance and an air gap resistance to heat transfer), and the introduction of a buffer volume (capacitance) and a connecting tube (resistance) into a pressure gauge. Most mechanical instruments, e.g. pressure gauges and manometers, have some mass that must move during operation. Damping may occur simply because of air resistance or viscous friction of the moving parts.

The equation describing the second-order model is given in Sec. 2.6.3. The solutions of this equation when subjected to some change in the measured variable (see Fig. 2.4) are well known, and can usually be found in textbooks concerned with either process control[4-7] or more advanced aspects of instrumentation (see Bibliography). Three particular cases are usually considered, whatever type of input is applied. These are an oscillatory response (roots of the auxiliary equation are conjugate complex with negative real parts), a critically damped response (negative roots, real and equal), and an overdamped response (negative roots, real and unequal).

The responses of a second-order instrument to a step change are shown in Fig. 2.7 for various values of the dimensionless parameters. The measuring lag is directly proportional to the process time constant ($\tau = \omega_n^{-1}$); a larger value of τ produces a greater measuring lag and a slower response. An increase in the damping ratio (ζ) or the damping number (v; if $\zeta > 1$) also increases the measuring lag and makes the response slower. Most pressure gauges and manometers have a slightly oscillatory response ($\zeta < 1$) with a damping ratio between 0.5 and 1.0, unless they are provided with a special device for damping. Most galvanometer-type electric instruments are nearly critically damped ($\zeta \simeq 1$), and the measuring lag can be specified by the process time constant. Temperature-measuring devices are generally overdamped (ζ or $v > 1$) and their measuring lag can be specified only by giving the characteristic time *and* the damping number. In general, an increase in either the resistance or the capacitance increases the characteristic time, however the damping number may increase or decrease.

Frequency response is also affected by the damping ratio. Resonance only occurs for

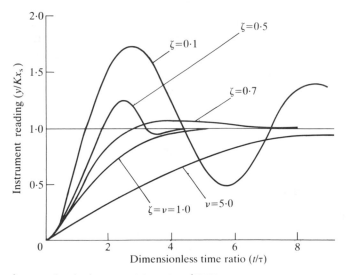

Figure 2.7 Response of a second order instrument to a step change.

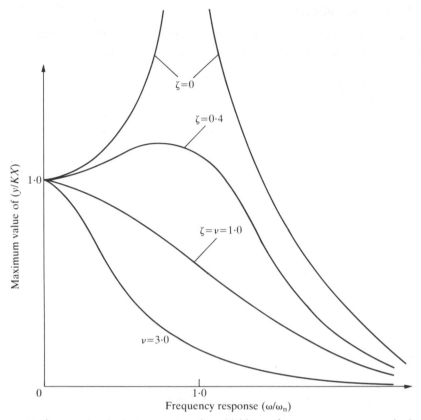

Figure 2.8 Response of a second-order instrument to a sinusoidal input (frequency response–magnitude ratio).

$0 \leqslant \zeta < 1$. The choice of damping ratio between 0.6 and 0.7 means that the value of (y/KX) is as close to unity as possible over the largest possible range of input frequencies; where X is the input amplitude of the sinusoidal function $X \sin \omega t$. The natural frequency (ω_n) also has a considerable effect on dynamic response. For a fixed value of ζ, an increase in ω_n decreases the time required for the oscillations to decay. A larger natural frequency also increases the range of input frequency that the instrument is capable of responding to adequately. The frequency response of a second-order instrument is shown in Fig. 2.8.

Unfortunately the dynamic response of all instruments cannot be described by first-order and second-order models, although for most practical purposes these are sufficient. However, some instrument mechanisms can be described only by third-order or higher-order differential equations. The student should refer to process control textbooks[4-7] for a more detailed analysis of the topics presented here, and for an analysis of more complex systems.

2.6.6 Instrument specifications

The terminology related to instruments in general was defined in Sec. 2.3; the static characteristics (Sec. 2.4) and dynamic characteristics (Sec. 2.5) have also been discussed. Thus far in this section, the performance and modelling of an instrumentation system have been considered. Our purpose now is to see how the principles that have been developed are applied to actual instruments.

Information regarding the static and dynamic characteristics of a particular instrument can usually be obtained from the manufacturer's catalogue. The values that are given are usually the *nominal* or *design values* of the parameters. The true values of these characteristics are slightly different from these nominal values because of variations in materials and manufacturing methods. In some cases the manufacturer measures certain values for each instrument and provides this specific information to the purchaser.

The characteristics specified by the manufacturer depend upon the particular type of instrument that is being supplied. An understanding of these specifications often requires some knowledge of the principle of operation of the particular instrument. For example, the static characteristics of a strain gauge usually include the gauge factor and its tolerance, the gauge resistance (ohm) and its tolerance, the maximum elongation of the gauge, and the effect of temperature changes upon the gauge factor. It is necessary to know that the gauge factor is defined as

$$\frac{\text{change in resistance}}{\text{unstrained resistance} \times \text{unit strain}}$$

This factor is usually measured at a particular temperature, e.g. 25°C. The gauge factor can also be obtained as the slope of the calibration curve, i.e. (change in resistance/unstrained resistance) against unit strain, which is sometimes provided by the manufacturer. The gauge factor in this case is equivalent to the static sensitivity. The tolerances are expressed as $\pm$ percentage values and are usually equivalent to three standard deviations from the mean value (see Sec. 7.2). The strain limits, e.g. ± 2 per cent, indicate the measuring range possible (± 0.02 cm per cm) without causing damage to the gauge. A graph is often provided showing the percentage variation in the gauge factor over a temperature range of (say) $-100°C$ to $+200°C$.

The specifications of the static characteristics of an instrument are intended to tell us how closely the output signal indicates a steady value of the input signal. However, if the input signal can vary, then the dynamic characteristics of the instrument also need to be specified. An accelerometer is used to measure absolute acceleration with respect to the earth's surface. For this type of instrument the manufacturer would specify both the static and dynamic characteristics; values are often given with reference to (i.e. as multiples of) the acceleration due to gravity (g). The static characteristics would include the range (e.g. $\pm 10g$) and static overload (e.g. $\pm 100g$). These values indicate the range of input readings to which the instrument is sensitive, and the maximum accelerations without damage occurring (although measurements may not be possible at these extreme values). The linearity and hysteresis may be combined, e.g., 0.5 per cent of full-scale deflection, as only the total deviation is of interest. A specification of infinite resolution actually means that the resolution (see Sec. 2.3) is infinitesimal, i.e. the output changes for any change of input. The static sensitivity may be obtained from the full-scale output, specified as ± 4 mV/V open circuit (or the excitation voltage, also specified). The output per unit change of input (e.g. $1g$) can then be calculated. Other information may include the operating temperature range, the thermal zero shift (e.g. <0.01 per cent of full-scale per °C) and thermal sensitivity shift (e.g. <0.01 per cent per °C).

The dynamic characteristics to be specified for this instrument would include the approximate natural frequency (Hz) and the damping ratio (e.g. 0.6 ± 0.10 of critical frequency at room temperature). Other information may include the allowable vibration, i.e. accelerations due to oscillatory motion, and the acceptable shock, e.g. $50g$ for up to 20 ms. The specifications for dynamic performance may be provided as a frequency response curve instead of values of natural frequency and damping ratio. Alternatively, the range in which the amplitude ratio lies for a given range of input frequency may be stated.

The general factors influencing the selection of an instrument are discussed in Sec. 2.2. If good estimates of the range of the expected inputs and the acceptable error are obtained, it should be possible to eliminate certain instruments from consideration. It should then be possible to select an instrument from those remaining by considering the static and dynamic characteristics that are specified. The instrument should be sensitive to the expected input and have the proper measurement range. Changes in the input can only be detected satisfactorily if the static sensitivity is sufficiently large and the resolution (discrimination) is sufficiently small. It will probably be possible to achieve the required accuracy from several instruments of a particular type. The maximum error (including static and dynamic errors) permissible by an instrument can be determined by the technique known as 'error analysis' (see Sec. 7.8). This procedure helps in the selection of an appropriate instrument, i.e. one that is adequate for the measurements required. An instrument possessing higher accuracy than is required will probably be more expensive, and may also require a skilled operator, lengthy setting-up procedures, and more costly servicing and repair. Such an instrument may also be less robust and unsuitable for the operating conditions. The dynamic characteristics must be adequate if the input is expected to change. For the best performance, the amplitude ratio should be constant over the expected range of input frequencies. The natural frequencies and the time constants should be properly selected, otherwise large amounts of useless data are obtained. The selection of an appropriate instrument requires a thorough knowledge of the physical system where measurements are to be made, a basic understanding of the principles of measurement, and extensive discussions with several instrument manufacturers. If an instrument is required to make measurements under conditions that are not 'standard', i.e. outside the range of the specifications, then advice should be obtained from the manufacturer. A written guarantee of performance under these unusual conditions may also be advisable.

2.7 CALIBRATION

Calibration can be defined as the procedure used to establish the relationship between the input and output of a device. Calibration is usually performed by applying inputs of different known magnitudes and measuring the output so produced.

Alternatively, calibration is the process of checking a measuring system (or instrument) against a standard, or checking a standard against a higher-grade standard. This process provides an opportunity to reduce the errors in accuracy. If errors are found during calibration, either they can be taken into account during subsequent use of the instrument or it may be possible to adjust the responses of the system (i.e. recalibration). For some instruments, e.g. a micrometer, recalibration is impracticable; other measuring devices may incorporate a number of calibration adjustments.

The International System of Units (SI) is discussed in Sec. 1.2. The SI system defines seven basic units and two supplementary units; all other units are derived from these. All units can be expressed in the dimensions of mass, length and time, but it is more convenient to define temperature and luminous intensity from readily reproduced physical phenomena. The International Prototype Standard of the kilogram is kept in Paris. The National Prototype Standards maintained by various countries are derived from the International Prototype Standards. These National Standards are held by the National Bureau of Standards (USA), the National Physical Laboratory (UK) and the British Calibration Service. The National Prototype Standard is protected from wear and corrosion and is not routinely used. A third level of standard is defined, the National Reference Standard, and subsequently the laboratory or working standards. The values of the standards obtained from the International Prototype Standard are not known exactly, but they are guaranteed to be within a specified range (to a high degree of precision). There is therefore a system of primary and secondary standards in existence.

Within the SI system there are fixed standards (e.g. the kilogram) and reproducible standards (e.g. temperature). A *fixed standard* is a unique physical object. A *reproducible standard* is one which, in principle, can be established locally as required. In practice, this may be possible only in a highly specialised laboratory because of the complex nature of the equipment and the expertise required. A system of secondary reproducible standards also exists, however in this case it is the equipment required to reproduce the standards that must be protected. Such equipment is complex and expensive, and requires properly trained operators. Unfortunately, the relationship between the terms used to describe the quality of the hierarchy of standards (both fixed and reproducible) is not precisely defined, except in certain specialized fields. The laboratory standard is the highest-grade standard in a particular laboratory, however this term does not define the absolute grade of the standard. Laboratory standards can vary in quality between different organizations. The principle of *traceability* means that any particular standard must have been checked against a standard of higher grade, and so on, being traceable back to the primary standard. The frequency of these checking processes depends upon the nature of the particular standard. In industrial situations, it is often more convenient to calibrate a working instrument (i.e. instruments actually carrying out operational measurements) against a reference instrument of a higher accuracy. The principle of traceability still applies.

An alternative method of calibration is by comparison of a particular instrument with a known input source; for example, the direct calibration of a flow-measuring device using a primary measurement, such as recording the time required for a measured quantity of water to flow through the device. Calibration is important because it establishes the accuracy of an instrument, and even a simple check is useful to confirm the validity of the measurements.

A manufacturer will provide instructions for the routine recalibration of an instrument. A zero adjustment and a sensitivity adjustment are almost always provided. *Zero adjustment* adds or subtracts an equal amount from all the indications of the measuring system; this is only approximately true if the relation between the input and the output is not linear. This type of adjustment is universal on devices having moving parts. The *sensitivity adjustment* (also called 'gain', 'range' or 'span adjustment') expands or contracts the response over the entire measuring range, i.e. it adds or subtracts an equal percentage to all measurements. This type of adjustment is usually provided where mechanical linkages or electronic amplifiers are incorporated. The zero and sensitivity adjustments are seldom completely independent; recalibration usually involves a series of alternative settings for each adjustment.

Static calibration is achieved by imposing constant values of known inputs upon an instrument and observing the resulting outputs. The problem of obtaining 'known constant values' of the input has already been discussed in relation to standards. However, two other problems are encountered and these will be considered now.

The instrument may be sensitive to inputs other than the desired input. It is necessary to control (or eliminate) these unwanted inputs so that changes in the measured output are due only to changes in the desired input. This may be achieved by careful control of the environmental conditions, or it may be necessary to perform several static calibrations—one for each unwanted input. It would be useful if all instruments exhibited a linear performance (calibration) curve, e.g. some thermocouples, but unfortunately this is often not the case. Where the calibration curve is non-linear or the input does not correspond to a known mathematical function of the output, it is necessary to obtain measurements over the expected range of the variables and to use either a tabular or graphical representation of these data for calibration purposes.

Most (or should this be all?) instruments do not have the same constant output for repeated

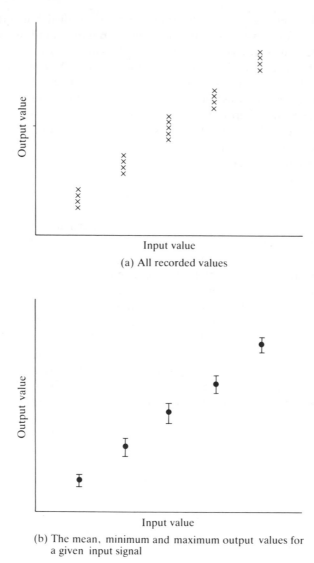

(a) All recorded values

(b) The mean, minimum and maximum output values for
a given input signal

Figure 2.9 Alternative presentations of calibration data.

applications of a particular input. The instrument is not perfect and errors can always occur in measurements, even calibrations. This problem (probably) cannot be eliminated; instead it is necessary to present results in a meaningful form, and that requires a knowledge of statistics. Basic statistical methods are described in Chapter 7. One possibility would be to represent the data for the output readings (for a particular input) by the mean value and its standard deviation. However, assuming that the output readings approximate to a normal distribution, the data can also be expressed as a confidence interval for a particular probability (see Sec. 7.3.4).

Two terms that are often used in relation to instrument calibration are 'bias' and 'precision'. The *bias* is defined as the difference between the mean value of all possible outputs and the input value. The bias can often be corrected, i.e. it can be eliminated by adjusting the instrument. *Precision* (i.e. repeatability) has been defined in Sec. 2.3 as the ability of an instrument to produce the same output for repeated applications of a given input. It is a statistical quantity and is usually expressed (for example) as ± 5 kPa (of the mean value of the output) with a probability of 0.95. A high-precision instrument produces outputs that are not widely scattered.

Calibration data are usually presented graphically. The data can be given as illustrated in Fig. 2.9, showing either all the points or the mean, minimum and maximum values of output for each input. Such a presentation is not particularly useful for interpolation between adjacent input points, or when only a very few values correspond to the maximum or minimum values which are far from the mean. A more convenient presentation is to draw the straight line that best fits the data. The procedure for calculating this line is given in Sec. 7.5.1; it is the regression line of x (input) on y (output). In order to indicate the scatter of the data points about this regression line, two lines can be drawn parallel to the calibration line at distances equal to $\pm$ one standard error of the mean. (The standard error of the mean is discussed in Sec. 7.3.2.) This type of presentation is shown in Fig. 2.10; it indicates that there is a probability of 0.683 that all outputs from the instrument will lie within this range. A range corresponding to any particular probability can be selected using a probability table. For example, 90 per cent of all outputs lie between the lines representing the *confidence limits* which are situated at a distance equal to $\pm 1.645\sigma_n$ from the calibration curve. The *confidence interval* is the distance between the lines representing the confidence limits, and the *confidence level* is 90 per cent.

This discussion relating to the presentation of the calibration data has been concerned with the prediction of the output, for a given input to an instrument. However, an instrument is normally used so that the output is known (measured) and the input needs to be determined. The input value can still be obtained from the regression line (best straight line fit). It can be shown that the confidence limits for determining an input from an output are not parallel with the calibration line. The confidence interval is narrowest at the mean of the input values ($\bar{x}$) and is broadest at the ends of the line.

If the confidence limits are parallel to the calibration line (Fig. 2.11a), this implies that the scatter in output is approximately the same for all values of input. Many instruments behave in

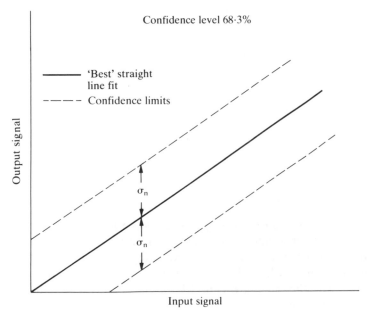

Figure 2.10 Limits of scatter as shown by the best straight line fit and the confidence limits. The confidence interval is the distance between the lines that indicate the confidence limits, i.e. $2\sigma_n$.

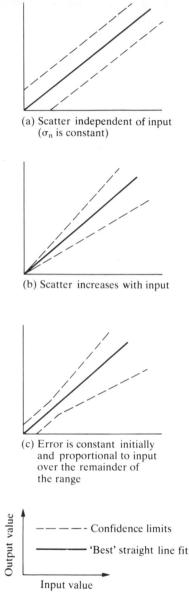

(a) Scatter independent of input
(σ_n is constant)

(b) Scatter increases with input

(c) Error is constant initially
and proportional to input
over the remainder of
the range

Output value

– – – – Confidence limits

———— 'Best' straight line fit

Input value

Figure 2.11 Variations in the confidence interval for a constant confidence level.

this manner, however others show increasing scatter as the input increases (Fig. 2.11b). In this second situation the confidence interval increases with input, although the confidence level is constant. The error of an instrument can then be expressed as $\pm x$ per cent of the indicated value. Another common situation is where the error remains constant over part of the range and is proportional to input over the remainder (Fig. 2.11c). The error can be stated as $\pm x$ units or $\pm y$ per cent of indication, whichever is larger.

Most calibration curves are presented as the range in which it is expected that some fraction of all outputs lie for a given input. This provides a measure of the precision of an instrument and is useful when selecting an instrument. Sometimes it is more useful to present calibration data as the range in which some fraction of all inputs lie for a given output, e.g. if the input value needs to be known.

2.8 TRANSDUCERS

2.8.1 Introduction

The term 'transducer' has already been defined (in a general sense) in Sec. 2.1 and used in the description of the basic elements of a measuring system. The operating characteristics of transducers are considered in more detail in this section, and particular examples of transducer systems are described.

Before discussing specific features of transducers, it is worth while reviewing the basic elements of an instrument system and their functions. An instrument system consists of a *primary sensing element* that senses the measured variable, and a *data presentation element* that provides the output signal or information. A *variable conversion element* is also required to convert the input signal into a different form, without altering the information. This variable conversion element is usually known by the general name of *transducer*, i.e. an energy transfer and conversion device. Many different conversions are possible; some of these are given as follows:

	input	*to*	*output*
(a)	pressure		displacement or force
(b)	temperature		displacement, voltage or resistance
(c)	voltage		current or frequency
(d)	displacement		voltage, current or force
(e)	rotation		translation

A mathematical operation must be performed upon the input signal without changing its physical form; this requires a *variable manipulation element*. In addition, a *data transmission element* is required to transfer the signal from the primary sensing element to the data presentation element. It should be possible to identify these elements in a simple instrument system, e.g. for pressure or temperature measurement (see Chapters 3 and 5). The equipment included in an instrument system (following the transducer) is discussed in more detail in Sec. 2.9.

An *analogue signal* is a continuous function of the input, i.e. the output from the transducer is proportional to the measured variable (or some other known function of it), and the signal has an infinite number of values. Each output value corresponds to a particular value of input; examples are a length measurement, a pen recorder and the movement of the pointer of a meter. A *digital signal* is discrete, having a finite number of values. Each value of the output signal may correspond to several values of input; the output may be a digit (pulse or step) for every successive increment of the input, or a coded discrete signal representative of the numerical value of the input. Examples are 'yes/no', 'on/off' and integer signals. Information can be carried on waves, e.g. a sine wave can give information through its amplitude, frequency, or phase. Transducers that produce a frequency output are analogue devices, although the frequency can easily be converted to digital form by counting over a known period. Analogue and digital signals each have certain abilities that are more appropriate to different measurement situations. Digital information is often more precise than analogue information (but not necessarily more accurate). However, analogue information is preferred in certain situations, e.g. for fast response to time-varying quantities. Decisions concerning the best form of information transport can be quite difficult in complex situations, however the output signal must be compatible with the rest of the system. For example, a control system utilizing a digital computer must provide input to the computer in a digital form.

The output from an instrument may be either a null (balanced) reading or a deflection (unbalanced) reading. Examples of instruments with a null reading are a weighing balance or a Wheatstone bridge, whereas a manometer or single pan (spring scale and pointer) weighing balance produces an unbalanced (deflection-type) output. In general, null devices are more accurate than deflection instruments and are often used to calibrate deflection systems, however only deflection instruments can be used to measure variables that change with time.

A transducer is an energy transfer device, and all measurement devices are transducers. A *passive transducer* obtains all of the energy transferred from the primary sensing element to the next element of the instrument, from the system upon which the measurement is performed. The output of passive elements is usually low in order to minimize the energy removed from the measured system. A passive transducer has two terminals, input and output. The following are examples showing the type of energy transfer:

(a) Tuning fork   —mechanical energy to acoustic energy.
(b) Compass   —magnetic to mechanical.
(c) Pyrometer   —thermal to optical.
(d) Microscope   —optical input and output.
(e) Spring balance—mechanical input and output.
(f) Galvanometer —electrical to optical.
(g) Thermometer —thermal to mechanical.
(h) Thermocouple—thermal to electrical.

An *active transducer* utilizes an auxiliary source of energy so that the output can be increased without significantly increasing the energy removed from the system, e.g. an electronic amplifier. This type of transducer has three terminals: major input, minor input and output. Each input and output can take the form of various types of energy, and each input and output transfers both energy and the required signal. Examples of active transducers are given as follows:

(a) Thermostat—temperature (major input) and voltage (minor input) produce electric current (output).
(b) Pressure transducer—pressure and voltage produce current.
(c) Oscilloscope—electrical signal and electric power produce an optical output.

The minor input represents an additional method of controlling the information, e.g. the information and energy may be changed in type, amplified, or controlled as to frequency. The minor input can also be used to discriminate between two or more ranges of information, or to filter out noise or static. The control afforded by the minor input may mean that an active transducer is the preferred instrument, although it comprises more equipment and hence higher costs and more sources of error.

Every transducer consists of several parts that can handle energy in one of three ways. They can either store potential energy (e.g. capacitors, springs, pressure chambers) or kinetic energy (e.g. flywheels, inductances), or they can dissipate power (e.g. dampers, resistors). The range of different types of transducer is very large if the types are based upon energy considerations, even more so if the information type is also considered. The study of a whole system to evaluate the overall performance may provide an easier means of definition, rather than identifying each individual transducer element.

2.8.2 Transducer characteristics

Most electrical instrumentation systems can be considered to consist of three parts; these are:

(a) *Input transducer*, converting a non-electrical quantity into an electrical signal.
(b) *Signal conditioner*, modifying an electrical signal, e.g. amplifier.
(c) *Output transducer*, converting an electrical signal into a non-electrical form, e.g. pen recorder.

This section will consider input transducers only. Other equipment comprising the instrument system is considered in Sec. 2.9.

The input, transfer and output characteristics all have to be considered when a transducer is chosen for a particular application. The type of input, i.e. any physical quantity, is usually a primary fixed specification. The type of transducer that can be used may be decided by the useful range of the input quantity. The lower limit of the useful range is normally fixed by the transducer error, or by the unavoidable noise originating in the transducer. *Noise* is a general term used to describe unwanted voltages consisting of a signal of random amplitude and random frequency. *Drift* is a slow change of zero level with time (temperature drift is due to a change in ambient temperature) having a frequency lower than the noise voltage. *Interference* is a general term used to describe any unwanted voltage that reduces measurement accuracy. The magnitude of the noise or drift is normally independent of the magnitude of the input signal.

The introduction of a transducer should not influence the measured quantity. Although this is impossible to achieve, it may be possible to reduce the magnitude of the effect the transducer produces so that in practice it is negligible. The magnitude of the effect can be expressed as the amount of energy or power required from the measured quantity in order to drive the transducer.

The *transfer function* is the relationship between the input and output quantities (P_i and P_o respectively) as given by

$$P_o = f(P_i)$$

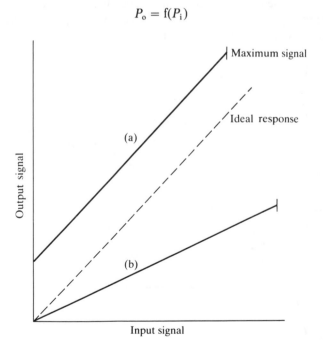

Figure 2.12 Performance curves for a practical transducer: (a) zero error; (b) sensitivity error.

and

$$\frac{dP_o}{dP_i} = K$$

where K is the sensitivity of the transducer. In general, the sensitivity is a function of P_i. If the transfer function is linear, the sensitivity is constant over the range of the transducer. The factor K is generally known as the *scale factor*, although some manufacturers designate the reciprocal $(1/K)$ as this quantity.

An error is introduced because the transducer does not exactly obey the relationship $P_o = f(P_i)$. The absolute error of the output is given by the difference between the actual output and its true value, although the error can be expressed in terms of either input or output quantities. The error is usually complex and consists of a combination of scale error, dynamic error and interference.

The *scale error* may be due to the occurrence of one or more of four different types of error. A *zero error* occurs when the observed output deviates from the correct value by a constant value over the entire range of the transducer. If the observed output deviates from the correct value by a constant factor, this is known as a *sensitivity error*. These types of errors are illustrated in Fig. 2.12; the ideal response of a transducer is also shown although this would be impossible to obtain in practice. Transducer *non-conformity* is the deviation of the transfer function obtained experimentally from the theoretical value; this is shown in Fig. 2.13. A scale error may also occur because of *hysteresis*, as shown in Fig. 2.14. With this type of response the output depends upon both the applied input and the input applied previously, i.e. the history of the input signal is important. A different output is obtained for the same input, depending on whether the input is increasing or decreasing.

A *dynamic error* only occurs when the input is varying with time. The output does not follow precisely the variations of the input with time, or it depends upon time functions of the input. The dynamic error depends upon the dynamic characteristics and performance of the instrument; these aspects were discussed in Secs 2.5 and 2.6.

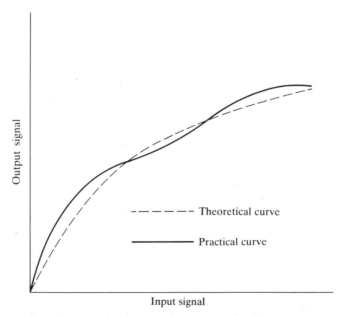

Figure 2.13 Non-linear performance curve showing transducer non-conformity.

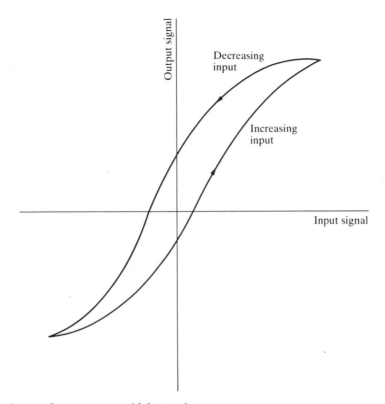

Figure 2.14 Transducer performance curve with hysteresis.

Any instrumentation system powered from a mains supply, or operating in a region where electrical power is being generated or transmitted, will suffer from *electrical interference*. One way in which noise affects the performance of a transducer is shown in Fig. 2.15. The level of interference depends upon the precautions taken when linking elements of the system, and on the design features of the elements themselves. Certain sophisticated instruments are designed specifically to operate in the presence of a high level of interference. Interference normally enters the instrument through either the input connections or the power supply. Additional interference may enter the system through the transmission cables connecting the transducer and the display or control elements.

Three types of interference may be identified. Alternating interference is the most common and is introduced by inductive coupling from a.c. power supplies. Stray magnetic and electric fields are usually present where instrumentation and control equipment are used. Steady interference takes the form of a d.c. voltage or current caused by thermo-electric or electro-chemical effects. Finally, random interference may be caused by intermittent operation of inductive loads. Any of these forms of interference may be present in an instrument system and may appear at the input terminals in one of two modes. Common (or parallel) mode interference is an interference voltage (usually at mains frequency) common to all terminals of a signal circuit with respect to a reference point (usually earth). Series (or differential) mode interference is an interference voltage that appears across the terminals of a signal circuit. This section only provides an introduction to the subject of signal interference; a more detailed explanation and analysis of this problem can usually be found in instrumentation textbooks (see Bibliography).

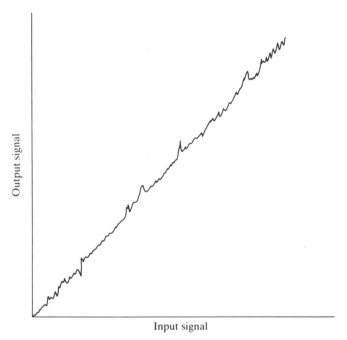

Figure 2.15 Noise affecting transducer performance.

2.8.3 Transducer performance

For most industrial applications, repeatability is of considerably more importance than absolute accuracy. Provided that the repeatability is small (see definitions in Sec. 2.3), very accurate measurements can be obtained even though the transducer may not be particularly accurate despite frequent calibration. The cross-sensitivity of a transducer needs to be considered when the actual variable is measured in one plane and the transducer is subjected to variations in other planes.

Effects caused by environmental changes can seriously affect the reliability of the measurements obtained from a transducer. In some cases there is a loss of accuracy; alternatively, too high an accuracy may be assigned to certain results. Errors may occur because of changes of temperature, pressure, supply voltages, magnetic or electric fields, etc.

The electrical output may be either voltage, current, impedance, or a time function of these magnitudes. It is important to consider the output characteristics of a transducer, including the useful output range and the output impedance. The useful output range is limited at the lower end by noise, and the upper limit is set by the maximum useful input level. The output impedance determines the amount of power that can be transferred from a transducer into successive stages of the instrumentation system at a given output signal level. The maximum power transfer is achieved when the output impedance is equal to the following input impedance; the impedances are then said to be 'matched'.

The interpretation of performance specifications for transducers requires careful reading in order to understand their significance fully. BS 4462[8] presents guidelines for manufacturers, although the literature often emphasizes the most impressive features of an instrument. The interpretation of typical static and dynamic characteristics has been presented in Sec. 2.6; some additional comments will be made here. Consider in general terms the specification for an instrument having a quoted accuracy of ± 0.01 per cent ± 1 digit. Suppose that two input ranges are possible for the instrument, namely 100 units and 1000 units, and the resolution is given as 10^{-2} units. On the 100 unit range, it is possible to measure the maximum input of 100 units ± 0.01 per cent $\pm 10^{-2}$ unit (i.e. ± 1 digit) the reading has an uncertainty of 0.02 per cent. However, at the lower end of the range, for an input of 1 unit the uncertainty is still ± 0.01 per cent $\pm 10^{-2}$ unit, i.e., an uncertainty of ± 1.01 per cent. The uncertainty will take decreasing values as the upper limit of the range is approached.

For the larger range of 1000 units, if an input amplifier and a $1:10$ input attenuator are used, then the resolution on this range becomes 10^{-1} (not 10^{-2}) units. All other factors bear the same relation to the full-scale reading, and the uncertainties are the same at corresponding percentages of full scale for both ranges.

The use of a transducer or measuring instrument on a process plant requires that many other factors are taken into account. These factors include any uncertainty due to the inherent design and manufacture of the instrument, and the zero stability as a function of temperature, e.g. 0.01 per cent of reading per °C (these factors should be quoted in the specification). Additional factors include vibrations, shock, humidity levels, electromagnetic interference, variations in power supply, etc. Once adequate information is available about the individual factors that contribute to inaccuracy, figures then have to be obtained for the operating conditions under which the instrument will have to perform. Error analysis (see Sec. 7.8) should be used to decide the contributions that individual errors make to the overall system error.

It is necessary to consider the alternative forms of transducer output that are often encountered in practice (e.g. d.c., a.c. or frequency). The criteria that should be applied can then be determined in order to avoid degradation of the transducer output signal by any subsequent processing or transmission media.

Most transducers are selected to measure a particular variable. Unfortunately, transducers are often sensitive to more than one physical quantity. If measurements are to be made where two variables are changing, a transducer should be chosen which is not sensitive to the change in the unwanted variable. Alternatively, compensation must be made for the changes produced by the unwanted variable.

2.8.4 Basic instruments

The concept of a transducer, as a device for converting the measured value into some convenient form of signal, has been used in this chapter to consider how instruments operate in general. In this section various transducers are described briefly, concentrating on how each instrument operates. There are many types of transducer, both electrical and non-electrical. Most of those described here convert a mechanical signal, e.g. force, strain or displacement, into an electrical signal of some kind. Other types of transducers are described in particular chapters for pressure, flow and temperature measurement, in the contexts of their particular applications. Transducers are named according to the nature of the input; a displacement transducer converts a displacement into a signal of some other kind, e.g. a voltage.

A *displacement transducer* is used to measure the position of one object relative to another, or the displacement of an object with respect to some reference point. The dial indicator is an example of this type of instrument. The linear motion of the spindle is translated into an angular displacement of the instrument pointer by means of the gear train as shown in Fig. 2.16. Small displacements of the spindle can result in a relatively large movement of the pointer by careful selection of the gear ratios. The spindle is sensitive to the input displacement and represents the primary sensing element. The gears transmit the input signal from the spindle to the pointer; they

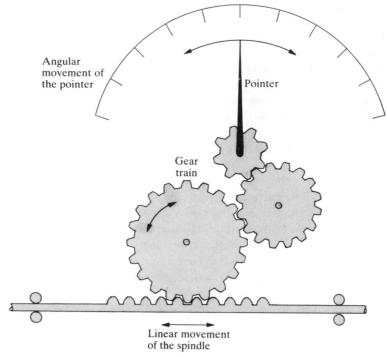

Figure 2.16 Displacement transducer (a dial indicator).

convert the signal from translation to rotation and amplify the input so that a large output displacement occurs. The pointer and scale comprize the data presentation element. As defined in Sec. 2.8.1, this transducer is a deflection-type device having an analogue output, and is classified as a passive transducer.

The *electrical-resistance strain gauge* converts strain, i.e., longitudinal displacement of one end of the gauge relative to the other, into a change in electrical resistance. The simple abbreviation 'strain gauge' is usually applied to this type of device. Bonded gauges are firmly cemented to a surface over their entire length; the common foil-type strain gauge is shown in Fig. 2.17. It consists of a strip of thin metallic foil that is bonded to the member in which the strain is to be measured; the bond ensures that both the gauge and the member experience the same strain. The bond also ensures that the gauge is electrically insulated from the material to which it is attached. A simple unbonded gauge consists of a wire stretched between two points as shown in Fig. 2.18. The wire must be subject to an initial tension if both positive and negative

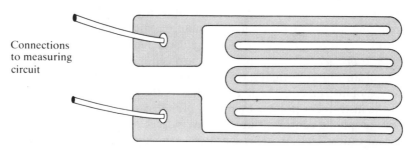

Figure 2.17 Foil-type bonded strain gauge.

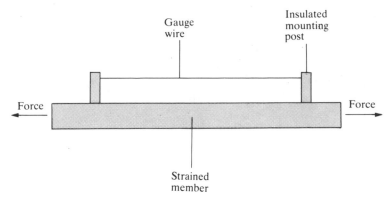

Figure 2.18 Simple unbonded strain gauge.

displacements are to be sensed. Unbonded gauges usually consist of four wires as shown in the simple arrangement of Fig. 2.19.

Usually an increase in resistance is obtained by straining the gauge (nickel gauges are an exception). The change in resistance is often very small and it is normally measured using a Wheatstone bridge circuit. A useful circuit is shown in Fig. 2.20 where any (or all) of the four resistances may be a strain gauge; this system operates as a deflection output device. Assume that resistance R_1 is a strain gauge and the resistances are initially adjusted so that the bridge is balanced. An output e.m.f. (e_o) is observed as the gauge is strained and the bridge remains unbalanced. The output voltage is used to calculate the change in resistance, assuming there is no current between points A and B (Fig. 2.20).

An alternative circuit is shown in Fig. 2.21 and includes a variable resistor R_2. Assume R_1 represents a strain gauge and the bridge is initially balanced so that the e.m.f. e_o is zero. As the gauge (R_1) is strained, the bridge becomes unbalanced and e_o is restored to zero by adjusting R_2.

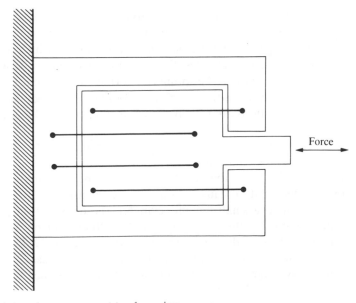

Figure 2.19 Unbonded strain gauge comprising four wires.

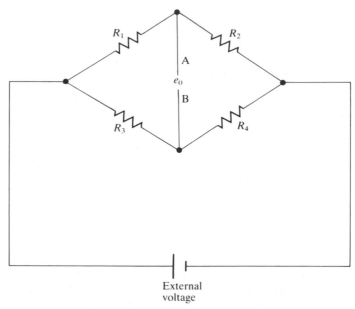

Figure 2.20 Wheatstone bridge circuit acting as a deflection output device (unbalanced). e_0 is the output voltage observed when resistance R_1 (strain gauge) changes.

The change in resistance of R_2 is equal to the change in resistance of the strain gauge. This is known as a null output device because the bridge output is always restored to zero.

The necessary design characteristics of a practical strain gauge include a small cross-sectional area so that it does not significantly affect the stress in the member, sufficiently sensitive to produce a measurable output, rigidly fixed to the strained member and electrically insulated from it. The gauge metal should have a high elastic limit, so that it can be used over a wide range of strain values without suffering permanent deformation.

The absolute velocity and acceleration (i.e. relative to the earth's surface) can be measured in several ways. A typical *accelerometer* (which can measure both variables) consists of a frame rigidly attached to an accelerating mass. The frame also contains a mass which is attached to it by a spring and damper. As the object accelerates, the mass also accelerates because of the force transmitted from the frame by the spring. The spring is deformed and the mass moves relative to the frame. A displacement-sensing transducer, e.g. an unbonded strain gauge, can be used to detect this displacement. Accelerations are usually stated as multiples of the acceleration due to gravity.

Force can be measured in several ways: an analytical balance compares the unknown force with the weights of objects of known mass; a platform-scale balance utilizes a system of levers so that the balance weight is much smaller than the force that is to be measured. Force can also be detected by observing the deformation caused to an elastic member. A spring scale uses this principle. If a force is applied to a small metallic cylinder, the deformation produced can be measured by strain gauges bonded to the side of the cylinder. Other elements are used to convert force (or pressure) into displacement; devices of this type are called *load cells*.

A *variable potential divider* converts displacement to voltage; it consists of a long resistor connected across a constant-voltage source. The output voltage is obtained between the sliding contact and one end of the resistor. This circuit is similar to that of a potentiometer, and the latter term is often used to describe both instruments.

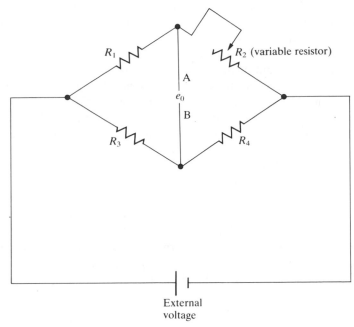

Figure 2.21 Wheatstone bridge circuit acting as a null output device. As resistance R_1 changes, the output voltage e_0 is restored to zero by adjusting R_2.

Piezo-electric transducers respond to pressure or force inputs by producing an electric charge. If the charge is removed, it is not regenerated by an existing steady load but only by a further change in the state of strain in the crystal. The charge gradually leaks away through the piezo crystals and through any insulation on connected equipment. Piezo transducers are therefore suitable only for dynamic measurements; they require sophisticated signal-conditioning equipment to measure the charge while also preventing its dissipation. The transducer action requires almost no displacement of mass, hence its very good response to dynamic inputs.

Inductive or capacitive transducers are used to measure small displacements; they do not suffer from the contact and resolution problems of the variable potential divider. Inductance or capacitance changes are usually measured in terms of the corresponding changes in reactance, hence a.c. measuring methods are required. The differential transformer is an inductive device used to measure a wide range of displacements and having good linearity. *Inductive* or *photo-electric pickoffs* are used to detect discrete events. They are digital devices used mainly for counting and timing applications.

2.9 DATA ACQUISITION SYSTEMS

This section is concerned with the general features of an instrument system following measurement of the basic primary data (i.e. the physical variable) by a transducer. These features include consideration of analogue and digital measurements, signal conditioning and conversion, computational aspects, data transmission and telemetry, signal recovery, data processing, display and recording. It is important that the preliminary general discussion relating to the elements of an instrument system (Sec. 2.1) has been studied, and also Sec. 2.8 concerning transducers. This section considers the specific aspects of an instrument system that are related to data processing, data transmission and recording. These aspects could be termed 'data acquisition' in its broadest sense, or more generally 'data handling'.

Data logging is one part of the data acquisition system; it is concerned with recording many channels of physical data on a routine time basis. It can also be used to provide automatic alarm signals in the event of a failure or an unacceptable system response. The major problem to be considered in the design of a data-logging facility is the communication of information between the measuring instrument (or process) and the data-collection point. Although this may present little difficulty in the laboratory, in industrial situations, e.g. pipeline measurements, a complex instrumentation system may be required to transmit data over very large distances or under adverse operating conditions. The final criterion of data measurement and transmission is often to provide the information necessary to identify and initiate control action. Therefore, the selection of a data transmission system to fulfil a specific function also depends upon the need to provide control actions, and whether a digital computer is also part of the control scheme.

Advances in microelectronics technology have made available smaller data acquisition systems and other instrumentation for control purposes. Electronic signal processing devices are now more compact, and are capable of performing complex operations, reliably, cheaply and on a commercial scale. The following notes describe the features of a data acquisition system in general terms; the choice of particular instruments or operational techniques in most cases depends upon the situation or application being considered.

2.9.1 Analogue and digital measurements

Most transducers produce analogue outputs which are continuous functions with time. The output after signal conditioning (see Sec. 2.9.2) is often displayed and recorded using analogue instrumentation. Digital instruments include those that count events and frequency, through to digital voltage measurement. Digital quantities are discrete and vary in equal steps, each digital number being a fixed sum of equal steps which is defined by that number. *Quantization* is the process of converting an analogue quantity to a digital number. The discrete digital output levels can be identified by a set of numbers such as a binary code. The two processes of quantization and coding represent the basic operations of analogue-to-digital conversion (see Sec. 2.9.2).

Many instrument systems consist only of a transducer which feeds directly to a local indicator. The drive can be electrical, pneumatic, mechanical or hydraulic. Amplification and other forms of signal processing (see Sec. 2.9.2) may be required for transmission to a remote indicator, recorder, computer interface or controller (see Sec. 2.9.6). Several standard instrument signal ranges are used so that equipment from different manufacturers can be directly linked, e.g. the electrical signal from a transducer being fed to a chart recorder.

Several standard ranges of electrical signals are used to represent analogue readings. The most common are 0–10 mA, 0–10 V and 4–20 mA. A range of 0–10 mA can be converted to 0–10 V by terminating the current signal in a precision resistance of 1 kohm. The range of 4–20 mA has now become almost a universal standard, and has the particular advantage of possessing a 'live' zero (equal to 20 per cent of the maximum value). A current of 4 mA flows when the transducer is reading zero, or the 'assumed' zero of the measurement scale, as shown below:

$$\ldots 50°C < \text{measurement range} > 150°C \ldots$$

$$0 \text{ mA} \ldots 4 \text{ mA} < \text{electrical analogue range} > 20 \text{ mA} \ldots$$

$$0 \text{ bar} \ldots 0.2 \text{ bar} < \text{pneumatic analogue range} > 1.0 \text{ bar} \ldots$$

Pneumatic transducers always possess a live zero, and the universal standard of 3–15 lb/in.^{2}

(p.s.i.g.) has even survived metrication (10^5 Pa = 1 bar = 14.5 lb/in.^{2}). An analogue signal can exceed maximum and also decrease below minimum measured values during operation, thus enabling a check that maximum or minimum readings on the measuring range are genuine. The absence of the 4 mA base signal can be used to indicate failure or alarm conditions.

Digital (or on/off) inputs are easier to generate from industrial equipment and easier to handle; also a signal is either present or not, and noise or earthing problems are less serious than for analogue inputs. Two types of signal can be used, either a voltage (e.g. +12 V or 0 V) to represent the equipment condition, or the more common contact closure, i.e. open-circuit or closed-circuit. Contact-closure signals allow 'multiplexing' of digital inputs (see Sec. 2.9.4) to be implemented more easily. There are several types of digital input signals including relay contacts, pushbuttons, limit switches, cams and transistor drives.

Analogue outputs are less common than analogue inputs; they are mainly used as setpoints to controllers or for drives to chart recorders. A 4–20 mA (or 0–10 V) signal range is normally used. Contact closure is the most common and convenient signal type for digital output. Digital outputs are commonly used for indicators and alarms, numerical displays, relay operation (e.g. motor start-up) and transistor drives (e.g. remote electronic equipment).

The interface between industrial equipment and a microcomputer could be a single analogue input plus a single digital input, or it may consist of several hundred analogue and digital inputs with several dozen outputs. However, more faults and inaccuracies are caused by the equipment–computer interface than the operation of the microcomputer equipment itself. Many problems that occur with an integrated computer–equipment logging and control system can be avoided if sufficient care is taken with the installation of cabling and equipment, and if correct screening and earthing arrangements are used.

2.9.2 Signal conditioning and conversion

The physical variables that are measured by a transducer in an industrial situation are usually analogue measurements. Almost all transducers are followed by signal-conditioning equipment. The signal-conditioning equipment is the *amplification system* for the transducer; it increases the transducer signal up to a sufficient value to make it useful for conversion, processing (including transmission), indicating and recording. Passive transducers (e.g. strain gauges or potentiometers) also require an excitation system because they do not generate their own voltage or current. In contrast, active transducers such as thermocouples or piezo-electric crystals produce their own output voltage which usually requires amplification. The excitation source can be alternating or direct voltage (or current).

In a data acquisition system, an amplifier is used to provide signal conditioning or amplification before the analogue-to-digital converter. The transducer signal usually comprises a low-level signal (due to the transducer input) and unwanted electrical noise. The amplifier should be designed or chosen to amplify the data signal and reduce the effect of the noise.

Pneumatic amplification of signals from pneumatic transducers (e.g. bellows, diaphragm) is achieved by the traditional techniques of flapper and nozzle and relay valve (see Fig. 3.11). The purpose of the electronic amplifier is to perform one or more of the following functions: boost the amplitude of the signal, buffer the signal, or convert a signal current into a voltage. The desired full-scale voltage level out of the amplifier is usually between 5 and 10 V, as this is the voltage level accepted by most analogue-to-digital converters (discussed later) and provides better accuracy. Electronic amplification techniques are based on the *operational amplifier* (*op-amp*) which amplifies d.c. and a.c. signals. It was often implemented as a discrete component but is now readily available in integrated circuit (IC) form. An IC op-amp chip is available for a fraction of the price of an equivalent discrete component. The advantages of the IC form include reliability, small size, low cost and less sensitivity to drift caused by temperature changes.

The particular features of the op-amp are:

(a) high input resistance;
(b) high gain;
(c) low output resistance;
(d) very little drift due to temperature changes;
(e) amplification of d.c. up to 1 MHz or more.

The main design feature is that it is a *differential amplifier*, i.e. it amplifies only the difference between the two voltages (V_1 and V_2) at its input terminals. Therefore, the output voltage (V_0) is given by

$$V_0 = A(V_1 - V_2)$$

where A is the intrinsic amplifier gain. If a common mode signal is present on *both* input connections, then this voltage component is not amplified and the amplifier is said to possess a high 'common mode rejection ratio'.

Signal conditioning transforms the measurements of physical quantities such as pressure, temperature, strain and position into an electrical voltage or current of sufficient level (e.g. 1–10 V) for further processing by electronic circuits. The data signal may be indicated and recorded at this stage, but often other operations are required. It may be necessary to perform some non-linear operation on the signal, e.g. squaring, linearizing, or multiplication by another function, before the indicating and recording stages. The signal may go to an *analogue-to-digital converter* where it is transformed from an analogue measurement into an alphanumeric (digital) output. The conversion may be from voltage to digital form, or from frequency (proportional to the input parameter) to digital form. The operation of a car speedometer provides a simple example of the analogue-to-digital conversion process. The movement of a pointer is proportional to the speed (i.e. rotation of the drive shaft) and this provides an analogue signal; conversion to the digital form is provided by the driver (eye and brain). The distance travelled is obtained in direct digital form by the movement of a number of rotating drums. Once in digital form, a signal may be fed to a variety of possible digital systems such as a computer, digital controller, digital data logger, or digital data transmitter.

The key elements of any analogue-to-digital conversion process are the input amplifier and the output counter. The technique used to achieve the conversion determines the intermediate stages and the control logic. Converters operate either by direct comparison with a reference voltage standard, or by indirect comparison (voltage–frequency or voltage–time conversion). The use of analogue-to-digital converters in data acquisition systems requires continuous operation, a convenient form of digital output and the ability to handle low-level signals in the presence of high-level noise. Most analogue-to-digital converters can now be purchased on a single chip.

There are two main analogue-to-digital conversion techniques. These are the direct measuring types including the servo-driven potentiometer (slow response) and the solid-state logic-controlled potentiometric divider, and the indirect measuring types which rely on frequency counting, e.g. ramp converter (voltage-to-time conversion), voltage–frequency converter (integrating type), etc. The most common analogue-to-digital converters use the *successive approximation method* (potentiometric type) where the input voltage is compared with accurately known fractions of full scale. The analogue voltage is balanced against a voltage which is obtained from a reference voltage. The balancing process is one of successive approximation. The input voltage is first compared with half of the reference voltage, and subsequently with other fractions of the reference voltage until the balance is achieved. These devices give good resolution at good speed, typically 20 μs conversion time.

The analogue-to-digital conversion technique known as the *integration method* (or 'double ramp' method) operates on the principle of counting pulses for a period which is proportional to the input voltage (i.e. voltage-to-frequency conversion). An integration counter gives high resolution and can correct for drift, but unfortunately it has a slow conversion time (typically less than 50 ms). The input voltage to any slow converter should be stabilized using a 'sample and hold' amplifier, i.e. an op-amp with a large capacitance across the input.

A *digital-to-analogue converter* is the essential element of an analogue output system. These converters are simpler than analogue-to-digital devices. Nearly all digital-to-analogue converters operate by using each bit of the digital signal in order to generate different binary-scaled voltages; these are summed to produce the final analogue voltage. Two methods are used to generate these binary multiples: the weighted-resistor method and the more common resistor ladder.

A variety of instruments can be classified as *electronic counters*; they all operate using the same basic technique but have widely different applications. Counting involves the serial addition of units of equal dimensions. The rate of operation can be obtained by relating the count to an increment of time. A cumulative numerical total, or an integral with respect to time, can be obtained by difference between the start and finish of a cycle.

2.9.3 Data acquisition computations

The use of the term 'computer' can have several different interpretations in relation to instrument systems, from a simple scaling operation to a complex control function. The automatic-sequence controller used on a manufacturing plant is also sometimes referred to as a computer. The 'on-line' computing requirements of a data acquisition system often involve simple arithmetic facilities, with occasional storage, to assist the interpretation of results. The data collection or data acquisition system will also require digital conversion in order to provide a useful output. Computing corrections in a data acquisition system are of two types; these are:

(a) Compensation of basic primary data (obtained from measurements of the physical variables) to account for scaling, zero offset, etc., and conversion of the data into a standard measurement unit.
(b) Arithmetic operations for two or more primary physical variables and their associated constants.

In its simplest form, a digital microcomputer may be used as the central controller for a data logger, and as a means of computing corrections to transducer outputs (which are seldom linear). The effects of sampling and continuous operation upon the computations used for data conversion, and subsequent control action, must be assessed, and in particular the costs involved. There is often far less flexibility in the choice and operation of the transducer and controller for an instrument system than in the range and performance of suitable microcomputers.

2.9.4 Data transmission

The terms *data transmission* and *telemetry* refer to the process by which a measurement is transferred to a remote location, typically to be processed, recorded and displayed. Measurements made in hazardous and remote locations can be transmitted to a central control room containing several instruments. The lengths of process instrumentation/analysis lines or cables can be minimized by converting the measured variable into a proportional air pressure or electrical quantity for transmission to a measuring instrument. A telemetering system usually consists of:

(a) A measuring instrument which may measure any variable, e.g. pressure, flow, etc.

(b) A conversion element which converts the measured variable into a proportional air pressure or electrical quantity.
(c) The pressure lines or connecting wires which carry the transmitted variable from transmitter to receiver.
(d) A receiver which indicates the size of the transmitted variable and may also record or control the measured variable.

A pneumatic system requires the following basic elements:

(a) A constant supply of air, usually at a pressure of 140 kPa gauge.
(b) A transmitter which reduces the pressure of the air supply so that the pressure of the air transmitted is directly proportional to the value of the measured variable.
(c) A suitable connecting pipe for transmitting the changes in air pressure.
(d) A receiver which translates the air pressure in terms of the original variable.

The time delay in a pneumatic system increases with the distance between transmitter and receiver, and is restricted to distances of about 500 m. However, the time lag in an electrical system is very small and signals can be transmitted over many kilometres using telephone circuits. The technique employed makes the signal independent of the line resistance and it is possible to transmit many signals over a single pair of lines simultaneously. Electrical systems can be classified as follows:

(a) Telemetering by variation of an electrical quantity, e.g. current, voltage or pulse duration.
(b) Balanced bridge systems, e.g. d.c. and a.c. bridges.
(c) Position systems.

Data transmission is an important aspect of an instrument system because degradation or even temporary loss of signal can have serious implications, especially for control action. The main problems associated with data transmission are:

(a) The information requirements, including volume of data, speed of transmission and reliability of the signal.
(b) Automatic detection and correction of errors.
(c) Coding techniques—the methods of presenting, storing and transmitting digital data.

The most fundamental aspects of the system are the modulation or encoding technique, and the type of transmission media used. Other important factors are the bandwidth, interference effects and the maximum amount of information that can be passed via a given data link, as well as methods of using a single link for many channels of data. The random noise that exists on the transmission channel can be described by statistical measurements.

Measured data may have to be transmitted over long distances (many kilometres) and the cost of pneumatic tubing (for a pneumatic signal) or single electric cable would be prohibitive. Also the pressure loss and voltage drop over large distances would be unacceptable. In the process industries, rapid transmission of data is often unnecessary and a simple method known as *pulse-duration modulation* can be used. An example of this method is the representation of the value of a measurement by the time that a tone is present on a telephone line. Another method known as *pulse-position modulation* uses the generation of a short-duration pulse to 'mark' the fraction of the full time interval; this represents the quantity of the measured variable compared with the full-scale value. These methods require only simple converters to generate the pulse signals, but they are only suitable for relatively slow data transmission over a few kilometres. The pulse energy reaches the receiving end of the transmission line more slowly than it is generated at the transmitting end. Errors can occur because it is difficult to determine exactly when the pulse is received. This problem is overcome by converting the analogue signal into a digital code so that the receiving equipment detects merely the presence or absence of a pulse.

The type of signal-modulation process that is selected also depends upon the transmission medium to be used. Direct transmission via cables (land-line telemetry) usually employs either current, voltage, frequency, position or impulses to convey the information. Radio-frequency telemetry employs either amplitude, frequency or phase modulation. Although cables and electromagnetic radiation (radio) links are the most common transmission media, optical, ultrasonic and magnetic induction data links are sometimes used.

The transmission to a computer of a large number of signals over a long distance can result in prohibitive cable costs. There are two solutions to this problem:

(a) Use distributed processing, i.e. install a signal-marshalling microcomputer adjacent to the plant connected by a serial link (three or four wires) to a centralized computer.
(b) Use a telemetry system with a master station sited at the computer end, and an outstation at the plant. They are connected by a single cable pair (normally a dedicated telephone pair). The outstation sends a bit pattern of 0s and 1s superimposed on a signal carrier. These 0s and 1s represent both digital inputs and the outputs of analogue-to-digital converters for analogue instrument signals. The fixed order in which the outstation scans and transmits the plant signals identifies the signals. Data (for digital outputs and analogue outputs) can similarly be transmitted in the opposite direction.

Multiplexing is a process of sharing a single transmission channel between more than one input. There are two main types of telemetry multiplexing transmission systems.

(a) *Frequency division multiplexing (FDM)*. For example, 24 different frequency carriers are scheduled over a frequency range (bandwidth) of 0 to 3000 Hz to handle 24 digital signals. A 0 or 1 is transmitted as a slightly higher or slightly lower frequency than the carrier for each signal.
(b) *Time division multiplexing (TDM)*. This technique involves simply transmitting voltage pulses *in a fixed sequence* (representing one variable after another), the level of each voltage pulse identifies the signal as a 0 or 1. Large numbers of signals (typically 120) can be transmitted using this method. It is possible to time-division-multiplex an FDM system, so that on each of 24 FDM frequency carriers, 120 signals can be transmitted sequentially by TDM. The maximum signal capacity becomes 2880 (i.e. 24×120).

A faster transmission rate is provided by TDM because the full 'bandwidth' is used (unacceptable pulse distortion is also avoided), rather than the narrow bandwidth channel in FDM. Systems using more than one channel of communication can use 'intelligent' TDM to assign data to a particular channel according to the data transmission loading. The data-gathering and communication 'channels' are separate, and any short-term data overloading can be temporarily stored in 'buffer' storage locations.

2.9.5 Signal recovery

Noise is the limiting factor in the measurement of small signals. It is often necessary to use a measurement technique which separates the noise from the data signal at the indicator or recorder, particularly if a data transmission link is used (see Sec. 2.9.4). Noise occurs throughout the frequency spectrum and the major portion is usually random, although fixed frequencies such as a.c. mains frequency can be picked up (usually called 'hum'). The 'hum' can be removed by correct grounding, using screens or guard shields or differential amplifiers (see Sec. 2.9.2). The measured data are usually contained in a signal of one frequency, or a narrow band of frequencies. The data signal can be identified from the noise by filtering, averaging, correlation, or coding.

Signal filtering is the most widely used method for separating data signals from noise. A filter is designed which attenuates the noise more than it attenuates the information signal (if the signal has a different spectrum from the noise), i.e. the signal-to-noise ratio is improved. The bandwidths of the measuring system and of the signal must be identical in order to minimize the effect of noise interference. This method is used if the signal is a repetitive symmetrical waveform of known frequency.

Signal averaging is applied when the data signal is a non-symmetrical waveform whose repetition rate may not be known. The system works on the principle that the standard deviation of random noise decreases as parts of it are added together and their average is calculated. It is sometimes called 'coherent' or 'synchronous detection', or simply 'averaging'. The measurement period must be finite and the observation time must be synchronized with the signal. The output signal from the system contains the desired response mixed with random noise. A series of responses are added and averaged so that the desired waveform is retained, and the noise tends to 'average out'. A typical signal averager incorporates a 'sample and hold' unit, an analogue-to-digital converter, digital accumulator, and output digital-to-analogue converter. Boxcar detectors also separate signal from noise using the averaging effect; however they only contain a single one-point analogue accumulator or memory. A boxcar requires more signal repetitions (and time) than an equivalent signal averager. However, the boxcar is cheaper and can be used to study much faster signals.

Signal correlation is used to detect a signal of known waveform which is transmitted through a medium, and is then received in the same form but containing a noise component. The transmitted signal is cross-correlated with the receiver output; the result consists of the autocorrelation function of the desired signal (common to both waveforms being correlated), and the cross-correlation of the desired signal with the unwanted noise. This second part tends to zero as there is usually no correlation between signal and noise. Therefore, only the signal remains in the form of its autocorrelation function, and the signal-to-noise ratio is substantially increased. Cross-correlation requires a reference signal which is usually similar to the signal to be recovered, and is therefore unsuitable for detection of unknown signals. Autocorrelation is uniquely successful for detection of unknown periodic signals, although it does not allow determination of the waveform.

Signal coding can involve carrier modulation and error-correcting codes. *Modulation* involves altering a signal to make it suitable for transmission over a 'noisy' channel where noise may be unavoidably introduced. The power of the signal at the source can be increased, although this is limited by the available power. The type of modulation method that is chosen depends upon the accuracy, reliability, available channel bandwidth, convenience, cost, etc. Each system has its own problems and requirements. The modulation of a carrier waveform in response to a measured signal usually involves either amplitude modulation, frequency modulation or pulse-modulation methods. *Coding* is a process of introducing error-correcting and error-detecting capabilities into a telemetry system by appropriate processing of the transmitted signal. This can be achieved by introducing additional information or by repeating the coded information (as a check).

2.9.6 Data processing, display and recording

It is often necessary to process, display and record the signals obtained from transducers. The required signal-conditioning, acquisition and conversion equipment may be provided in a single instrument known as a *data logger*. The output from a data logger is in a convenient digital form to operate a printer or paper punch, for recording by magnetic tape, or direct feed to data processing equipment. The data processing may involve modification of the data, analysis, or generation of control signals. Data modification includes correction of measurements of physical variables, e.g. for non-linearity, scaling, offset, etc., and arithmetic calculations involving two or more variables or operations such as averaging, squaring, etc. Particular equipment may incorporate analogue-to-digital converters and accept analogue inputs.

The particular situation or application determines the type of display or recording device which is appropriate. Visual displays include pointer-scale indicators, alphanumeric devices, graphic and pictorial displays. Digital displays can usually be read more accurately by unskilled personnel. Most devices operate as visual displays, although audible signals are also important in particular situations, e.g. alarms.

Recording devices assist in the presentation of data so that they are preserved in a manageable and usable form. Particular problems include the identification and measurement of rapid changes or long-term trends, and the recording and analysis of large amounts of data. An instrumentation signal may be analogue or digital, and a record is possible in both cases. Galvanometric pen recorders account for the majority of all direct-write recorders. Ultraviolet recorders are useful for high-frequency measurements. Servo-recorders provide high accuracy at low frequency; they do not require load-damping adjustment, unlike recorders incorporating moving-coil elements. Direct recording is employed with magnetic-tape recorders to give wide bandwidth. Frequency-modulation recording is used for d.c. or very low-frequency signals. Storage oscilloscopes, and transient and fibre-optic recorders, provide solutions to particular recording problems.

GLOSSARY OF TERMS

(Additional terms that are not defined in Sec. 2.3)

Active arm A term used in conjunction with strain-gauge or resistance-thermometer bridges referring to the arm of the bridge subject to resistance variation.

Amplifier A device that increases the value of current, voltage, or power supplied to it.

Analogue A term used to describe a quantity whose value is represented by a single continuous variable, e.g. voltage or current.

Analogue-to-digital converter Equipment that converts data in analogue form, e.g. an amplified output voltage from a temperature, pressure, or strain-measuring instrument, into digital form for processing by a digital computer.

Auto ranging A facility on a digital voltmeter that automatically selects the optimum range (sensitivity) to suit the voltage level being measured.

Backing off A voltage that reduces the magnitude of another voltage, usually by opposing it.

Bias The average d.c. voltage existing between specific elements of a circuit.

Block diagram Pictorial representation of a data processing system; it is a functional representation and may therefore differ in detail from the actual system (see Sec. 9.5.1).

Bus A common conductor used for carrying signals from several sources to several destinations.

Channel Data input path of a scanner or data-logging system.

Characterize To change deliberately the linearity of an instrument to compensate for any non-linearity of the measured variable.

Coupler A unit that transfers data from a measuring device to a recording device.

Damping Preventing rapid changes or excessive instability.

Data   General term used to denote the information that describes an object, condition, situation, etc. (plural of the term *datum*, although *data point* is often used as the singular term).

Data logging   Automatic recording of multi-channel data in digital form.

Dead band   A range of values over which the signal can be altered without changing the value of the associated output.

Digitize   Conversion of an analogue quantity (usually voltage) to its digital equivalent.

Gain   Ratio of output signal to input signal of an amplifier.

Gauge factor   A factor allowing for dimensional changes in a strain gauge, in addition to changes in resistivity.

Linearize   To correct a measurement for the non-linearity of the transducer characteristic.

Loop (*closed*)   A control system incorporating feedback of information from the output in order to modify the input.

Loop (*open*)   A term relating to a system having no corrective action supplied by feedback.

Mean time between failures   Statistically, the average time between random failures, usually expressed in thousands of hours.

Module   A self-contained functioning unit of a system, usually a standard item.

Non-linearity   The deviation from the theoretical straight-line law of a practical device, usually expressed as a percentage of full scale.

On-line   Descriptive of a process or action carried out as part of a complete sequence of operations where the timing is fixed, and can be shown on a timing diagram as forming an intrinsic part of the system operation.

Over-range   The overload capacity of a voltmeter or counter which allows it to measure (without additional error) an input in excess of full scale.

Pulse   A significant and sudden short-duration change of a quantity, usually voltage.

Real time   The processing of information in such a manner as to generate results in a brief interval.

Reliability   The ability to work to a specification for protracted periods.

Scaling   The linear attenuation of amplification of a signal carried out to provide a reading in engineering units.

Scan   A sequential sampling of all selected inputs in which each channel is sampled once.

Scanner   A device that selects each input for presentation to the digital voltmeter, etc.

Sensitivity   Ratio of cause to effect, e.g. 1 mV per deg C. Sometimes used incorrectly as a synonym for resolution.

Sensor   General term used to describe a measuring instrument with an electrical output, i.e. it senses rather than measures.

Signal   A voltage (or current) that conveys data or an instruction.

Stability   Degree of constancy of an output for a sustained constant input.

Systems analysis   A critical examination of procedures in order to ascertain the best way of making them achieve desired objectives.

Timing diagram   Pictorial representation of the operating sequences and time periods in a system. Only drawn up when the timing is critical.

Zero suppression   The changing of the bottom point of a range.

EXERCISES

2.1   Describe the basic functional elements comprising an instrument system.

2.2   Select some common measuring/control instruments, e.g. thermometer, pressure gauge, level

QUALITIES OF MEASUREMENTS **69**

controller, flowmeter, etc., and identify the basic functional elements described in Exercise **2.1**.

2.3 What is the difference between a digital and an analogue display instrument?

2.4 What is a transducer? Describe some common (and simple) examples.

2.5 List the important factors that need to be considered when selecting an instrument for a particular application.

2.6 Describe (briefly) the main types of errors that can occur when using a measuring instrument, and their possible sources.

2.7 Describe (*in your own words*) the meaning of the terms:

Accuracy
Repeatability
Discrimination
Sensitivity

What other characteristics are important when considering the application of an instrument system?

2.8 Distinguish between the static and dynamic characteristics of instruments.

2.9 Discuss the desirable and undesirable static characteristics that an instrument may possess.

2.10 Discuss the desirable and undesirable dynamic characteristics that an instrument may possess.

2.11 Explain what is meant by 'dynamic performance'.

2.12 Explain what is meant by a 'mathematical model'. Describe how such a model may be useful when studying the performance of an instrument.

2.13 Describe zero-order, first-order and higher-order models that are used to describe the dynamic response of instruments.

2.14 Explain what is meant by:
(a) forcing functions;
(b) measuring lag;
(c) dead time;
(d) time delay;
(e) time constant;
(f) natural frequency;
(g) damping ratio.

2.15 Classify some common measuring instruments, e.g. thermometers, pressure gauges, flowmeters, according to the type of mathematical model that can be used to describe their performance. Explain the physical significance of some of the characteristics of these instruments, e.g. time constant, damping ratio, etc.

2.16 Describe typical specifications for particular static and dynamic measuring instruments.

2.17 Explain why instrument calibration is important.

2.18 Describe the hierarchical system of standards that is used for the calibration of measuring instruments.

2.19 Explain what is meant by:
(a) a fixed standard;
(b) a reproducible standard;
(c) traceability;
(d) zero adjustment;
(e) sensitivity adjustment;
(f) bias;
(g) precision.

2.20 How is calibration data presented? Give examples of typical calibration curves.

2.21 What is a transducer (see Exercise **2.4**)? Explain the difference between an active and a passive transducer, and give examples of each type.

2.22 Discuss the importance of transducer characteristics with reference to (and explaining the meaning of) the terms:
   (a) noise;
   (b) drift;
   (c) interference;
   (d) transfer function;
   (e) scale factor.

2.23 Describe the four different types of errors that may give rise to a scale error.

2.24 When is a dynamic error important in relation to transducer performance?

2.25 Describe some basic measuring instruments that are commonly encountered in the laboratory.

2.26 Explain the meaning of the terms:
   (a) signal conditioning and conversion;
   (b) data transmission and telemetry;
   (c) signal recovery;
   (d) data processing, display and recording.
   Discuss the ways in which these operations are incorporated into an instrumentation system.

2.27 Describe the operation and function of an amplifier and an analogue-to-digital converter in an instrument system.

2.28 What is multiplexing and how is it implemented?

2.29 Specify the equipment comprising a particular measuring/instrumentation system, e.g. for pressure measurement (Chapter 3), flow measurement (Chapter 4), temperature measurement (Chapter 5), or any other physical variable that you have measured in the laboratory.

2.30 Discuss the problems associated with the transmission of data over distances of several kilometres or more.

REFERENCES

1. *PD 6461: Vocabulary of legal metrology. Fundamental terms (1980).* (2nd edn of 1978 version published by OIML.) *Part 1 (1985): Basic and general terms (international).*
2. *BS 2643: Glossary of terms relating to the performance of measuring instruments.*
3. *BS 5233: Glossary of terms used in metrology.*
4. Douglas, J. M., *Process Dynamics and Control,* Volume 1, Prentice-Hall, Englewood Cliffs, New Jersey, 1972.
5. Friendly, J. C., *Dynamic Behaviour of Processes,* Prentice-Hall, Englewood Cliffs, New Jersey, 1972.
6. Luyben, W. L., *Process Modelling, Simulation, and Control for Chemical Engineers,* McGraw-Hill, New York, 1973.
7. Stephanopoulos, G., *Chemical Process Control: An Introduction to Theory and Practice,* Prentice-Hall, Englewood Cliffs, New Jersey, 1984.
8. *BS 4462: Guide for the preparation of technical sales literature for measuring instruments and process control equipment.*

BIBLIOGRAPHY

British Standards

BS 89: Specification for direct acting indicating electrical recording instruments and their accessories.
BS 1986: Design and dimensional features of measuring and control instruments for industrial processes.

BS 4308: Specification for documentation to be supplied with electronic measuring apparatus.
BS 5164: Indirect acting electrical indicating and recording instruments and their accessories.
BS 5781: Measurement and calibration systems. Part 1: Specification for systems requirements. Part 2: Guide to the use of
    BS 5781: Part 1.
BS 6348: Method for expression of the properties of signal generators.

US Standards

Abbreviations used to designate standards organizations are listed at the beginning of the book (page xiii).

ASHRAE SI Ch 13–85: Measurement and Instruments.
ASME PTC 19.1–85: Part 1, Measurement Uncertainty—Instruments and Apparatus.
IEEE 196–51: Transducers, Definitions of Terms.
ISA Directory of Instrumentation, 1982–83.
ISA S5.1–84: Instrumentation Symbols and Identification.
TAPPI TI 406–15–80: Instrument Application Check Sheet Factors and Considerations for Instrument Application.
TAPPI TI 414–01–81: Instrument Symbols and Nomenclature.

Other sources

Abernethy, R. B., *Measurement Uncertainty Handbook*, Instrument Society of America, Research Triangle Park, NC, 1980.
Anderson, N. A., *Instrumentation for Process Measurement and Control*, Chilton Book Co., Radnor, PA, 1980.
Barford, N. C., *Experimental Measurements: Precision, Error and Truth*, 2nd edn, Wiley, New York, 1985.
Barry, B. A., *Errors in Practical Measurement in Science, Engineering and Technology*, Wiley, New York, 1978.
Bell, E. C. and R. W. Whitehead, *Basic Electrical Engineering and Instrumentation for Engineers*, 2nd edn, Sheridan House, New York, 1981.
Daly, J. W. and W. F. Riley, *Instrumentation for Engineering Measurements*, Wiley, New York, 1984.
Doebelin, E. O., *Measurement Systems: Applications and Design*, McGraw-Hill, New York, 1982.
Galyer, J. and C. Shotbolt, *Metrology for Engineers*, 4th edn, Cassell, London, 1980.
Gray, B. F., *Measurements, Instrumentation and Data Transmission*, Longman, London, 1977.
Haslam, J. A. and G. R. Summers, *Engineering Instrumentation and Control*, Edward Arnold, London, 1981.
Johnson, C. D., *Process Control Instrumentation Technology*, 2nd edn, Wiley, New York, 1982.
Kirk, F. W., and N. R. Rimboi, *Instrumentation*, 3rd edn, American Technical Publishers, Alsip, IL, 1974.
Norton, H. N. (ed.), *Sensor Selection Guide*, Elsevier Sequoia, Switzerland, 1984. (Contains data sheets, specifications and manufacturers' data concerning sensors for solid-mechanical quantities, fluid-mechanical quantities and temperature sensors.)
O'Higgins, P. J., *Basic Instrumentation, Industrial Measurement*, McGraw-Hill, New York, 1966.
Pease, B. F., *Basic Instrumental Analysis*, Robert E. Krieger Publishing Co., Melbourne, Fla, 1980.
Ramalingom, T., *Dictionary of Instrument Science*, Wiley, New York, 1982.
Sirohi, R. S. and R. Krishna, *Mechanical Measurements*, Wiley, New York, 1983.
Sydenham, P. H., *Measuring Instruments: Tools of Knowledge and Control* (History of Technology Series, Number 1), Peregrinus, Hitchin, 1980.

TWO

MEASURING INSTRUMENTS

SCOPE

Part Two is concerned with the measurement of pressure (Chapter 3), flow (Chapter 4) and temperature (Chapter 5). These quantities were selected because they are frequently measured by engineers of all disciplines. Other measurements such as level, chemical composition, force, motion, etc., were considered outside the scope of this book as a general text on engineering experimentation.

Rapid changes are occurring in instrument technology as a result of the development and use of microelectronics in instrument systems. However, the basic principles of measurement and operation remain largely unchanged and these three chapters concentrate upon aspects of the measurement techniques used by certain types of instruments rather than detailed descriptions of particular instruments.

All engineering students (and graduate engineers) should be encouraged to keep abreast of new developments in instrument technology, and to obtain manufacturers' catalogues for a wide variety of measuring instruments. More detailed information regarding the measurement of particular quantities and measurement methods can be found in specialist instrument handbooks. Some of these books are listed in the Combined Bibliography to Chapters 3, 4 and 5 at the end of Chapter 5; the reader is particularly recommended to refer to Jones's *Instrument Technology* (several volumes; Volume 1 reissued, B. E. Noltingk (ed.), Butterworths, 1985). I acknowledge frequent reference to this source when preparing these chapters.

THREE

MEASUREMENT OF PRESSURE

CHAPTER OBJECTIVES

1. To explain the basic principles associated with pressure measurement.
2. To describe the different methods of pressure measurement.
3. To provide details of the principles of operation and descriptions of some common pressure-measuring instruments.
4. To consider the operation of pressure transmitters.
5. To describe some instruments used for measurement of vacuum.

QUESTIONS

- What is pressure?
- How is pressure measured?
- What is vacuum?
- How is vacuum measured?
- What are the units of pressure?
- What are the common pressure-measuring instruments?
- What are the principles of operation of these instruments?

3.1 DEFINITIONS

Density Mass of unit volume of a substance.
Units: kg/m^3, lb_m/ft^3; symbol: ρ.

Specific gravity $\dfrac{\text{Mass of any volume of substance}}{\text{Mass of the same volume of water}}$
measured under identical conditions and both masses being measured in the

same units. Alternatively, $\dfrac{\text{density of substance}}{\text{density of water}}$ measured in the same units and under identical conditions.

Units: dimensionless ratio; abbreviation: SG.

Gravitational acceleration The International Standard value of the acceleration due to gravity is $9.806\,65$ m/s^2, measured at sea level at latitude $45°$.

Units: m/s^2; symbol: g.

The value of g at any location is given by:

$$g = g_e(1 + \beta \sin^2 \theta - 5.9 \times 10^{-6} \sin^2 \theta) - 3.086 \times 10^{-4} H$$

where $g_e = 9.780\,49$ m/s^2 (at the equator);

$\beta = 5.288 \times 10^{-3}$;

θ = geographical latitude;

H = height above sea level (m).

Pressure Force per unit area.

Units: N/m^2; Pa; lb$_f$/in.^{2}; symbol: P.

Several units have been used for pressure measurements; the relation between some of these is shown in Table 3.1.

Pressure can also be expressed as the height of a column of liquid, e.g. feet of water or mm of mercury. The pressure due to a column of liquid 1 mm high is $\rho g \times 10^{-3}$ N/m^2. The *standard mm water gauge* is defined as the pressure due to a 1 mm column of water of maximum density, i.e. 1000 kg/m^3 at $3.98°$C, where $g = 9.806\,65$ m/s^2.

Therefore, the standard mm water gauge is

$$1000 \times 9.806\,65 \times 10^{-3} \text{ N/m}^2$$

Table 3.1 Comparison of different units used for pressure measurement

	Pa	mbar	mm H$_2$O	atm	in. WG	lb$_f$/in.^{2}
Pa	1	10^{-2}	0.102	9.869×10^{-6}	4.02×10^{-3}	1.4504×10^{-4}
mbar	100	1	10.197	9.869×10^{-4}	0.402	1.4504×10^{-2}
mm H$_2$O	9.807	9.807×10^{-2}	1	9.678×10^{-5}	3.937×10^{-2}	1.4223×10^{-3}
atm	1.013×10^5	1013	1.0332×10^4	1	406.77	14.696
in. WG	249.10	2.491	25.4	2.453×10^{-3}	1	3.613×10^{-2}
lb$_f$/in.^{2}	6894.8	68.948	703.07	6.805×10^{-2}	27.68	1

	MN/m^2	dyn/cm^2	lb$_f$/in.^{2}	kg$_f$/mm^2	bar	ton$_f$/in.^{2}
MN/m^2	1	10^7	1.45×10^2	0.102	10	6.48×10^{-2}
dyn/cm^2	10^{-7}	1	1.45×10^{-5}	1.02×10^{-8}	10^{-6}	6.48×10^{-9}
lb$_f$/in.^{2}	6.89×10^{-3}	6.89×10^4	1	7.03×10^{-4}	6.89×10^{-2}	4.46×10^{-4}
kg$_f$/mm^2	9.81	9.81×10^7	1.42×10^3	1	98.1	6.35×10^{-1}
bar	0.10	10^6	14.50	1.02×10^{-2}	1	6.48×10^{-3}
ton$_f$/in.^{2}	15.44	1.54×10^8	2.24×10^3	1.57	1.54×10^2	1

Note: 1 Pa = 1 N/m^2
1 bar = 10^5 Pa
1 atm = 760 mm Hg at $0°$C
1 mm Hg ($0°$C) = $1.333\,224 \times 10^2$ N/m^2.

At 20°C, $\rho = 998.203 \text{ kg/m}^3$ and the pressure is $9.789\,03 \text{ N/m}^2$.

Pressures are often measured as mm mercury; comparable values for 1 mm column of mercury are:

0°C (SG = 13.5955), $P = 133.326 \text{ N/m}^2$

20°C (SG = 13.5462), $P = 132.843 \text{ N/m}^2$

For other temperatures the actual density of the water or mercury is required. The pressure at any point in a liquid (due to the presence of the liquid) is proportional to the depth of the point below the liquid surface. The pressure is therefore the same at all points on the same horizontal level. The pressure at any point in a liquid is the same in all directions.

3.2 BASIC PRINCIPLES

The pressure in a static fluid is exerted equally in all directions and is referred to as the *static pressure*. In a moving fluid, the static pressure is exerted on any plane parallel to the direction of motion. The pressure exerted on any plane surface at right angles to the direction of flow is greater than the static pressure because of the additional force required to bring the fluid to rest. This additional pressure is proportional to the kinetic energy of the fluid.

To measure the static pressure in a moving fluid, the surface where the measurement is taken must be parallel to the direction of flow. This ensures that no kinetic energy is converted into pressure energy at the surface. For a circular pipe, the measuring surface must be perpendicular to the radial direction. The pressure connection is known as a *piezometer tube*; its end should be flush with the pipe wall so that the flow is not disturbed. The fluid velocity is a minimum near the walls and any error in the reading, because the surface is not exactly parallel to the direction of flow, will be minimized.

If eddies or cross-currents are likely to occur in the fluid, a piezometer ring should be used. This consists of four pressure tappings (holes), equally spaced at 90° intervals round the circumference of the piezometer tube. These tappings are joined by a circular tube that is connected to the pressure-measuring device. Using the ring, a mean pressure value is obtained. The cross-section of the piezometer tubes and ring should be small. The static pressure should be measured at a distance equal to at least fifty pipe diameters from any bends or obstructions. This ensures that the fluid flow lines are almost parallel to the pipe walls.

3.2.1 Absolute and differential pressure measurements

It is important to understand the difference between absolute and differential pressure before considering the methods and instruments that are used to measure pressure.

The *absolute pressure* is the difference between the pressure at a point in the fluid and the 'absolute zero' of pressure, i.e. a complete vacuum. A mercury barometer is an example of an absolute pressure gauge. The height of the mercury column measures the difference between the atmospheric pressure and the 'zero' pressure of the Torricellian vacuum above the mercury column. It may be difficult to obtain a complete vacuum in certain instruments, however a very low pressure may be sufficient for practical purposes.

Most pressure-measuring instruments measure the *difference* between the absolute pressure of the fluid and the atmospheric pressure. The reading obtained is called the *gauge pressure* and is an example of a *differential pressure measurement.* Therefore

$$\text{gauge pressure} = \text{absolute pressure} - \text{atmospheric pressure}$$

or

$$\text{absolute pressure} = \text{gauge pressure} + \text{atmospheric pressure}$$

Vacuum or suction gauges indicate the amount by which the absolute pressure of the fluid is below atmospheric pressure; therefore

$$\text{gauge pressure (vacuum)} = \text{atmospheric pressure} - \text{absolute fluid pressure}$$

Differential pressure-measuring devices are also used to measure the difference between two unknown pressures (neither of which is atmospheric). These absolute pressures can be obtained either at different locations for the same fluid or between two different fluids. The principles of absolute and differential pressure measurements are shown in Fig. 3.1.

3.2.2 Pressure-measurement methods

There are two methods that can be used to measure pressure directly: first, balancing the unknown pressure against the pressure produced by a column of liquid of known density; second, allowing the unknown pressure to act on a known area and measuring either the resultant force, the stress, or the strain produced in an elastic medium.

The instruments commonly used for pressure measurement can be classified as follows:

(a) *Balancing a column of liquid*
   Simple U-tube with vertical or inclined limb
   Well-type manometer
   Closed-limb U-tubes
   McLeod gauge
   Calibrating manometer
(b) *Balancing the resultant force on a known area*
   Piston-type pressure gauge or dead-weight tester
   Ring balance-type pressure gauge
   Bell-type pressure gauge
(c) *Balancing against the stress in an elastic member*
   Bourdon tubes
   Diaphragm types
(d) *Indirect methods*
   Katharometer type
   Piezo-resistive pressure sensor
   Quartz sensor
   Strain gauge

Some of the more common pressure-measuring instruments are described in the following sections, while other devices are mentioned briefly. This chapter is intended to provide an introduction to pressure-measuring instruments commonly encountered by students in their experimental work. For this reason, attention is focused upon the more traditional and simpler methods of pressure measurement, e.g. U-tube manometers and Bourdon gauges, rather than the extensive range of modern sophisticated and often specialist instruments. For more detailed descriptions of the principles of operation, and a comprehensive coverage of the instruments available, the reader is referred to an instrument handbook or to a particular manufacturer, e.g. Foxboro Company or GEC Ltd. These comments also apply to the measurement of flow and temperature covered in Chapters 4 and 5.

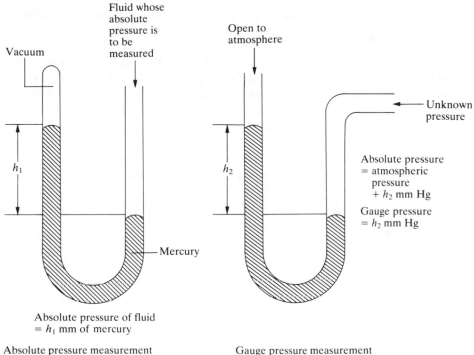

Vacuum

Fluid whose absolute pressure is to be measured

h_1

Mercury

Absolute pressure of fluid
$= h_1$ mm of mercury

Absolute pressure measurement

Open to atmosphere

Unknown pressure

h_2

Absolute pressure
$=$ atmospheric pressure
$+ h_2$ mm Hg

Gauge pressure
$= h_2$ mm Hg

Gauge pressure measurement

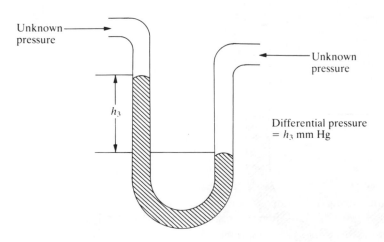

Unknown pressure

Unknown pressure

h_3

Differential pressure
$= h_3$ mm Hg

Differential pressure measurement

Figure 3.1 Types of pressure measurements.

3.3 U-TUBE MANOMETER

A simple U-tube containing a liquid of density ρ is shown in Fig. 3.2. Points A and B are at the same horizontal level and the pressure is the same at each point. Therefore

pressure at A = atmospheric pressure (at C) + pressure due to column of liquid (BC)

= atmospheric pressure + $h\rho$

The units of pressure depend upon whether the liquid in the U-tube is water (mm H_2O) or mercury (mm Hg). The pressure, $h\rho$ mm of liquid, is the gauge pressure. This system assumes that the density of the fluid above the manometer liquid is negligible compared with that of the liquid. However if this is not the case, then a correction must be applied. This is known as a wet-leg connection and is shown in Fig. 3.3. Therefore

pressure at A = pressure at B

$$h_1\rho_1 + P = h\rho + \text{atmospheric pressure}$$

$$P = h\rho - h_1\rho_1 + \text{atmospheric pressure}$$

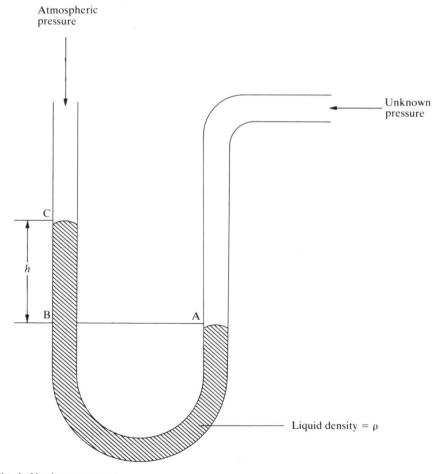

Figure 3.2 Simple U-tube manometer.

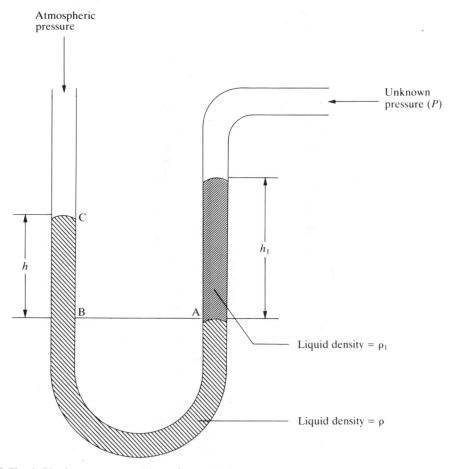

Figure 3.3 Simple U-tube manometer with wet-leg connection.

or

$$P = (h\rho - h_1\rho_1) \text{ gauge pressure}$$

If both limbs are filled with the same liquid (density ρ_1) above the manometer fluid, and to the same horizontal level as shown in Fig. 3.4, then

$$\text{pressure at A} = \text{pressure at B}$$

$$h_1\rho_1 + P_1 = (h_1 - h)\rho_1 + h\rho + P_2$$

Therefore, differential pressure is

$$P_1 - P_2 = (h_1 - h)\rho_1 + h\rho - h_1\rho_1$$

$$= h\rho - h\rho_1$$

$$= h(\rho - \rho_1)$$

This is the correction to be applied if the *difference* in levels (h) is measured. If the rise of the manometer fluid above a *fixed level* is measured as shown in Fig. 3.5, then a different correction is required. Assuming that the change in level in each limb (h_m) is equal, i.e. uniform tube

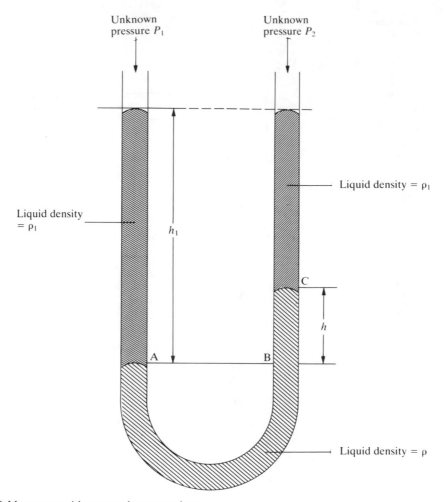

Figure 3.4 Manometer with two wet-leg connections.

cross-section, then

$$\text{pressure at A} = \text{pressure at B}$$

$$(h_1 + h_m)\rho_1 + P_1 = (h_1 - h_m)\rho_1 + 2h_m\rho + P_2$$

Therefore

$$P_1 - P_2 = 2h_m\rho - 2h_m\rho_1$$

$$= 2h_m(\rho - \rho_1)$$

A well-type manometer is shown in Fig. 3.6. The limbs have different diameters and the rise in one limb does not equal the fall in the other. However, the loss of liquid in one limb must equal the gain of liquid in the other. Therefore

$$h_m A = h_2 a$$

or

$$h_2 = \frac{h_m A}{a}$$

(notation as given in Fig. 3.6).

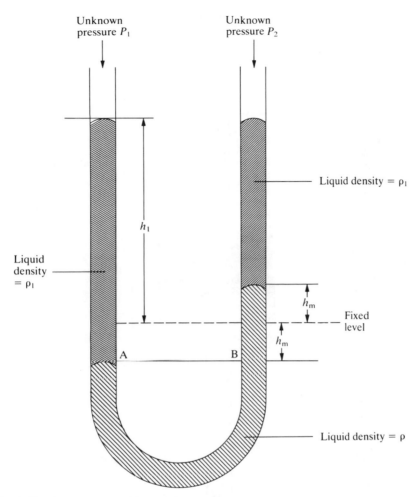

Figure 3.5 Simple U-tube manometer with wet-leg connections.

For a simple well-type U-tube *without* wet legs:

$$P_1 = (h_2 + h_m)\rho + P_2$$

If the left-hand limb *only* becomes a wet leg (fluid density ρ_1), then

$$P_1 + (h_1 + h_2)\rho_1 = (h_2 + h_m)\rho + P_2$$

Therefore

$$P_1 - P_2 = (h_2 + h_m)\rho - (h_1 + h_2)\rho_1$$

If both legs are wet, then

$$P_1 + (h_1 + h_2)\rho_1 = (h_2 + h_m)\rho + (h_1 - h_m)\rho_1 + P_2$$
$$P_1 - P_2 = (h_2 + h_m)\rho + (h_1 - h_m)\rho_1 - (h_1 + h_2)\rho_1$$
$$= (h_2 + h_m)\rho - (h_2 + h_m)\rho_1$$
$$= (h_2 + h_m)(\rho - \rho_1)$$

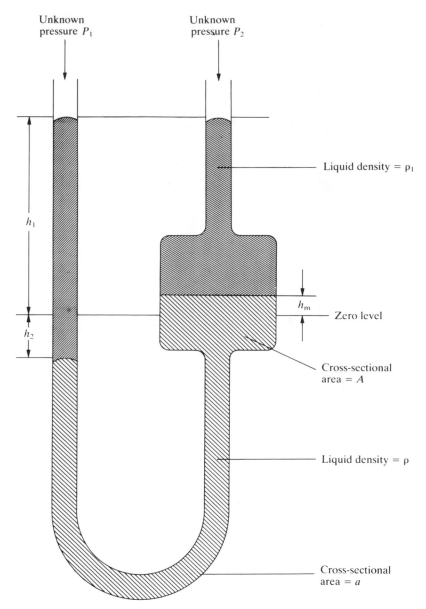

Figure 3.6 Well-type U-tube manometer.

$$P_1 - P_2 = h_m \left[\frac{A}{a} + 1 \right] (\rho - \rho_1)$$

For accurate work, the effect of any variation in temperature upon the readings (e.g. change in fluid density or expansion of the scale) must be considered. If the conditions change from density ρ_0 at temperature T_0 to ρ_1 at T_1, then

$$\rho_1 = \frac{\rho_0}{1 + \beta(T - T_0)}$$

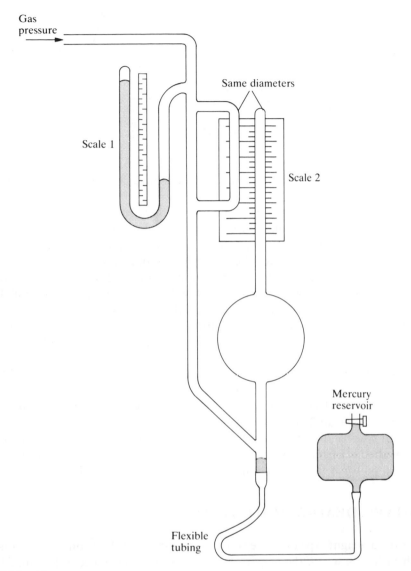

Figure 3.7 McLeod low-pressure gauge.

where β is the coefficient of volume expansion. The expansion of the gauge tube does not affect the reading because the pressure in the liquid depends upon the depth and density only.

The liquid used in a manometer depends upon the pressure, or pressure differential, to be measured. For a small pressure difference between the limbs, a low-density liquid will be required to give the maximum reading. Water (often coloured) is the most common liquid; other alternatives include transformer oil (SG 0.86), carbon tetrachloride (SG 1.61), tetrabromoethane (SG 2.96) and several others.

To measure high pressures, one limb of the manometer must be very long. However, if this limb is closed so that it contains dry air, then the pressure is measured by the amount the air is compressed. Volumes of compressed gas can be regarded as proportional to the length if the closed limb has a uniform bore. This type of manometer is suitable for pressure measurements of

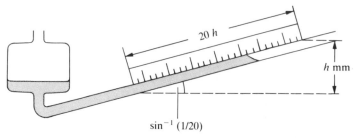

Figure 3.8 Inclined-limb manometer.

several atmospheres, because at high pressures a small change of pressure will not produce an appreciable change in the air volume.

A mercury vacuum gauge is used to measure very small absolute pressures. One limb of the manometer is closed and *completely* filled with mercury. The low-pressure source is connected to the open limb and the mercury falls, leaving a vacuum in the closed limb. The difference in height of the mercury levels is equal to the pressure acting on the open limb.

The *McLeod gauge* is used for measuring very low pressures, as low as 2.5×10^{-4} mm Hg. A particular example of this type of gauge is shown in Fig. 3.7. Gas is trapped in the bulb by raising the mercury reservoir, but details of the operation and theory are not given here. Pressures from 100 mm Hg down to 12 mm Hg can be read from scale 1, and from 12 mm Hg down to 2.5×10^{-4} mm Hg on scale 2. At very low pressures, allowance must be made for the vapour pressure of mercury.

When a small difference in level must be measured, the accuracy can be improved by using a U-tube with an inclined limb. The other limb has a large bulb as shown in Fig. 3.8. If the tube is inclined at an angle of $\sin^{-1}(1/20)$ to the horizontal, then a vertical rise of h mm in liquid level will correspond to a movement of $20h$ mm along the tube.

Many different types of industrial manometers are available, too numerous to describe here. However, their method of operation should be easily ascertained if the descriptions in this section have been understood (or if the manufacturers' instructions are available!).

3.4 PISTON-TYPE DEAD-WEIGHT TESTER

A free-piston or dead-weight type of gauge is used for calibrating Bourdon-type gauges, and for measuring high pressures where the height of a column of mercury becomes impractical. The principle of operation involves allowing the unknown pressure to act on a piston of known area and measuring the resultant force that is generated. This force is measured directly by the mass it will support. The dead-weight pressure tester is illustrated in Fig. 3.9. The accuracy depends upon the precision obtained in the manufacture of the piston and its cylinder, and also upon reducing the effect of friction. For low-pressure measurements (up to 500 bar) the masses are placed directly above the piston; for higher pressures (up to 8000 bar) an overhang design is used. The tester is fitted with a pump for priming and a screw press to produce the pressure; lubrication is required between the cylinder and the piston (*not* oil for oxygen gauges!). Neglecting frictional forces, the pressure (P, N/m^2) acting on the piston (area A, m^2) is given by

$$PA = Mg$$

where M is the mass (kg) supported by the piston. Various types of dead-weight testers are available from instrument manufacturers.

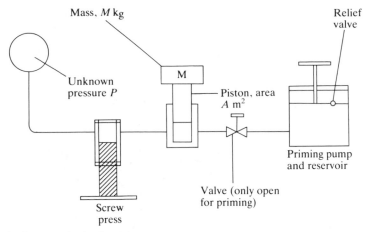

Figure 3.9 Schematic diagram of a dead-weight tester.

3.5 BOURDON GAUGE

The *Bourdon gauge* is a simple and versatile pressure-measuring instrument; it is widely used and is available with many modifications and improvements. The simplest type consists of a curved tube of oval cross-section that is bent into a circular arc. The applied pressure produces a deflection at the end of the tube, and this movement is communicated through a system of levers to a recording needle.

This simple form of the gauge is shown in Fig. 3.10; the sealed end of the tube is attached to the lower end of a pivoted quadrant (via an adjustable connecting link). The play between the quadrant and pinion is taken up by a fine phosphor bronze hair-spring. Tubes are available in phosphor bronze, beryllium–copper and stainless steel, and in K-Monel if corrosion is a problem. The performance of Bourdon pressure gauges depends upon their basic design and the materials of construction, and also the conditions under which they are used. The main sources of measurement error are hysteresis in the tube, temperature changes affecting the sensitivity, frictional effects and backlash in the pointer mechanism. A typical accuracy is ± 2 per cent of the span.

The sensitivity of the gauge can be improved by using a spiral (for low pressures) or a helix (for high pressures) configuration of the tube. These arrangements avoid the necessity for the toothed quadrant and hence reduce the backlash and frictional errors. The spiral or helix increases the effective angular length of the tube; the movement of the free end is also increased and the need for greater magnification is avoided.

3.6 DIAPHRAGM-TYPE GAUGES

The diaphragm elements may be of two types: stiff metallic diaphragms, or bellows and slack diaphragms with drive plates.

The *Schaffer diaphragm gauge* consists of a stainless steel corrugated diaphragm held between two flanges. Pressure is applied beneath the diaphragm and the movement at the centre is transmitted through a linkage and is used to drive a pointer. This gauge can be used to measure pressures above and below atmospheric; it is more sensitive than a Bourdon gauge for low-pressure measurements and it is suitable for measuring fluctuating pressures.

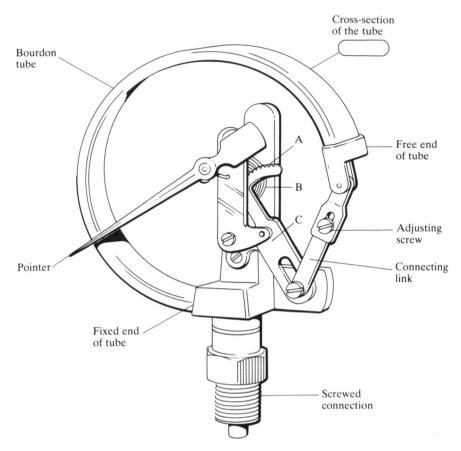

A — Toothed segment; B — Hair-spring; C — Pivoted quadrant

Figure 3.10 The Bourdon pressure gauge.

An *aneroid barometer* uses an evacuated circular capsule (supported by a spring) which acts as a stiff diaphragm. Changes in atmospheric pressure cause movement of the capsule surface, and this movement is transmitted to an indicating lever. If the instrument is calibrated as height above sea level (which is a function of the change in air pressure), it is known as an *altimeter*. Several high-precision capsule-type pressure indicators are available, and also instruments for differential-pressure measurement incorporating elements built up of several diaphragms of various diameters.

Improved manufacturing methods have led to the replacement of many diaphragm capsules by bellows elements. A typical bellows unit can now be used to withstand static pressures as high as 400 bar while measuring a wide range of differential pressures from 20 mm Hg gauge to 200 bar. The unit must be protected from distortion caused by excessive pressure. Absolute pressure measurements can be obtained if one of the bellows elements is evacuated to approximately 0.05 mm Hg, and sealed.

Measurement of small pressures requires a large diaphragm area; this is a slack diaphragm made from a thin non-porous material and supported on both sides (over most of its area) by drive plates. Diaphragm movement is opposed by a flat beryllium–copper spring; the drive plates also provide overload protection.

3.7 INDIRECT MEASUREMENT METHODS

Lack of space prohibits detailed descriptions of the many alternative methods of pressure measurement. Only very short descriptions of the principles of operation of selected methods are included; further details can be obtained from the manufacturers' or specialist instrument handbooks.

A *capacitance manometer* uses an electrode assembly, located in the sensor body, to measure the movement of a tensioned metal diaphragm. Pressures as low as 10^{-3} Pa can be measured reliably for a wide range of process fluids, including those which are highly corrosive.

The operation of a *quartz electrostatic pressure sensor* utilizes the transverse piezo-electric effect in which the applied force causes an electrostatic charge to be developed across the quartz crystal. This charge is measured using a charge amplifier and the resultant signal is used to provide an indication of the applied force. Sensors possessing high stability, wide dynamic range, good temperature stability, good linearity and low hysteresis are available. The typical pressure range is from 200 kPa to 100 MPa.

A *quartz resonant pressure sensor* utilizes the piezo-electric effect to observe the change in resonant frequency of the sensor due to the force developed on a flexible diaphragm.

Strain gauge-pressure sensors have resistance-type strain sensors connected in a Wheatstone bridge network.

3.8 PRESSURE TRANSMITTERS

The control of a process usually requires that a pressure signal is transmitted from the actual measuring instrument to a central operations room; these are often separated by a substantial distance. The first transmission systems were pneumatic because of their inherent safety aspects; also the diaphragm actuator provided a powerful and fast-acting device for driving the final operator. The original *pneumatic motion-balance pressure transmitters* were particularly sensitive to vibration and have now been mainly superseded by force-balance systems. The motion-balance system uses a primary element, e.g. Bourdon tube, to produce a movement that is proportional to the pressure.

Pneumatic force-balance transmitters convert a force applied to the input point into a proportional pneumatic signal. The force may be generated by a Bourdon tube, bellows or diaphragm, and it is applied to the free end of the force bar. A differential-pressure cell that utilizes this principle is illustrated in Fig. 3.11. The force is produced in the bar B by the difference between the high and low pressures fed to either side of the diaphragm A. The force bar is located by the closure at E which acts as a fulcrum so that the flapper nozzle separation at H is modified, thus causing a change in the air pressure from the relay I. The output air pressure is fed to the flexible bellows G, producing a force on the bar B in such a direction as to restore the system to equilibrium. Thus, the air pressure at G is proportional to the differential pressure on the diaphragm A, and can be transmitted over several hundred metres.

An equivalent electronic system has been developed such that a current (range 4–20 mA d.c.) is generated by the sensor and transmitted. The current is proportional to the span of the measured quantity. This system has negligible delay and response lag, and there is sufficient power below the live zero (i.e. 4 mA) to operate the sensing device. Electronic force-balance transmitters are available in many different designs. However, the basic principle is to measure the force produced in a primary element, e.g. a force bar, resulting from the applied pressure to be measured.

Force-measuring pressure transmitters have also been developed. They measure pressure by

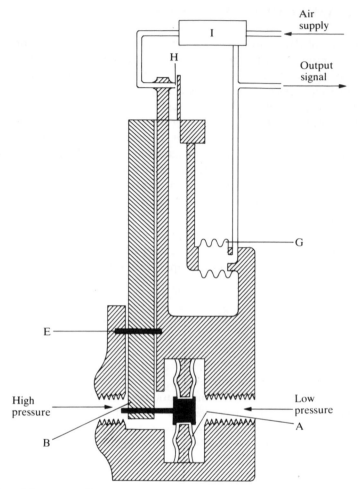

Figure 3.11 Differential-pressure cell.

measuring the deflection of an elastic member. One example is the use of a pre-stressed wire situated in the field of a permanent magnet. A force is applied to the wire which oscillates at its resonant (or natural) frequency. The output voltage is proportional to the tension in the wire, and hence proportional to the measured pressure.

3.9 VACUUM MEASUREMENT

Vacuum is the range of pressures below atmospheric. Atmospheric pressure is often used as a reference and lower pressures are expressed as *mm Hg vacuum* (for example). However, this is often inconvenient because of the changes in atmospheric pressure that occur; the use of zero pressure as the reference is preferable. Pressures expressed this way are referred to as *absolute* pressures. Many of the pressure gauges already described can also be used to measure vacuum pressures. *Absolute pressure gauges* measure the pressure by knowing the force per unit area, and the physical quantities associated with the gauge, e.g. the length. At very low pressures the force exerted by the gas is too small to be measured accurately and *non-absolute gauges* are used. These gauges must always be calibrated against an absolute gauge for each gas to be used. Non-absolute gauges measure the pressure indirectly by measuring a pressure-dependent physical property of the gas, e.g. thermal conductivity, ionization, viscosity, etc.

Absolute pressure (vacuum) gauges include mechanical gauges such as the Bourdon tube gauge and diaphragm gauge, liquid manometers and the McLeod gauge. Several of these instruments have been described previously in this chapter for the measurement of pressure.

Different types of thermal conductivity and ionization gauges are mentioned briefly here as examples of non-absolute vacuum gauges. The change in thermal conductivity of a gas or vapour can be used to measure its pressure in the range 10^{-1} Pa to 1000 Pa. These gauges are known as *hot-wire gauges* because an electrically heated wire is used as the sensitive element. A *thermocouple gauge* has a thermocouple attached to the centre of an electrically heated wire mounted inside a metal or glass envelope that is connected to a vacuum. The wires are welded so that one dissimilar pair form the hot wire and the other pair form the thermocouple. The *Pirani gauge* consists of an electrically heated wire (Pt or Ti) situated along the axis of a glass or metal tube that is connected to a vacuum. Pressure variations cause changes in the temperature of the wire, and hence changes in its electrical resistance which are measured. A dummy head gauge (or coil of wire) is used to compensate for changes in room temperature. The dummy head is sealed at low pressure. The *thermistor gauge* has a similar construction to the Pirani gauge except that the sensitive element is a small bead of semiconducting material (a mixture of metallic oxides) mounted on two platinum wires. The thermistor gauge has greater sensitivity and quicker response; it is smaller and requires less power.

Ionization gauges operate in the range 10^{-8}–10^3 Pa; they utilize the current carried by ions that are formed in a gas because of the impact of electrons. The *discharge-tube gauge* is a simple cold-cathode ionization gauge where the electrons are released from the cathode by the impact of ions. The gauge head is a glass tube (approximately 1 cm diameter and 15 cm long) containing a metal electrode (preferably aluminium) at each sealed end; it is connected to a vacuum source. A stable power supply is connected across the electrodes in series with a resistor (used to limit the current). During operation, a positive ion strikes the cathode and releases an electron which is attracted to the anode. This electron subsequently produces ionization when it contacts a gas molecule. Hence, more ions are attracted to the cathode. The range of operation is between 10^{-1} Pa (too few gas molecules to sustain the process) and 10^3 Pa (too many gas molecules). The *Penning ionization gauge* is simple, robust and sensitive; it operates in the range 10^{-5}–1 Pa. One form of this gauge has a cylindrical cold-cathode enclosed in a glass envelope (connected to vacuum) with a stiff wire anode along its axis. An axial magnetic-flux density is produced by a cylindrical permanent magnet or solenoid.

The *hot-cathode ionization gauge* is extremely sensitive and its reading is directly proportional to the pressure over the range 10^{-6}–100 Pa. A special triode valve contains a heated tungsten or thorium-coated iridium filament; this is enclosed by a molybdenum grid surrounded by a cylindrical nickel ion-collector. At pressures below 10^{-6} Pa, soft X-rays are generated when electrons strike the grid, and they produce a spurious current in the ion-collector circuit. The *Bayard–Alpert ionization gauge* is a modified design such that the ion-collector area (and hence the spurious current) is reduced by a factor of 10^3. It can, therefore, be used for pressures down to 10^{-9} Pa. The filament is mounted outside the cylindrical grid, having a fine tungsten-wire collector mounted along its axis. The glass envelope has a transparent conducting coating.

EXERCISES

3.1 Define the following terms:
(a) force;
(b) stress;
(c) pressure.

3.2 Explain what is meant by, and distinguish between:
(a) absolute pressure;
(b) gauge pressure;
(c) atmospheric pressure;
(d) differential pressure;
(e) vacuum.

3.3 Obtain values for a pressure of 1 atmosphere in units of:
(a) Pa
(b) N/m^2
(c) kN/mm^2
(d) mm Hg
(e) in. H_2O
(f) bar
(g) p.s.i.

3.4 What units are used for the measurement of vacuum pressure? Explain clearly how vacuum readings are measured.

3.5 List the different methods that are commonly used for the measurement of pressure. Give examples of instruments that use each method.

3.6 Describe the use of a simple U-tube manometer, a manometer with one and two wet-leg connections, and a well-type manometer.

3.7 What are the advantages of a McLeod gauge for pressure measurement? Describe the method of operation.

3.8 When is an inclined-limb manometer used?

3.9 Describe the operation of a piston-type dead-weight tester.

3.10 Describe the principle of operation of a Bourdon gauge. State the advantages of this type of pressure-measuring instrument.

3.11 Explain how pressure signals are transmitted.

3.12 Describe some devices for vacuum measurement.

3.13 Examine the pressure-measuring and vacuum-measuring instruments that are used in your laboratory; identify their method of operation.

3.14 Obtain a range of catalogues and technical literature related to pressure measurement from several instrument manufacturers. Study this material and the descriptions given in instrument handbooks. Become aware of new developments in this field.

BIBLIOGRAPHY

See also the Combined Bibliography to Chapters 3, 4 and 5 which follows Chapter 5.
Abbreviations used to designate standards organizations are listed at the beginning of the book (page xiii).

British Standard

BS 3127: Specification for ferrous and non-ferrous Bourdon tubing.

US Standards

Pressure gauges

ASME B40.1–85: Gauges—Pressure Indicating Dial Type—Elastic Element.
ASME PTC 19.2–64, Part 2: Pressure Measurement Instruments and Apparatus (Performance Test Code).
SAE ARP 427–58: Pressure Ratio Instruments.
SAE AS 411A–63: Manifold Pressure Indicating Instruments.
UL 404–79: Gauges, Indicating Pressure, for Compressed Gas Service (30 July 1979).

Pressure regulators

AGA Z21.18–81: Gas Appliance Pressure Regulators.
CGA E–7–83: Flowmeters, Pressure Reducing Regulators, Regulator/Flowmeter and Regulator/Flowgauge Combinations for the Administration of Medical Gases.
NFP(A) B93.13M–81: Pneumatic Fluid Power-Pressure Regulators—Industrial Type.
NFP(A) T3.12.3 R1–78: Fluid Power Industrial Type Air Line Pressure Regulators.

Pressure transducers

ASME MC88.1–72: Guide for Dynamic Calibration of Pressure Transducers.
ISA S37.3–75: Specifications and Tests for Strain Gauge Pressure Transducers.
ISA S37.6–76: Specifications and Tests for Potentiometric Pressure Transducers.
ISA S37.10–75: Specifications and Tests for Piezoelectric Pressure and Sound-Pressure Transducers.

Other sources

Carpenter, L. G., *Vacuum Technology*, Adam Hilger, Bristol, 1970.
Lyons, J. L., *The Designer's Handbook of Pressure-Sensing Devices*, Van Nostrand Reinhold, New York, 1980.

FOUR

MEASUREMENT OF FLOW

CHAPTER OBJECTIVES

1. To explain the basic principles related to flow measurement.
2. To describe the principles of operation of some common types of flow-measuring instruments:

   Differential-pressure instruments, e.g. orifice plate, venturi, rotameter (variable-area meter), pitot
   Quantity meters
   Velocity meters
   Electronic flowmeters

3. To describe flow measurement in open channels.
4. To describe methods of calibrating flow-measuring instruments.

QUESTIONS

- What are the units of flow?
- What is the Reynolds number?
- What instruments are used to measure fluid flow?
- What are the principles of operation of these instruments?

4.1 DEFINITIONS AND BASIC PRINCIPLES

It is assumed that the reader has previously studied basic fluid mechanics. If not, or if a refresher is needed, reference should be made to one of the very many introductory texts on the subject (see Bibliography). This section states, with a minimum of explanation, some of the relevant background principles.

Flow can be measured either as an instantaneous velocity or a measured quantity over a period of time.

$$\text{Volumetric flow rate, } Q = \frac{\text{quantity}}{\text{time}} \; \frac{\text{m}^3}{\text{s}}$$

$$\text{Mass flow rate, } G = \frac{\text{mass}}{\text{time} \times \text{cross-sectional area } (A)} \; \frac{\text{kg}}{\text{s m}^2}$$

$$\text{Therefore, } G = \frac{Q\rho}{A}$$

$$\text{Also, } Q = \text{average velocity} \times \text{cross-sectional area}$$

$$Q = \bar{V} A$$

If the total quantity is required, and the flow rate is subject to fluctuations, an electronic or mechanical integrator can be incorporated in the measuring instrument (i.e. integration of the area under the graph of flow rate against time).

The *Reynolds number* (Re) is a dimensionless group used to characterize the flow of fluids in a particular channel. The Reynolds number is discussed in more detail in Chapter 8. At low flow rates, i.e. corresponding to $Re < 2000$, the flow is streamlined or laminar and all the fluid particles move parallel to the wall. Above a critical fluid velocity ($Re > 4000$ for flow in a pipe), the fluid particles also have a transverse velocity and mixing occurs. This is known as turbulent flow. At intermediate Reynolds numbers, there is a critical transition region.

For flow in a pipe, the fluid has a parabolic velocity profile for laminar flow and a flat velocity profile (velocity at centre approximately 1.2 times the mean velocity) for turbulent flow.

Viscosity is the frictional resistance in a flowing fluid such that the fluid at the pipe walls is stationary, and the fluid at the centre of the pipe has maximum velocity. In the SI system of units, the *dynamic viscosity* is expressed in units of $N \, s/m^2$ and the *kinematic viscosity* (defined as dynamic viscosity/density, both measured at the same temperature) is expressed in m^2/s. The corresponding c.g.s. units are the poise (0.1 kg/m s) or the poiseuille (1 $N \, s/m^2$), and the stokes (1 St $= 10^{-4} \, m^2/s$).

A moving fluid may possess the following types of energy:

(a) Pressure energy due to the pressure of the fluid.
(b) Kinetic energy due to the motion of the fluid.
(c) Potential energy due to the position or height of the fluid above a fixed reference level.
(d) Internal energy due to the temperature, i.e. heat energy; if there is a frictional resistance to flow, other forms of energy will be converted into heat energy.

The total energy possessed by a moving fluid is the sum of these four types of energy. For laminar flow in a closed channel as shown in Fig. 4.1, the total energy at any two sections must be equal. Therefore, for 1 kg of fluid:

$$Z_1 g + \tfrac{1}{2} V_1^2 + P_1 v_1 + I_1 = Z_2 g + \tfrac{1}{2} V_2^2 + P_2 v_2 + I_2$$

If the temperature is unchanged, $I_1 = I_2$; then

$$Z_1 g + \tfrac{1}{2} V_1^2 + P_1 v_1 = Z_2 g + \tfrac{1}{2} V_2^2 + P_2 v_2$$

This equation is known as *Bernoulli's theorem* and applies to ideal gases and liquids.

For *liquids only*, which may be regarded as incompressible, the specific volume (v) and

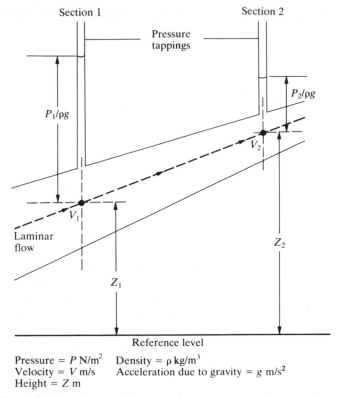

Pressure = P N/m^2 Density = ρ kg/m^3
Velocity = V m/s   Acceleration due to gravity = g m/s^2
Height = Z m

Figure 4.1 Laminar flow of fluid in a pipe.

density (ρ) are unchanged:

$$v_1 = v_2 = \frac{1}{\rho_1} = \frac{1}{\rho_2} = \frac{1}{\rho}$$

Therefore

$$Z_1 g + \tfrac{1}{2}V_1^2 + \frac{P_1}{\rho_1} = Z_2 g + \tfrac{1}{2}V_2^2 + \frac{P_2}{\rho_2}$$

or

$$Z_1 + \frac{1}{2}\frac{V_1^2}{g} + \frac{P_1}{\rho g} = Z_2 + \frac{1}{2}\frac{V_2^2}{g} + \frac{P_2}{\rho g}$$

As shown in Fig. 4.1, if a hole is drilled in the channel at section 1 and a gauge tube attached, then liquid will rise in the tube to a height that balances the static pressure P_1. The liquid column will be a distance $[(P_1/\rho g) + Z_1]$ above the reference plane. Similarly at section 2, the liquid column will rise to a height of $[(P_2/\rho g) + Z_2]$ above the reference plane. The differential head (h) is the difference in level between the gauge tubes, and is given by

$$h = \left(\frac{P_1}{\rho g} + Z_1\right) - \left(\frac{P_2}{\rho g} + Z_2\right)$$

Substituting from Bernoulli's theorem:

$$h = \frac{V_2^2}{2g} - \frac{V_1^2}{2g}$$

or

$$V_2^2 - V_1^2 = 2gh$$

The volume of *liquid* (Q) flowing along the channel is given by

$$Q = A_1 V_1 = A_2 V_2$$

where A is the cross-sectional area. Therefore

$$V_1 = \frac{A_2 V_2}{A_1}$$

Substituting into the above equation:

$$V_2^2 - \frac{V_2^2 A_2^2}{A_1^2} = 2gh$$

or

$$V_2^2 \left[1 - \frac{A_2^2}{A_1^2} \right] = 2gh$$

$$V_2 = \frac{\sqrt{(2gh)}}{\sqrt{\left(1 - \dfrac{A_2^2}{A_1^2} \right)}}$$

The ratio (A_2/A_1) is often represented by the symbol m; the expression

$$\sqrt{\left(1 - \frac{A_2^2}{A_1^2} \right)} = \sqrt{(1 - m^2)}$$

is known as the *velocity of approach factor* (symbol E). Therefore

$$V_2 = E\sqrt{(2gh)}$$

and

$$Q = A_2 V_2 = A_2 E\sqrt{(2gh)}$$

In terms of the differential pressure ($\Delta P = h\rho$):

$$Q = A_2 E\sqrt{(2g\,\Delta P/\rho)}$$

and the mass flow ($\dot{m}$, kg/s) is

$$\dot{m} = Q\rho = A_2 E\sqrt{(2g\rho\,\Delta P)}$$

These equations only apply to laminar (streamlined) flow. In order to take into account the effects of turbulence and viscosity, another factor is introduced into the flow equations. This factor is the *discharge coefficient* (C), defined by

$$C = \frac{\text{actual mass flow rate}}{\text{theoretical mass flow rate}}$$

The coefficient can be defined in terms of volume flows if the density, temperature, etc., are constant. The discharge coefficient is a function of pipe size, type of pressure tappings and Reynolds number; its value is determined experimentally. The modified flow equation ($Re > 20\,000$) becomes

$$Q = CA_2E\sqrt{(2g\,\Delta P/\rho)}$$

For lower Reynolds numbers and for very small or rough pipes, a correction factor (Z) must also be applied. The value of this factor also depends upon the area ratio. Values of C and Z and other relevant data are given in British Standard BS 1042: Part 1. The corrected flow equation becomes

$$Q = CZA_2E\sqrt{(2g\,\Delta P/\rho)}$$

Gases (unlike liquids) are compressible and in order to modify the flow equations to apply to gases, certain gas laws are applied. These laws apply to ideal gases (most gases not near their critical temperatures and pressures), and are stated as follows for dry gases.

Boyle's law: the volume (v) of a given mass of gas is inversely proportional to the absolute pressure (at constant temperature). Therefore

$$P_1v_1 = P_2v_2$$

Charles' law: the volume of a given mass of gas is directly proportional to the absolute temperature (T, kelvin), that is

$$\frac{v_1}{T_1} = \frac{v_2}{T_2}$$

The ideal gas law combines Boyle's law and Charles' law (for a given mass of gas):

$$\frac{P_1v_1}{T_1} = \frac{P_2v_2}{T_2}$$

or

$$\frac{Pv}{T} = \text{constant}$$

For n moles of gas, this constant is the *universal gas constant* (R) and the equation becomes

$$Pv = nRT$$

where P is N/m^2, v is m^3, and $R = 8.314\,J/mol\,K$.

Adiabatic expansion occurs due to a rapid pressure change and a change in temperature; in this situation Boyle's law does not apply. The appropriate equation is

$$P_1v_1^\gamma = P_2v_2^\gamma$$

or

$$Pv^\gamma = \text{constant}$$

where γ is the ratio of the specific heats (heat capacities) of the gas at constant pressure (C_p) and constant volume (C_v), and is defined by

$$\gamma = C_p/C_v$$

Values of γ are 1.40 for dry air and other diatomic gases, 1.66 for monatomic gases (e.g. helium) and approximately 1.33 for triatomic gases (e.g. carbon dioxide).

For compressible fluids, the change in volume due to expansion through a restriction requires use of the *expansibility factor* (ε) in the flow equations. Details are given in BS 1042: Part 1, including values of the pressure ratio (r) for flow through a convergent tube, e.g. a nozzle. The maximum flow rate occurs when the value of the critical pressure ratio (r_c) is approximately 0.5.

Carbon dioxide does not behave as an ideal gas, unlike most other gases at absolute pressures less than 10 bar. For departure from the ideal gas laws, a *deviation coefficient* (K) is used to calculate gas densities (see BS 1042: Part 1).

All of these equations apply to dry gases and a correction factor must be applied to account for the presence of any water vapour. This correction is necessary because the partial pressure due to saturated water vapour does not obey Boyle's law.

4.2 CLASSIFICATION OF MEASUREMENT METHODS

The instruments and techniques used for the measurement of fluid flow are classified here in the following categories.

(a) *Differential-pressure devices*
   (i) Orifice plate, venturi meter, nozzles and Dall tube—all of which operate on the principle of measuring the pressure drop across a flow restriction in a pipe (fixed-orifice meters).
   (ii) Rotameter, gate meter, Gilflo element and target meter—examples of variable-orifice (or variable-area) meters.
   (iii) Point-velocity measurement, e.g. pitot tube, hot-wire anemometer, etc.

(b) *Quantity meters*
   A flowmeter of this type would measure either the mass flow rate or volumetric flow rate (for gases and liquids), e.g. collecting tanks, displacement meters, rotating impellers.

(c) *Velocity meters*
   These devices are also referred to as 'rate-of-flow' meters; they are used to measure the velocity of a flowing liquid or gas. Most instruments of this type operate with a rotating vane inside the meter through which the fluid flows.

(d) *Electronic flowmeters*
   Either the principle of operation is electronically based or the primary sensor uses an electronic device.
   (i) Electromagnetic flowmeters.
   (ii) Ultrasonic flowmeters.
   (iii) Oscillatory flowmeters.

(e) *Flow in open channels*
   (i) Head–area methods, e.g. weirs, hydraulic flume.
   (ii) Velocity–area methods.

There are many different types of flow-measuring instruments available; a few of the more common devices are described here. Only basic details are given and the reader should refer to either specialist instrument handbooks, or manufacturers' literature, or appropriate standards for additional information.

4.3 DIFFERENTIAL-PRESSURE INSTRUMENTS

Measurement of the pressure drop in a pipe (caused by a constriction) is the most common method of obtaining the fluid flow rate. The differential pressure depends upon the fluid density and the flow velocity (a square root relationship as shown previously in Sec. 4.1). This type of flowmeter consists of an element that causes the pressure drop and a pressure-measuring device (e.g. a pressure transducer, see Chapter 3).

4.3.1 Fixed-orifice meters

The *orifice plate* is a thin steel plate with a square-edged circular orifice located in the centre of the plate. The plate is clamped between adjacent flange fittings in the pipeline. The build-up of solids and formation of gas pockets can be avoided by including a vent hole and a drain hole in the

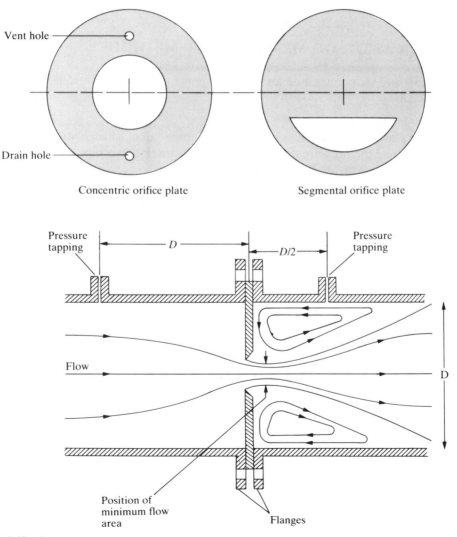

Figure 4.2 Orifice flowmeter.

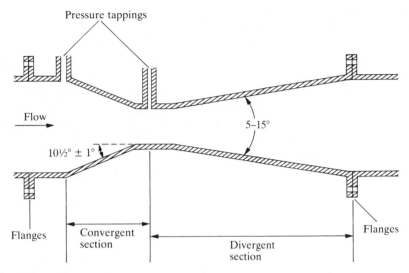

Figure 4.3 Venturi flowmeter.

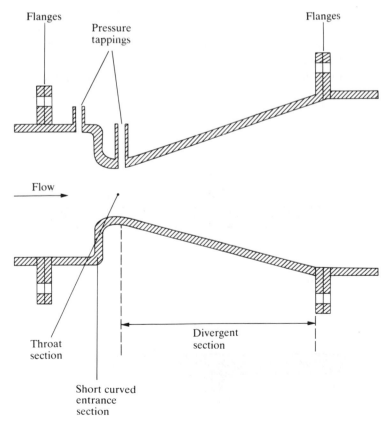

Figure 4.4 Venturi nozzle.

plate. The device is shown in Fig. 4.2. Pressure tappings are located on the pipeline at each side of the plate; the position of these tappings (e.g. D or $D/2$) depends upon the particular application (refer to BS 1042: Part 1). A segmental orifice (Fig. 4.2) is used for liquids having solids in suspension. The opening is positioned in the lower portion of the pipe to allow passage of the solids. An eccentric orifice is used for measurements with liquids containing undissolved gases, and for gases containing liquid condensate. This type of orifice is useful for pipeline drainage.

The *venturi tube* (shown in Fig. 4.3) consists of a cylindrical inlet, a convergent section, a cylindrical throat section and a divergent outlet section. Complete details of the construction and operation of a venturi tube are given in BS 1042, Part 1. This standard includes details of the dimensions of each section expressed in terms of the throat diameter and the entrance diameter, and the angle of each tapered section. The velocity of the flowing fluid increases in the convergent section and the differential pressure between the inlet and the throat is measured. The measurement of pressure and the flow equation to be used are as described for the orifice plate. The locations of the pressure tappings are given in BS 1042 Part 1. The pressure holes should be large enough to prevent blockages occurring; several equally spaced holes connected together as an annular piezometer ring may be used to provide a true mean pressure reading at each section. The divergent outlet section provides for a high pressure recovery, i.e. a low head loss, and the venturi is particularly suitable for flow measurements when there is a high solids content.

Nozzles are shorter (and cheaper) versions of the venturi tube. The *venturi nozzle* has a shortened convergent entrance section with a curved profile as shown in Fig. 4.4. The flow nozzle is further reduced in length, having a bell-shaped entrance section and no exit cone, as shown in Fig. 4.5. The nozzle is preferable to the orifice plate for high-velocity fluids, as it produces a lower pressure drop for the same area ratio. There is less flow resistance due to the smooth entrance cone, although the nozzle is not suitable for use with viscous liquids. The main advantages of nozzles are their compact size and reduced cost.

The *Dall* tube shown in Fig. 4.6 is also a shortened version of the venturi tube, being

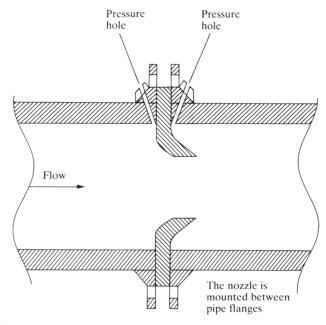

Figure 4.5 Flow nozzle.

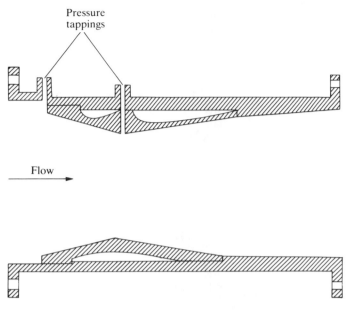

Figure 4.6 Dall tube.

approximately two pipe diameters long. A *Dall orifice* (*or insert*) is only about 0.3 pipe diameters long, as shown in Fig. 4.7.

Summary
The advantages of the orifice plate and venturi tube are simple operation, long-term reliability and no moving parts; the disadvantages are the square root pressure–velocity relationship, poor turn-down ratio and critical installation requirements. The differences between the devices are the low head loss of the venturi and the cheapness of the orifice plate. All the devices described so far suffer from some degree of pressure loss which should be minimized. This factor often influences the selection of a particular instrument, venturi tubes and nozzles having superior pressure recovery. Correct installation of these instruments is critical and they should be located as far downstream as possible from bends, valves and reducers.

4.3.2 Variable-area meters

A *rotameter* is an example of a variable-area meter; it maintains a nominally constant differential pressure by allowing the flow area to increase with flow rate. This type of instrument is sometimes referred to as a 'variable-orifice' meter (and may be classified as a variable-area–constant-head method of flow measurement). A rotameter consists of a long graduated tube having a uniform taper (usually the narrowest section at the bottom), with the tube axis vertical. The float moves freely within the tube; it is maintained centrally within the tube by a series of guides. A schematic diagram is shown in Fig. 4.8.

As the rate of flow through the instrument increases, the float (or plug) rises in the tube and the annular flow area increases; the differential pressure across the float remains constant. When the flow rate is steady, the float is in equilibrium under the action of three forces: the force of gravity due to the float mass, the upthrust of the fluid, and the difference between the forces on the upper and lower surfaces of the float. This assumes that the forces due to the viscosity of the fluid,

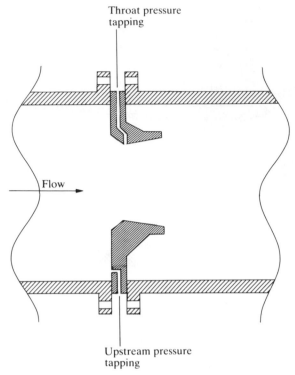

Throat pressure
tapping

Flow

Upstream pressure
tapping

Figure 4.7 Dall orifice (or insert).

and the tendency of the fluid to adhere to the float surfaces which are parallel to the stream, are small and can be neglected. Let

$$\rho_s = \text{density of float (kg/m}^3)$$

$$\rho' = \text{density of fluid (kg/m}^3)$$

$$v_s = \text{volume of float (m}^3)$$

$$A_2 = \text{orifice area (m}^2)$$

$$A_e = \text{effective area of float perpendicular to the fluid flow (m}^2)$$

$$P_1 = \text{impact pressure on lower surface (N/m}^2)$$

$$P_2 = \text{downstream static pressure on upper surface (N/m}^2)$$

then

$$\begin{array}{c} \text{force due to} \\ \text{the float} \end{array} = \left[\begin{array}{c} \text{pressure} \\ \text{difference} \end{array} \times \begin{array}{c} \text{effective} \\ \text{area} \end{array} \right] + \begin{array}{c} \text{upthrust} \\ \text{on float} \end{array}$$

$$v_s \rho_s g = (P_1 - P_2)A_e + v_s \rho' g$$

Rearranging:

$$(P_1 - P_2)A_e = v_s(\rho_s - \rho')g$$

$$(P_1 - P_2) = \frac{v_s}{A_e}(\rho_s - \rho')g$$

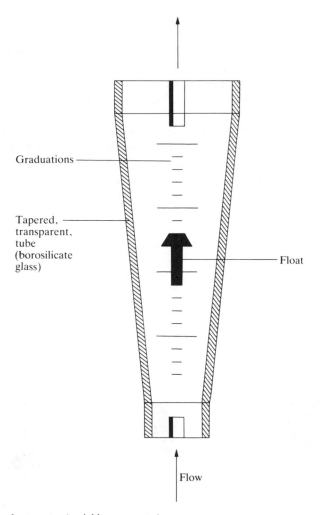

Figure 4.8 Schematic of rotameter (variable-area meter).

The velocity (V_2) of the fluid through the annular space is given by

$$\tfrac{1}{2}V_2^2 = \frac{(P_1 - P_2)}{\rho'}$$

therefore

$$V_2^2 = 2g\,\frac{v_s}{A_e}\,\frac{(\rho_s - \rho')}{\rho'}$$

The volumetric flow (Q) through the annular space is

$$Q = CA_2 V_2$$

$$= CA_2 \sqrt{\left(2g\,\frac{v_s}{A_e}\,\frac{(\rho_s - \rho')}{\rho'}\right)}$$

The discharge coefficient (C) depends upon the flow pattern within the tube, which is influenced by the fluid viscosity and flow rate. The effect of the float shape upon the flow pattern in the fluid stream is illustrated in Fig. 4.9. For a simple float (plumb bob type), the value of C increases rapidly as the Reynolds number (Re) increases up to about 7000; it then remains constant. The change in flow pattern due to increased velocity for this type of float is shown in Fig. 4.9(a). The liquid tends to carry the float along because of the action of the viscous forces; this effect can be reduced by using the float shown in Fig. 4.9(b). For this type of float, C is constant for $Re > 300$.

The float shown in Fig. 4.9(c) produces a constant value of C and constant flow pattern for $Re > 40$. This instrument is capable of giving a constant calibration for a very large range of flows and for a large range of fluid viscosities. It is possible to produce a series of instruments having a wide variety of ranges without the necessity of individual calibration (constant value of $C = 0.61$). In Fig. 4.9(c), the body of this float (which provides the downward force as a result of its mass) is outside the flowing liquid.

A rotameter can be arranged and give an indication of the flow rate (in mass units) which is independent of small changes in the fluid density; this is useful for fluids such as petrol. The mass flow rate ($\dot{m}$, kg/s) is given by

$$\dot{m} = CA_2 \sqrt{\left(2g \frac{v_s}{A_e} (\rho_s - \rho')\rho' \right)}$$

Hence, the mass flow rate depends upon $\sqrt{[(\rho_s - \rho')\rho']}$. If the float material is selected so that

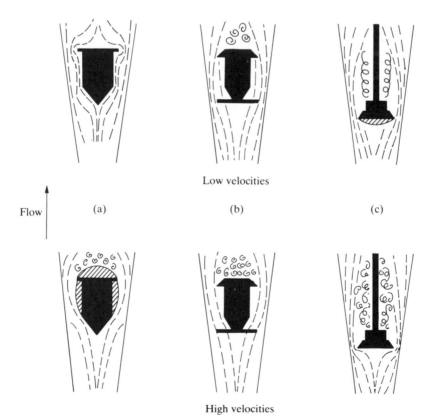

Figure 4.9 Effect of float shape upon flow pattern in the fluid stream.

$\rho_s = 2\rho'$, then an increase in ρ' will be compensated by a similar decrease in $(\rho_s - \rho')$. To make the flow measurement independent of small changes in ρ', the fluid must flow downwards. The term $(\rho_s - \rho')$ then becomes $(\rho' - \rho_s)$ and if the float is made of a very light material, $[(\rho' - \rho_s)\rho']$ remains approximately constant for small changes in ρ'.

Rotameters are available in a wide range of materials depending upon the corrosive nature of the metered fluid. The tube is often made of glass and the flow is measured by reading directly the position of the float in the tube. When a metal tube is used, e.g. non-magnetic stainless steel, a float magnet and an external follower magnet are used to transmit the movement of the float to a pointer. A wide range of tube sizes is available for a large variety of pressure ratings. The accuracy of a rotameter is normally ± 2 per cent of the full-scale flow over a 10:1 range of flows. Other rangeabilities are available and instruments having an accuracy of ± 2 per cent of the indicated flow are available. The rotameter requires little servicing except for the removal of deposits; this is achieved either by using a wash liquid or by dismantling. The sensitivity can be checked by using a process flow-control valve to produce small flow changes. The rotameter should be examined regularly for corrosion and the mass of the float checked for wear by abrasive action.

Other instruments that operate on the principle of a variable flow-area are the gate meter, orifice and plug meter, the Gilflo and the target meter.

In a *gate meter*, the orifice area is varied by raising or lowering a gate (either manually or by automatic control) to maintain a constant pressure drop across the orifice. The position of the gate is indicated by a scale; the positions of the pressure tappings are shown in Fig. 4.10(a). The flow increases more rapidly than the area and if the vertical movement of the gate is to be directly proportional to the rate of flow, the width of the opening must decrease towards the top as shown in Fig. 4.10(a). The flow can be made to depend directly upon the orifice area if the *impact pressure* is measured at the upstream tapping, as shown in Fig. 4.10(b). The open end of the pressure tapping faces directly into the flow, i.e. it acts as a pitot tube (described in Sec. 4.3.4). The flow equations used for other orifices also apply to this meter (Fig. 4.10b) but the velocity of approach factor is unity and the flow is directly proportional to the orifice area. The gate opening can therefore be rectangular, and the vertical movement is directly proportional to the flow rate.

A hinged gate meter uses a weighted gate situated in the flow stream; the deflection of the gate is proportional to the flow rate. The flow is indicated on a recorder using mechanical linkage between the gate and the recorder head. The main application of this device is in water mains where step changes are of more interest than the absolute accuracy of the flow measurement.

A simple *orifice-and-plug meter* is shown schematically in Fig. 4.11 with a tapered plug fitting into a circular orifice. The plug movement is a measure of the flow rate. The effective area of the plug changes as the area of the annular space increases, and the flow is not directly proportional to the annular area. However, the shape of the plug can be chosen so that the flow rate is directly proportional to the lift of the plug. The cone-and-disc type of meter is an alternative form having a cone or disc which moves in a conical chamber. The mass of the disc is balanced by the differential pressure across it. The chamber is shaped so that the flow is proportional to the rise of the disc.

The *Gilflo meter* is a variable area–variable head meter. It was developed to avoid the restrictions due to the square-root law for fixed-orifice meters $(Q \propto \sqrt{(\Delta P)})$. The Gilflo 'A' meter is shown in Fig. 4.12(a), it is available in sizes from 10 mm to 40 mm and has a fixed cone. A movable orifice is mounted on a strong linear bellows which is fixed at one end. When gas or liquid flows, the orifice moves axially along the cone thus creating a variable annulus. The differential pressure is proportional to the flow rate, producing a measurable range of 100:1. The Gilflo 'B' meter shown in Fig. 4.12(b) (sizes 40 mm to 300 mm) has a fixed orifice and a movable cone. The cone moves against the resistance of a spring; it also produces a linear differential

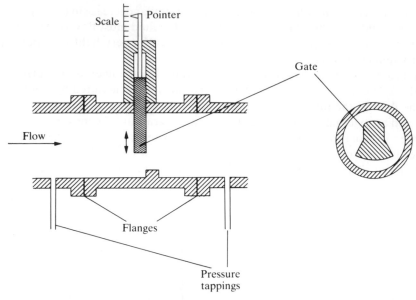

(a) Shape of the gate makes vertical movement proportional to flow

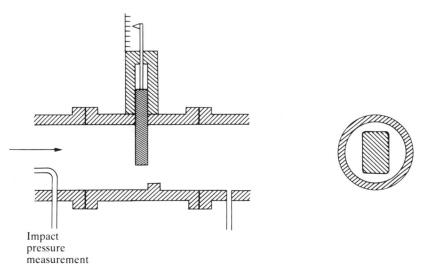

(b) Impact pressure measurement makes the flow reading proportional to orifice area
(hence a rectangular shaped gate)

Figure 4.10 Gate-type area meter.

pressure. Gilflo meters are used for saturated and superheated steam systems, with pressures up to 200 bar and temperatures of 500°C.

4.3.3 Target flowmeters

Target flowmeters are not strictly differential pressure meters although they are usually categorized as such. The primary device and the responsive element form an integral unit and

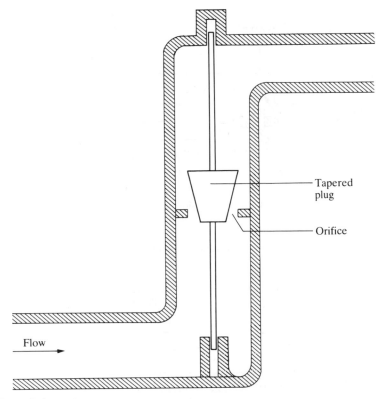

Figure 4.11 Orifice-and-plug meter.

pressure tappings are not required. These meters are used for high viscosity liquids, hot oils and slurries, for pressures up to 100 bar and Reynolds numbers as low as 2000. The principle of operation of a target flowmeter is shown in Fig. 4.13. The liquid impinging on the target is brought to rest, so that the pressure increases by $(V^2/2g)$ in terms of the head of liquid. The force on the target is balanced through a force bar by the air pressure in the bellows; a 0.2 bar to 1 bar signal is obtained (proportional to $\sqrt{Q}$). The circular square-edged target is positioned exactly concentric with the pipe, thus forming an annular orifice. A leakproof seal isolates the process fluid from the upper section of the meter. The instrument requires the same length of straight pipe upstream and downstream as an orifice plate-type instrument having a large d/D ratio. Flow ranges of 0–120 litres/minute (up to 400°C) and 0–2250 litres/minute (up to 250°C) can be measured; meters are also available for measuring gas flows. The overall accuracy of the meter is ± 0.5 per cent with repeatability of ± 0.1 per cent.

4.3.4 Point-velocity measurement

The *pitot static tube* is used to measure the fluid velocity at a point; the average fluid velocity can be obtained by taking measurements at several points and hence the average flow rate can be calculated ($\bar{Q} = \bar{V} \times$ cross-sectional area). A pitot tube can be used to measure the total pressure in a fluid, comprising the static pressure and the impact (or velocity) pressure. The elementary type of pitot tube, as shown in Fig. 4.14, consists of a single-hole pitot tube and a separate static-pressure tapping. The pitot tube must withstand the impact of the fluid and any vibrations. An ellipsoidal-nosed pitot tube consists of two concentric tubes, the inner tube transmits the total

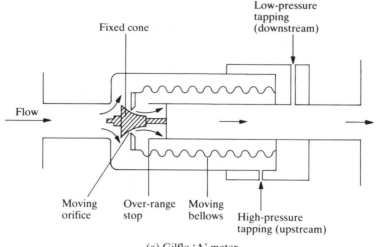

(a) Gilflo 'A' meter

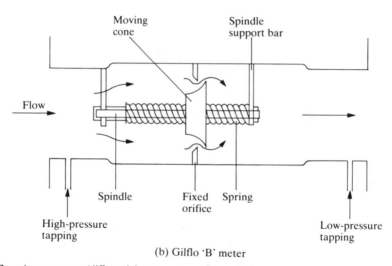

(b) Gilflo 'B' meter

Figure 4.12 Gilflo primary sensor (differential pressure-type flowmeter).

pressure (static and dynamic) and the annular space between the tubes transmits the static pressure. Several static pressure holes are equally spaced around the head.

The pitot tube is positioned with its open end facing the flowing fluid as shown in Fig. 4.14. The fluid impinging on the open end is brought to rest and its kinetic energy is converted to pressure energy. The pressure measured by the tube is greater than the static pressure, the difference between these measurements being a measure of the 'impact' pressure and, therefore, the velocity of the stream. The pressure differential (h) is given by

$$h = \frac{V_2^2}{2g} - \frac{V_1^2}{2g}$$

where V_1 and V_2 are the initial and final velocities of the small fluid stream impinging on the tube. Since $V_2 = 0$: $h = -V_1^2/2g$ (negative sign indicates a pressure increase). Therefore, $V_1 = \sqrt{(2gh)}$. Part of the fluid stream may be deflected and not brought to rest. The value of V_1 may not be the

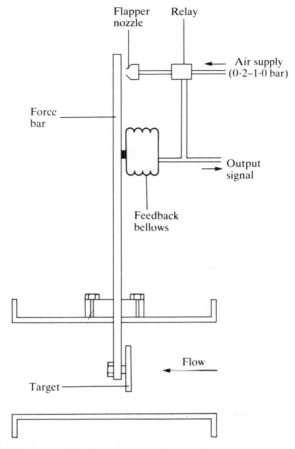

Figure 4.13 Principle of operation of a target flowmeter.

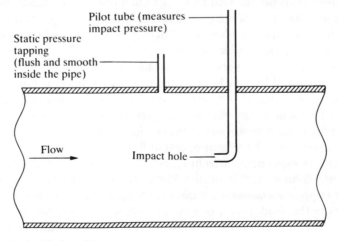

Figure 4.14 Simple pitot tube (single hole)

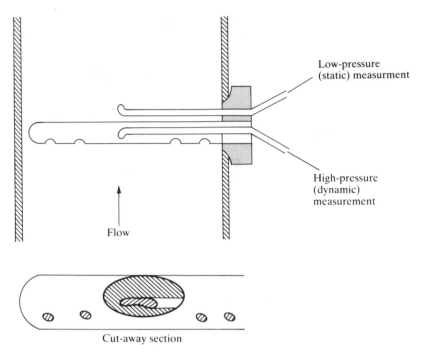

Low-pressure
(static) measurment

High-pressure
(dynamic)
measurement

Flow

Cut-away section

Figure 4.15 The annubar.

true velocity and a pitot tube coefficient (C) is introduced to compensate for this variation. The coefficient is unity for a correctly designed tube. If an overall accuracy of ± 1 per cent is required, the measurement conditions should conform to BS 1042: Part 2.

To find the average velocity across a pipe, it is assumed that the velocity profile follows a logarithmic law. Local velocities are measured on a ten-point (log–linear) traverse; the average velocity is the sum of the individual velocities divided by 10. Good results can be obtained even if the velocity profile does not obey the logarithmic law, provided a large number of diameters are traversed. Readings should be obtained for at least two perpendicular diameters. Several practical precautions must be observed, including single-phase flow, the use of two diametrically opposed holes so that the pitot tube is not inserted by more than half a pipe diameter, and the diameter of the pitot tube head less than 1/25 of the pipe diameter. The advantages of the pitot tube include low cost and negligible pressure loss. However, it may be difficult to obtain accurate readings and the differential pressure produced is usually small.

The *annubar* is often used for permanent installations and is illustrated in Fig. 4.15. The dynamic pressure is measured by four holes in the tube facing into the fluid stream; these holes measure the representative dynamic pressure of equal annuli. The inner tube is connected to the high-pressure side of a manometer and the low-pressure side is connected to a downstream element (measuring the static pressure minus the suction pressure). This instrument can measure the flow with an accuracy of ± 1 per cent of actual flow.

Other point-velocity measurement techniques are available for measuring the velocity of a fluid at a point or establishing the flow profile. Methods have been developed for pipes and open channels. The *laser Doppler anemometer* utilizes the Doppler shift of light that has been scattered by moving particles in the fluid. It is a non-contact technique used for gases and liquids, with particular application for flow around propellers and in turbines. The ultrasonic Doppler velocity probe is an adaptation of this method for open-channel flow. The *hot-wire anemometer* consists of a small electrically heated element placed in the fluid stream (gas or liquid). The cooling of the element as the flow increases causes a change in resistance which is proportional to the fluid velocity. The *electromagnetic velocity probe* consists of a field coil (which generates an electromagnetic field) and two electrodes. These electrodes detect the voltage generated which is proportional to the point velocity. The point velocity can also be measured by inserting either a small turbine, a five-bladed rotor or propeller, or a vortex-shedding bluff body into the fluid stream.

4.4 QUANTITY METERS

Quantity meters usually produce a reading proportional to the total quantity that has passed in a known time. The fluid passes through the primary element in successive and (almost) completely isolated quantities by alternately filling and emptying a fixed (known) volume. The secondary element indicates the number of times this procedure occurs, usually on a counter or a suitably calibrated dial. Numerous quantity-measuring instruments are available; their classifications and some brief notes are presented here, but detailed descriptions of the different meters are not given. Most quantity meters are provided as 'closed' instruments and the reading is obtained directly from a counter or a dial. The student concerned with engineering experimentation usually needs to know the type of meter being used and how to convert the reading into a flow rate (and a suitable calibration method to check the accuracy and repeatability of the readings). Additional information concerning installation, correct operation, appropriate operating conditions and relevant precautions is best obtained from the manufacturer's handbook for a particular instrument. The reader concerned with selection of a meter should consult a specialist instrument handbook, and then obtain professional advice from a manufacturer.

The following classification can be applied to quantity meters. *Weighing meters*—either a container of known volume is alternately emptied and filled (with calibrations for temperature and density changes) or a counterbalance scale beam is used to indicate when a known mass of liquid has been collected. *Volumetric meters* are described in Secs 4.4.1 and 4.4.2.

4.4.1 Volumetric meters for liquids

(a) *Simple tank system*, float operated.

(b) *Positive-displacement meters* operate on the principle that as the liquid flows through the meter it moves a measuring element; this isolates the measuring chamber into a series of measuring compartments each of known volume. As the measuring element moves, these compartments are successively filled and emptied. For each complete cycle of the measuring element, a fixed quantity of liquid passes through the meter. These meters provide a high degree of accuracy and good repeatability. The main sources of error are due to differences in the values of temperature, density and viscosity of the liquid at the working conditions from the values that apply at the calibration conditions. The most common forms of positive-displacement meters are: rotary piston, reciprocating piston, nutating disc, fluted-spiral rotor, sliding vane, rotating vane, and oval gear.

4.4.2 Volumetric meters for gases

The volume of gas is usually measured in terms of the standard cubic metre, i.e. dry at 15°C and 101.325 kPa. A correction must be applied for other conditions

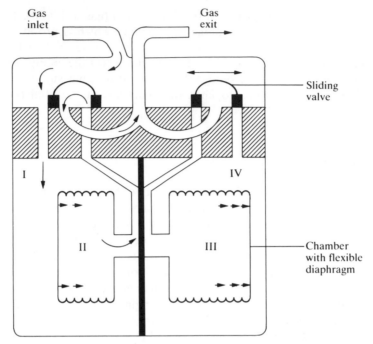

Figure 4.16 Operation of a diaphragm meter. Chamber I is filling; II is emptying; III is full; IV is empty.

(a) *The diaphragm meter* (*bellows type*) is used for metering the supply of gas to domestic and commercial users, and is also known as a *dry gas meter*. The meter consists of a metal case having an upper and lower section. The lower section consists of four chambers; two chambers are enclosed by flexible diaphragms that expand and contract as they are charged and discharged with gas. This type of meter is illustrated in Fig. 4.16. The quantity of gas is measured by the number of times the chambers are filled and emptied, and is recorded by counting the number of

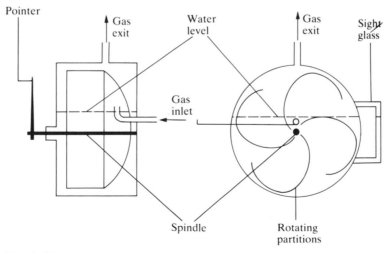

Figure 4.17 Liquid-sealed drum meter.

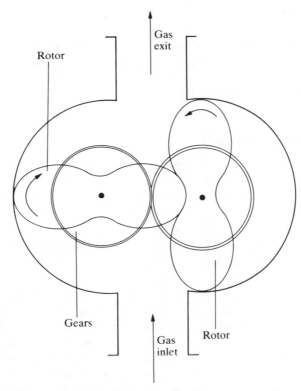

Figure 4.18 Rotating impeller-type meter (rotary displacement meter).

horizontal movements of the two diaphragms. This type of meter is highly accurate with trouble-free operation.

(b) *The liquid-sealed drum meter* differs from the dry gas meter ((a) above) by using water or another suitable liquid as the sealing medium, rather than a solid diaphragm. The instrument is shown in Fig. 4.17. The rotating part consists of shaped partitions forming four measuring chambers; they are balanced about a centre spindle that rotates freely. The measuring chambers are sealed off by water which fills the outer chamber to just above the centre line. The water level is critical and is adjusted so that when one chamber becomes open to the gas outlet, the partition (between it and the next chamber) isolates it from the gas inlet. This type of meter is unsuitable for high pressure or for gases that dissolve in water; the exit gas becomes saturated with water vapour.

(c) *The rotating impeller-type meter* could also be described as a two-toothed gear pump; the principle of operation is shown in Fig. 4.18. Rotation of the impellers is caused by the pressure drop at the outlet as the gas is used; the impellers are timed relative to each other by gears fitted to one or both ends of the impeller shafts. As an impeller passes through the vertical position, a volume of gas is trapped between the impeller and the casing. A counter records the rotation of the impellers, reading directly in cubic metres or cubic feet. The volume indicated by the meter must be corrected for pressure, temperature and compressibility. The leakage rate is very small (<1 per cent) and a correction can be applied. These meters are available for pressures up to 60 bar and for flow rates from 10 m³/hour to 10 000 m³/hour. The pressure drop across the meter is very small and an accuracy of ±1 per cent over a range from 5 to 100 per cent of maximum capacity can be achieved.

4.5 VELOCITY METERS

The velocity of a fluid moving through a pipe is greater at the centre and decreases near the pipe walls. The rate-of-flow meter is based upon the principle of measuring a velocity that has a constant relationship to the mean velocity across that section. The volume flow is $KAV\,\mathrm{m^3/s}$, where K is a constant for a particular pipe.

4.5.1 Rate-of-flow meters for liquids

(a) *The deflecting vane meter* uses a rectangular vane that is freely pivoted about its upper edge and is allowed to hang in a liquid stream. The angle through which the vane is deflected by the impinging liquid depends upon the liquid velocity; although this is not a linear relationship, it can be found experimentally. There is a significant head loss within this type of meter.

(b) *The rotating vane meter* uses a number of wings arranged around the circumference of a disc, or attached radially at intervals around a pivoted spindle. At least one wing is in the liquid stream at any time; the flowing liquid produces continuous rotation and the rate is a measure of the liquid velocity.

(c) *The helical vane meter* uses a hollow cylindrical vane (with accurately formed wings); it is mounted centrally in the body of the meter with its axis along the direction of flow. The meter is used for large flow rates in closed channels; it is available in sizes from 40 mm to 150 mm, having a maximum continuous flow rate from $15\,\mathrm{m^3/hour}$ to $200\,\mathrm{m^3/hour}$ respectively.

(d) *The turbine meter* consists of an almost friction-free rotor pivoted along the axis of the meter tube, and designed so that the rate of rotation is proportional to the flow rate of fluid. The speed of rotation is sensed by an electrical pick-up coil mounted in the meter housing.

(e) *The by-pass meter* (or shunt meter or combination meter) is used where widely fluctuating flows are encountered. A large meter such as a helical vane is used in the flow main, and a small rotary meter in the by-pass line. An automatic valve is used to direct the flow through the appropriate meter, depending upon the flow rate. This type of combination can provide reasonable accuracy over a 5:1 turn-down ratio, and is suitable for liquids, gases and steam.

4.5.2 Rate-of-flow meters for gases

(a) *The deflecting vane-type meter* (*velometer*) is similar to the instrument used for liquid measurement (described in Sec. 4.5.1(a)) except that the vane is either very much larger or very much lighter when used for gases. This is because gas densities are considerably less than for liquids, and the higher gas velocities (6 to 10 times) are insufficient to compensate for this effect.

(b) *Rotating vane-type meters.* An *anemometer* is used to measure wind speeds; it uses either light or large multiple vanes or cups. The rotor must be accurately balanced and the bearings nearly friction-free. The speed of rotation is proportional to the air speed.

The *rotary gas meter* is a development of the air anemometer. The measuring element comprises an internal tubular body that directs the gas through a series of circular ports on to a vaned anemometer. The meter also includes a multi-point index driven through the intergearing. It is used in industrial and commercial applications at pressures up to 1.5 bar and flows up to $200\,\mathrm{m^3/hour}$, giving an accuracy of ± 2 per cent over a flow range of 10:1.

(c) *Turbine meters* for gases operate on the same principle as for liquids (described in Sec. 4.5.1(d)) except that high gas velocities are required to turn the rotor blades.

4.6 ELECTRONIC FLOWMETERS

The instruments considered in this category have been the subject of major innovations and developments in recent years. Some improvements in flow-metering equipment are due to the development of more sophisticated and more manageable signal-detection and read-out equipment. It is likely that new methods, and adaptations of existing techniques, will continue to be developed. Electronic flowmeters either use an electronic device as the primary sensor or their principle of operation is electronically based. The techniques that are mentioned briefly here are electromagnetic, ultrasonic and oscillatory.

The principle of operation of an *electromagnetic flowmeter* is based upon Faraday's law of electromagnetic induction which states that: an electric conductor (i.e. the flowing liquid) moving in a magnetic field induces an electromotive force (e.m.f.) whose amplitude is dependent upon the force of the magnetic field, the velocity and the length of the conductor. An electromagnetic flowmeter consists of a primary device (also containing the flow channel), the measurement electrodes and the magnetic field coils, and a secondary device. The secondary device provides the field-coil excitation and amplifies the output of the primary device, converting it into a suitable form. The use of standard 50 Hz mains voltage as an excitation source for the field coils in the early flowmeters resulted in disadvantages such as interfering voltages, high power consumption and zero drift. Most of these problems have been overcome by using a low frequency (2–7 Hz) system. These flowmeters are suitable for a wide variety of liquids and the accuracy is unaffected by changes in the temperature, pressure, density, viscosity or conductivity of the liquid. The accuracy is affected by the flow profile, and a minimum of 10 straight pipe diameters upstream and 5 straight pipe diameters downstream of the primary element are required. Electromagnetic flowmeters are available in sizes from 32 mm to 1200 mm for flow velocities from 0–0.5 m/s to 0–10 m/s respectively, having an accuracy of ±1 per cent over a 10:1 turn-down ratio.

Ultrasonic flowmeters operate by monitoring the interaction between a flowing liquid and an ultrasonic sound wave; they measure the velocity of the fluid. Many techniques have been developed, the most common being the Doppler system and the pulse transmission system. The *Doppler flowmeter* comprises a unit containing two piezo-electric crystals (a transmitter and a receiver) that are located on the pipe wall. Ultrasonic waves are transmitted into the fluid at an angle and are reflected back by any discontinuities, e.g. solid particles or gas bubbles. The magnitude of the frequency change (between the transmitted and reflected waves) is proportional to the velocity of the fluid. The meter is inexpensive and can be fixed to the pipe wall (which must be acoustically transmissive); however, the fluid must contain some discontinuities and the accuracy and repeatability of measurements are doubtful. *In situ* accuracies of ±5 per cent should be possible, although the meter is more suitable for flow indication where absolute accuracy is not required.

Transmissive flowmetering devices operate by transmitting an ultrasonic pulse sound wave across the flowing fluid (at an angle) in each direction. The difference between the flight times is proportional to the fluid velocity. Discontinuities in the fluid are not required. The piezo-electric ceramic transducers on each side of the pipe act as both transmitter and receiver. Measurements are influenced by velocity profile effects across the pipe and suitable upstream and downstream straight pipe lengths are required. This flowmeter can be used with liquids and gases for pipe sizes from 75 mm to 1500 mm; an accuracy of ±1 per cent of the flow rate can be achieved over a flow range of 0.2–12 m/s.

Oscillatory flowmeters operate on the principle that an obstruction causes the fluid to oscillate in a predictable manner; the degree of oscillation is related to the flow area. The *vortex*

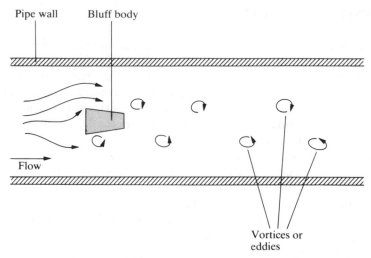

Figure 4.19 Principle of operation of the vortex flowmeter.

flowmeter uses a symmetrical bluff (non-streamlined) body to generate regular vortices (circular eddies) in the fluid stream as shown in Fig. 4.19. Under appropriate conditions, a regular vortex-shedding pattern is obtained alternately from each side of the body (situated centrally in the pipe and producing a pressure-feedback system). The vortex-shedding frequency provides a measure of the flow rate. Various sensing methods can be used to detect the vortices; these include ultrasonic, heated thermistor, oscillating disc, capacitance, and internal strain gauge methods. The output from the primary sensor is a low-frequency signal that is dependent upon the flow velocity. The meter output is independent of the pressure, temperature and density of the fluid and can be used for gases ($2 \times 10^3 < Re < 10^5$) and liquids ($4 \times 10^3 < Re < 1.4 \times 10^5$). An accuracy of ± 1 per cent of full scale can be achieved for turn-down ratios of 20:1.

The *swirlmeter* uses curved inlet blades to impart a swirl to the fluid (i.e. a tangential flow component) as it enters a converging section. As the rotating fluid enters an enlargement, the region of highest velocity rotates about the meter axis. The frequency of the oscillation (or precession) produced is proportional to the volumetric flow rate. A heated bead thermistor is used as the sensing element. The meter is available in different sizes for a range of flow rates (depending upon the specific application) for gases and liquids. An accuracy of ± 1 per cent of the flow rate can be achieved, with repeatability of ± 0.25 per cent of flow rate.

4.7 FLOW IN OPEN CHANNELS

Flow measurement in open channels, e.g. rivers and sewers (part-filled pipes), can be obtained by:

(a) *Head–area methods*, e.g. weirs and flumes, where a device is incorporated in the flow stream to develop a unique head–flow relationship.
(b) *Velocity–area methods* where both head and velocity are measured; the flow can be determined if the geometry of the structure is known. Examples are the turbine current meter, and electromagnetic and ultrasonic techniques.
(c) *Dilution gauging* is a technique of injecting a tracer element, e.g. salt solution or radioactive solution, and estimating the degree of dilution caused by the flowing liquid.

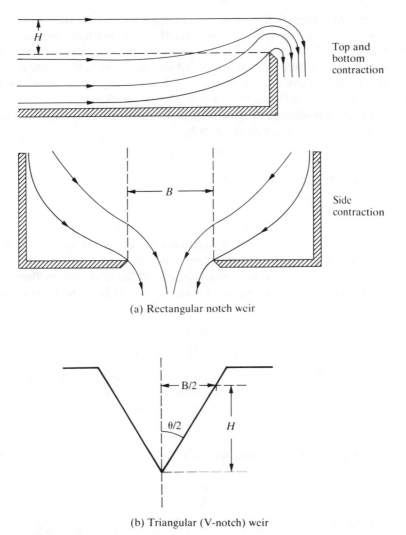

Top and
bottom
contraction

Side
contraction

(a) Rectangular notch weir

(b) Triangular (V-notch) weir

Figure 4.20 Flow in open channels.

Weirs are classified according to the shape of the notch or opening; the rectangular notch and the V-shaped (or triangular) notch are the most common. The weir is a dam over which liquid flows; the depth of liquid above the sill is a measure of the flow rate. The term 'head of a weir' refers to the height of the liquid above the sill, measured just upstream of where it begins to curve over the weir (denoted by H in Fig. 4.20a).

For a *rectangular notch weir*, if the cross-section of the stream approaching the weir is large compared with the steam area over the weir, then the velocity over the weir is given by $V = \sqrt{(2gh)}$. This assumes that the upstream velocity is negligible in comparison with V. The area of the stream is BH (Fig. 4.20a) and it can be shown (by calculus) that the flow (Q; m^3/s) over the weir is given by:

$$Q = \tfrac{2}{3}B\sqrt{(2gH^3)}$$

The actual flow is less because the stream contracts at both the top and bottom as it flows over the weir, i.e. H is reduced. Also there is friction between the liquid and the sides of the channel; this effect may be reduced by making the notch narrower than the width of the stream (Fig. 4.20a), although side contraction then occurs. A discharge coefficient (as defined in Sec. 4.1) must be included in the equation to account for the actual reduced flow; the value of the coefficient varies with H. If the upstream velocity cannot be neglected, a velocity of approach factor must be included; this will also influence the value of the discharge coefficient.

A *triangular notch weir* is shown in Fig. 4.20(b), and based upon this notation it can be shown (by calculus) that the flow over the weir is given by

$$Q = \tfrac{4}{15}B\sqrt{(2gH^3)}$$

Since $B = 2H \tan(\theta/2)$, the equation becomes

$$Q = \tfrac{8}{15} \tan(\theta/2)\sqrt{(2gH^5)}$$

A discharge coefficient should also be included in this equation. For satisfactory operation, the value of θ should be between $35°$ and $120°$. The triangular notch weir is used where low flow rates may occur; the head is greater than for a rectangular notch.

A *hydraulic flume* is used if only a very small head is available. A venturi flume (many other types exist) is shown in Fig. 4.21. Assuming that the channel is flat bottomed, the volume flow rate is given by

$$Q = Bh_2\sqrt{\left(\frac{2g(h_1 - h_2)}{1 - (Bh_2/B_1h_1)^2}\right)}$$

A discharge coefficient should also be included; its value depends upon the shape of the channel and the flow pattern. A carefully designed flume can be made to function as a free-discharge outlet so that the throat depth is maintained at a certain critical value. The flow rate is then given by

$$Q = kh_1^{3/2}$$

where k is a constant for a particular installation. In this situation only the upstream depth of liquid needs to be measured.

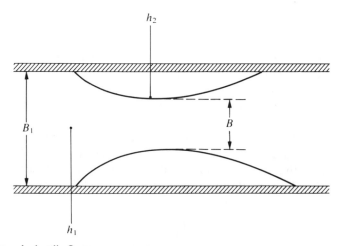

Figure 4.21 Venturi-type hydraulic flume.

4.8 FLOWMETER CALIBRATION METHODS

Many methods exist and these can be classified as *in situ* or laboratory methods. *In situ* methods used for liquids and gases include insertion-point velocity and dilution gauging. Laboratory methods for liquids include the master meter, volumetric or gravimetric methods, and the pipe prover; for gases the soap-film burette and water-displacement methods are used. Details of these methods are given in more detailed instrument handbooks.

EXERCISES

4.1 How are the velocity, volumetric flow rate and mass flow rate of a fluid related?

4.2 Explain the significance of the Reynolds number.

4.3 Why is the viscosity of a fluid important? What are the different units used for the measurement of viscosity?

4.4 List the types of energy that a moving fluid may possess.

4.5 Write down an equation for Bernoulli's theorem, and explain the meaning/significance of each term. Why and when is Bernoulli's theorem useful?

4.6 Explain the meaning of the terms:
(a) velocity of approach factor;
(b) discharge coefficient;
(c) laminar flow;
(d) turbulent flow;
(e) transition flow.

4.7 Write down equations describing:
(a) Boyle's law;
(b) Charles' law;
(c) the ideal gas law.

4.8 Define the universal gas constant (R).

4.9 Explain what is meant by adiabatic expansion.

4.10 List the different methods of measuring fluid flow, and give examples of instruments based upon each method.

4.11 Describe on an annotated diagram, for each type, the principle of operation of an orifice plate and a venturi meter. Compare and contrast the advantages and disadvantages of each meter. Describe the operation, and the advantages, of flow nozzles and the Dall tube.

4.12 Draw a schematic diagram showing the operation of a rotameter. What factors influence the selection of a particular float shape?

4.13 Describe the principles of operation of:
(a) a gate meter;
(b) an orifice and plug meter;
(c) Gilflo meters;
(d) the target meter.

4.14 Distinguish between the static and impact pressures of a fluid.

4.15 Describe the operation of a pitot static tube; list the advantages and disadvantages of this type of flow-measuring instrument.

4.16 List the different types of quantity meters that are available for gases and liquids. Describe the operation of a diaphragm meter, a liquid-sealed drum meter and a rotating impeller-type meter.

4.17 Describe briefly the principles of operation of several types of rate-of-flow meter.

4.18 Obtain manufacturers' information related to modern electronic flowmeters. Study this information and be aware of new developments in this field.

4.19 Explain the principles of measurement of flow in open channels.

4.20 How are flowmeters calibrated?

4.21 Examine the flow-measuring instruments that are used in your laboratory; identify their method of operation.

4.22 Obtain a range of catalogues and technical literature related to flow measurement from several instrument manufacturers. Study this material and the descriptions given in instrument handbooks. Become aware of new developments in this field.

BIBLIOGRAPHY

See also the Combined Bibliography to Chapters 3, 4 and 5 which follows Chapter 5.

British Standards

BS 1042: Methods of measurement of fluid flow in closed conduits.
   Part 1: Section 1.1: Orifice plates, nozzles and Venturi tubes inserted in circular cross-section conduits running full. Section 1.2: Specification for square-edged orifice plates and nozzle (with drain holes, in pipes below 50 mm diameter, as inlet and outlet devices) and other orifice plates and Borda inlets. Section 1.4: Guide to the use of devices specified in Sections 1.1 and 1.2. Part 2: Section 2.1: Method using Pitot-static tubes. Section 2.2: Method of measurement of velocity at one point of a conduit of circular cross section. Section 2.3: Method of flow measurement in swirling or asymmetric flow conditions in circular ducts by means of current-meters or pitot static tubes. Part 3: Guide to the effects of departure from the methods in Part 1.
BS 3680: Methods of measurement of liquid flow in open channels. Part 1: Glossary of terms. Part 2: Dilution methods. Part 3: Stream flow measurement. Part 4: Weirs and flumes. Part 5: Slope area method of estimation. Part 6: The measurement of flow in tidal channels. Part 7: The measurement of liquid level (stage). Part 8: Measuring instruments and equipment. Part 9: Water level instruments. Part 10: Sediment transport.
BS 5792: Specification for electromagnetic flow-meters.
BS 5844: Methods of measurement of fluid flow: Estimation of uncertainty of a flowrate measurement.
BS 5857: Methods for measurement of fluid flow in closed conduits using tracers. Part 1: Measurement of water flow. Part 2: Measurement of gas flow.
BS 5875: Glossary of terms and symbols for measurement of fluid flow in closed conduits.
BS 6199: Methods of measurement of liquid flow in closed conduits using weighing and volumetric methods. Part 1: Weighing method.

US Standards

Abbreviations used to designate the standards organizations are listed at the beginning of the book (page xiii).

Flow measurement

ASHRAE Handbook, 1985: Fundamentals; Chapter 2—Fluid Flow; Chapter 4—Two-Phase Flow Behavior.
ASHRAE 41.7–84: Method for Measurement of Flow of Gas.
ASHRAE 41.8–78: Standard Methods of Measurement of Flow of Fluid—Liquids.
ASME MFC–1M–79: Glossary of Terms Used in the Measurement of Fluid Flow in Pipes (R 1986).
ASME MFC–2M–83: Measurement Uncertainty for Fluid Flow in Closed Conduits.
ASME MFC–3M–85: Measurement of Fluid Flow in Pipes Using Orifice, Nozzle, and Venturi.
ASME MFC–5M–85: Measurement of Liquid Flow in Closed Conduits Using Transit-Time Ultrasonic Flowmeters.
ASTM C518–85: Test Method for Steady-State Heat Flux Measurements and Thermal Transmission Properties by Means of the Heat Flow Meter Apparatus.
ASTM D1941–67: Method for Open Channel Flow Measurement of Water and Waste Water by the Parshall Flume (R 1975).

ASTM D2458–69: Method of Flow Measurement of Water by the Venturi Meter Tube (R 1975).
ASTM D3857–79: Practice for Measurement of Water Velocity in Open Channels by Acoustic Means (R 1984).
ASTM D3858–79: Practice for Open-Channel Flow Measurement of Water by Velocity–Area Method (R 1984).
ASTM D4408–84: Practice for Open Channel Flow Measurement by Acoustic Means.
ISA RP3.2–60: Recommended Practice for Flange Mounted Sharp Edged Orifice Plates for Flow Measurement (R 1978).
ISA RP31.1–72: Recommended Practice for Specification, Installation, and Calibration of Turbine Flowmeters (R 1977).
SAE AS 407B–60: Fuel Flowmeters.
SAE AS 431A–62: True Mass Fuel Flow Instruments.

Gas meters

AGA B109.1–73: Gas Displacement Meters (500 Cubic Feet per Hour Capacity and Under).
AGA B109.2–80: Diaphragm Type Gas Displacement Meters (Over 500 Cubic Feet per Hour Capacity).
AGA Report No. 3 (1985): Orifice Metering of Natural Gas and Other Related Hydrocarbon Fluids.
ASME MFC–4M–86: Measurement of Gas Flow by Turbine Meters.

Pitot tubes

ASTM D3796–79: Practice for Calibration of Type S Pitot Tubes.
SAE ARP 920–68: Design and Installation of Pitot–Static Systems for Transport Aircraft.
SAE AS 390–63: Pitot or Pitot–Static Pressure Tubes, Electrically Heated (Turbine Powered Sub-Sonic Aircraft).

Rotameters

ASTM D3195–73: Practice for Rotameter Calibration.
ISA RP 16.1, 16.2, 16.3–59: Terminology, Dimensions and Safety Practices for Indicating Variable Area Meters (Rotameters).
ISA RP 16.1: Glass Tube.
ISA RP 16.2: Metal Tube.
ISA RP 16.3: Extension Type Glass Tube.
ISA RP 16.4–60: Recommended Practice for Nomenclature and Terminology for Extension Type Variable Area Meters (Rotameters).
ISA RP 16.5–61: Recommended Practice for Installation, Operation, Maintenance Instructions for Glass Tube Variable Area Meters (Rotameters).
ISA RP 16.6–61: Recommended Practice for Methods and Equipment for Calibration of Variable Area Meters (Rotameters).

Velocity measurement

ASTM D3154–72: Test Method for Average Velocity in a Duct (Pitot Tube Method).
ASTM D3464–75: Test Method for Average Velocity in a Duct Using a Thermal Anemometer.
ASTM D3857–79: Practice for Measurement of Water Velocity in Open Channels by Acoustic Means.
ASTM D3858–79: Practice for Open-Channel Flow Measurement of Water by Velocity–Area Method.
ASTM D4409–84: Practice for Velocity Measurements with Rotating Element Current Meters.

Water meters

AWWA C700–77: Cold Water Meters—Displacement Type.
AWWA C701–78: Cold Water Meters—Turbine Type for Customer Service.
AWWA C704–70: Cold Water Meters—Propeller Type for Main Line Applications.
AWWA C708–82: Cold Water Meters—Multi Jet Type.
AWWA M6–73: Water Meters—Selection, Installation, Testing and Maintenance.

Other sources

Cheremisinoff, N. P., *Applied Fluid Flow Measurement: Fundamentals and Technology*, Marcel Dekker, New York, 1979.
Cheremisinoff, N. P. and D. S. Azbel, *Fluid Mechanics and Unit Operations*, Butterworth, London, 1983.
Haywood, A. T. J., *Flowmeters: A Basic Guide and Sourcebook for Users*, Macmillan, London, 1979.
Kinsky, R., *Applied Fluid Mechanics*, McGraw-Hill, Sydney, 1982.
Miller, R. W., *Flow Measurement Engineering Handbook*, McGraw-Hill, New York, 1982.
Patterson, A. R., *A First Course in Fluid Dynamics*, Cambridge University Press, 1983.
Robertson, J. A. and C. T. Crowe, *Engineering Fluid Mechanics*, 2nd edn, Houghton Mifflin, Boston, MA, 1980.
Vennard, J. K. and R. L. Street, *Elementary Fluid Mechanics*, 6th edn, Wiley, New York, 1982.

MEASUREMENT OF TEMPERATURE

CHAPTER OBJECTIVES

1. To explain the basic principles related to temperature measurement.
2. To describe the principles of operation of some common temperature-measuring instruments.
3. To consider possible sources of measurement errors.
4. To consider the calibration of temperature-measuring instruments.

QUESTIONS

- What is temperature?
- What types of instruments are used to measure temperature?
- What are the principles of operation of these instruments?
- What errors can occur in temperature measurements?

5.1 BASIC PRINCIPLES

Temperature is probably the most frequently measured variable in industrial situations, and in the world in general. However, temperature is not readily defined in simple terms. It is very important to distinguish between temperature and heat. A substance possesses internal energy due to the motion of its molecules, and this energy is manifested in the temperature of the body. Consider two bodies that are in contact or in view of each other. If these bodies are at different temperatures, a transfer of internal energy occurs from the body at the higher temperature to the other at a lower temperature. The transferred energy is referred to as 'heat'.

The quantity of heat that a body contains depends upon its temperature, the mass of the body, and the nature of the material from which it is made. Heat flows from a body at a higher temperature to one at a lower temperature, even though the low-temperature body may contain more heat. This transfer of heat will continue until both bodies are at the same temperature. Changes in the temperature of a body can cause a wide variety of effects which may be categorized as physical, chemical, electrical and optical. Many of these effects are used as transducing effects in temperature-sensing devices.

In order to be able to compare the temperatures measured by different devices, some recognized fixed temperature points are required. These points must be constant in temperature and easily reproduced. Two suitable points are the melting point of ice and the temperature of condensing steam at a pressure of one standard atmosphere (101.325 kN/m^2). The temperature interval between these two fixed points is known as the *fundamental interval*; it is divided into a number of equal parts. On the Celsius scale (formerly called the Centigrade scale) the interval is divided into 100 parts, and the melting point of ice is designated 0°C.

Note: The boiling point of water is very dependent upon the applied pressure; its variation is given by the formula

$$T_s = 100 + 3.67 \times 10^{-2}(P - 760) - 2.3 \times 10^{-5}(P - 760)^2$$

where T_s is the boiling point (°C) and P is the total pressure (mm Hg).

5.2 TEMPERATURE SCALES

The basic unit of temperature is the kelvin (K); however, it cannot be represented by a single primary standard of either the 'fixed' or 'reproducible' type. The nature of temperature requires that the unit should be defined in terms of the *difference* between two standard temperatures.

In the 19th century, Lord Kelvin defined a temperature scale in terms of the mechanical work obtained from a reversible heat engine working between two temperatures. Unlike previous scales, this scale did not depend upon the properties of a particular substance. Kelvin also divided the interval between the ice and steam points into 100 parts, so that one Kelvin degree represented the same temperature interval as one Celsius degree. The only difference between the Celsius and Kelvin scales is that the zero point on the Kelvin scale is the *absolute zero of temperature* (-273.15°C, 0 kelvin), and on the Celsius scale it is 0°C (273.15 K). The Kelvin scale is known as the *absolute thermodynamic temperature scale*.

The unit of thermodynamic temperature is the kelvin (K), it is defined as 1/273.16 of the temperature interval between absolute zero (0 K) and the triple point of water (i.e. equilibrium between ice, water and steam, at 0.01°C).

The use of an engine to define temperatures is impractical, and thermodynamic temperatures are realized by using the ideal gas law: $PV = nRT$, where P is the absolute pressure (N/m^2), V is the volume (m^3), n is the number of moles of gases, R is the universal gas constant (J/mol K) and T is the thermodynamic temperature (K). 'Ideal' gases do not exist but the permanent gases (hydrogen, nitrogen, oxygen, helium) provide a close approximation at low pressures. For other gases (and for high pressures), a correction can be applied. Absolute thermodynamic temperatures can be measured by observing either the change in pressure of a given mass of gas at constant volume, or the change in volume of the gas at constant pressure.

The constant-volume gas thermometer is easier to use and simpler in construction than a constant-pressure instrument. Helium at a very low pressure is used to measure very low temperatures, hydrogen is used up to 500°C and nitrogen for measurement between 500°C and 1500°C. The constant-volume gas thermometer is used to establish thermodynamic temperature values over a range of approximately 20–1000 K, and comparisons are made with values on the practical temperature scale.

Note: Temperatures below 0°C are usually expressed in kelvin, and above 0°C in degrees Celsius.

Although the thermodynamic temperature scale is fundamental and necessary, the gas thermometer (the final standard of reference) is unsuitable for industrial use. However, temperature-measuring instruments capable of a very high degree of reproducibility are available, although the actual value of the thermodynamic temperature is not known with the same degree of accuracy. The temperature scales can be reproduced to a much higher degree of accuracy than they can be defined. The *International Practical Temperature Scale (IPTS)* provides a more

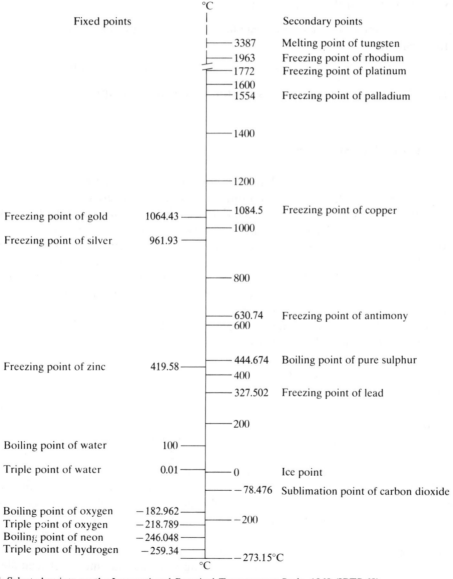

Figure 5.1 Selected points on the International Practical Temperature Scale, 1968 (IPTS 68).

readily available means of reference and calibration. This scale was adopted in 1929, and revised in 1948 and 1968 (*IPTS 68*).

Two standard temperatures would be sufficient to define the temperature scale if suitable methods existed for extrapolation beyond these temperatures. Because no satisfactory methods are available, eleven *primary fixed points* are defined. These points include the temperatures of the triple points of hydrogen, oxygen, and water; the boiling points of some pure liquids; and the freezing points of some pure metals. (These points should be referred to as 'temperatures of equilibrium' between the phases of the pure substances.) The boiling points and freezing points (with one exception) are all measured at the standard atmospheric pressure of 101.325 kN/m². *Secondary fixed temperature points* are also given in *IPTS 68*, and provide convenient laboratory calibration points for temperature-measuring devices. Some of the primary and secondary fixed points are given in Fig. 5.1.

Having specified certain fixed points, it is necessary to specify the method of measuring temperature values between the fixed points. It can be shown in the laboratory that some temperature-measuring methods have a more linear response over a particular temperature range, when compared with the thermodynamic scale. However the stability of the method is more important than its linearity for obtaining good repeatability at intermediate temperatures. The following methods are used for interpolating temperatures between (and above) the fixed points.

(a) For the temperature range 13.81 K to 630.74°C, the standard instrument specified (*IPTS 68*) is the platinum resistance thermometer. Below 0°C, the resistance–temperature relationship is defined by a reference function and specified deviation equations. From 0°C to 630.74°C, two polynomial equations define the resistance–temperature relationship. The constants in these equations are found by making resistance measurements at specified fixed points.

(b) From 630.74°C to 1064.43°C, a standard thermocouple of platinum–10 per cent rhodium/ platinum is used. The Celsius temperature (T_c) is defined by the equation

$$e = a + bT_c + cT_c^2$$

where e is the e.m.f. The constants (a, b, c) are calculated from the values of e at 630.74°C $\pm$ 0.2°C (determined by a platinum resistance thermometer) and also at the freezing points of silver and gold.

(c) Above 1064.43°C, the thermodynamic temperature scale is defined by Planck's law of radiation. This law relates the radiated energies of a black body at a particular temperature and at 1064.43°C, at the same wavelength of radiation (see Sec. 5.8.1).

5.3 METHODS OF MEASUREMENT

The effects caused by the change in temperature of a body, e.g. expansion, change of state, radiation emission, etc., and certain electrical phenomena are used as the bases for temperature measurement. The most common practical methods may be classified as follows.

(a) *Expansion* of solid, liquid or gaseous materials.
(b) *Chemical and physical changes.*
(c) *Electrical methods* including use of the thermo-electric effect and the change in electrical resistance of metals and semiconductors.
(d) *Radiation* and optical pyrometry.

The principles of these four classes of measurement are discussed in the following sections. Certain common instruments are also described.

5.4 EXPANSION THERMOMETERS

An increase in temperature causes most materials, whether solid, liquid, or gaseous, to expand. Many thermometers indicate the temperature of a body by direct observation of its increase in size, or by the signal from a secondary transducer used to detect this increase. It is necessary to provide some means of magnifying the small expansion of most solids and liquids so that the temperature can be measured accurately. This is usually achieved by mechanical indicating devices.

5.4.1 Expansion of solids

The expansion of a single metal, or a metal rod encased in a concentric metal tube, is used in a solid rod thermostat. This is a temperature-controlling device rather than a temperature-indicating thermometer. A typical instrument is shown diagrammatically in Fig. 5.2; it is widely used with water heaters and also to control the temperature of domestic ovens. The expansion of the frame or the support for the rod must also be taken into account.

Most thermometers using the principle of solid expansion employ a *bimetal strip*. This consists of two strips of different metals that are bonded together to prevent relative motion between the strips. The two metals, e.g. Invar and brass, have different coefficients of linear thermal expansion and an increase in temperature causes movement of the free end of the strip. A simple bimetal strip and the bending due to an increase in temperature are shown in Fig. 5.3 (assuming metal A has the higher coefficient of expansion). This form of the bimetal element is used in thermostatic switches, and as a temperature-compensating device in other instruments. For use as a measuring device, the sensitivity can be improved by incorporating a longer strip coiled in a helical form as shown in Fig. 5.4. A temperature change produces a twisting of one end relative to the other, and this moving end may be connected directly to an indicating pointer on a scale.

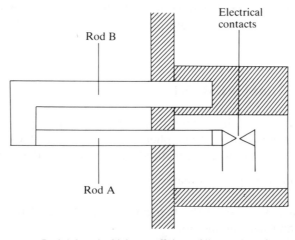

Figure 5.2 Solid rod thermostat. Rod A has the higher coefficient of linear expansion.

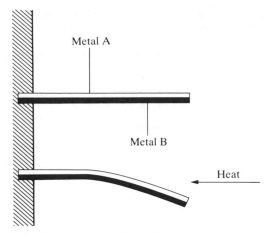

Figure 5.3 Deflection of a bimetallic strip (metal A has the higher coefficient of linear expansion).

5.4.2 Expansion of liquids

The common mercury-in-glass thermometer is widely used because of its repeatability, reliability and cheapness. The coefficient of volumetric (or cubical) expansion of mercury is approximately eight times that of glass. The temperature-measuring range is from freezing at $-35°C$ to vaporizing at $375°C$, and can be extended to $510°C$ by pressurizing the capillary tube with an inert gas such as nitrogen. Other liquids and the use of metal tubes extend the range of liquid-filled thermometers from $-200°C$ to $650°C$, as shown in Fig. 5.5. The most common thermometer consists of a glass stem having a very small (but uniform) bore, with a thin-walled glass bulb at the end. The bulb and bore are completely filled with mercury and the open end is sealed either at high temperature, or under vacuum so that no air is included in the system. Various grades of liquid-in-glass thermometers are available depending upon the accuracy required.

The mercury-in-glass thermometer is used mainly in the laboratory; its fragility (even when protected by a metal sheath) and the closeness of the actual reading to the measuring point make it unsuitable for industrial applications. The mercury-in-steel thermometer shown in Fig. 5.6 is more robust and can be arranged for remote reading, and hence it is of more importance industrially. The steel bulb (the whole of which should be at the measured temperature) and the stainless steel capillary tube are completely filled with mercury under pressure. The steel capillary

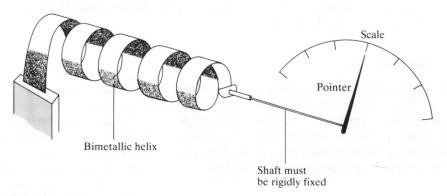

Figure 5.4 Helical bimetal thermometer.

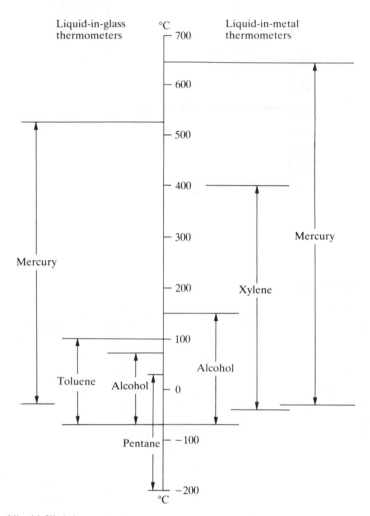

Figure 5.5 Ranges of liquid-filled thermometers.

tube is connected to a Bourdon tube (see Sec. 3.5); this responds to pressure changes in the system caused by temperature changes at the bulb (and hence volume changes of the mercury). The gauge is calibrated in temperature units and has a nearly linear scale. The mercury-in-steel thermometer can be arranged for remote readings.

5.4.3 Gas thermometers

The construction of a constant-volume gas thermometer is identical to the mercury-in-steel thermometer (Fig. 5.6), however the filling material (an inert gas, usually nitrogen) and the principle of operation are different. The gas thermometer uses the change in pressure of the gas contained within a constant volume to provide an indication of the temperature change. The volume increases as the temperature and pressure increase; a linear temperature scale is obtained provided that the volume increase is very small compared with the total gas volume. However, the pressure developed for a given temperature change is usually small, and compensation for the ambient temperature of both the capillary and the Bourdon tube is difficult.

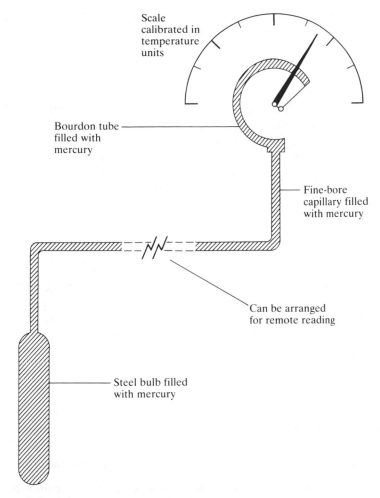

Scale
calibrated in
temperature
units

Bourdon tube
filled with
mercury

Fine-bore
capillary filled
with mercury

Can be arranged
for remote reading

Steel bulb filled
with mercury

Figure 5.6 Mercury-in-steel thermometer.

5.5 CHEMICAL-CHANGE AND PHYSICAL-CHANGE METHODS

5.5.1 Vapour-pressure thermometer

The construction of a vapour-pressure thermometer is identical to that of a mercury-in-steel thermometer (Fig. 5.6). However, the bulb is only partially filled with liquid and the space above the liquid is initially completely evacuated. Liquid evaporates until equilibrium is achieved between the vapour–liquid interface in the bulb. The saturated vapour pressure depends only upon the temperature; it is independent of the container size if *both* liquid and vapour are present. The vapour pressure measured by the Bourdon gauge gives an indication of the temperature. Although the scale is not linear, it can be calibrated accordingly or compensation may be achieved by suitable design of the gauge linkage. The pressure depends only upon the interface temperature; the reading is virtually unaffected by ambient temperature changes. However, measurements near ambient temperature are inaccurate because of vapour condensation in the capillary and a consequent change in the static head. Only a limited number of liquids are suitable (those having an appropriate temperature–saturated vapour pressure relationship), e.g. toluene, methyl chloride, etc., giving instrument ranges of approximately 100 degrees between 0°C and 250°C.

5.5.2 Pyrometric cones

Mixtures of solid substances which melt at predictable temperature values can be produced. These mixtures consist of silicate minerals such as aluminium silicate (china clay) and magnesium silicate (talc), and they are manufactured in the form of cones (6 cm high) known as Seger cones. By varying the composition of the cones, a range of temperatures between 600°C and 2000°C may be covered in convenient steps. When a cone bends over (collapses) so that its tip touches the base level, the stated temperature has been reached. The heating must be at a controlled rate and the collapse temperature is time dependent. The conductivity of the cones is poor and their precision is only about 10°C.

5.5.3 Temperature-sensitive coatings

Chemical and physical changes in certain substances can cause distinct colour changes at reasonably well-defined temperature values. Paints and crayons that incorporate such substances are available, and provide a positive colour indication of temperature in the range 40°C to 1400°C. Temperatures can be measured in steps of 5°C at the lower end and 20°C or more at the upper end, with a precision of about 5 per cent. Some coatings provide two, three or four different temperature changes. The colour changes are usually non-reversible and are time dependent, hence there is a recommended 30 minute heating period.

5.6 SOURCES OF ERRORS

Two types of error should be considered when measuring temperature; these are:

(a) *Static errors* which occur when reading a steady temperature. (Note that the term 'static' refers to the temperature; the measured fluid itself may be moving.)
(b) *Dynamic errors* which occur when a fluctuating temperature is being measured, or when the thermometer is brought suddenly into contact with the measured temperature.

The method of measuring temperature must be chosen so that the indication obtained for a particular temperature is always the same, i.e. the results must be consistent. This is usually more important than the absolute accuracy of the results. There must be no chemical action between a thermometer and the surrounding substance, vapour must not condense on the thermometer (or latent heat will be taken up by the thermometer) and liquid must not evaporate from the thermometer.

5.6.1 Static errors

Manufacturing imperfections in the instrument or the system may cause measurement errors. The effects of some of these errors may be removed by calibration.
Design faults in the instrument, such as stress on metal springs, frictional forces, etc., may also cause measurement errors.
Operating errors occur for a variety of reasons, e.g. parallax error for scale readings, misalignment

when using a micrometer, deformation caused by contact between the component and the instrument.

Environmental errors occur because of variations in the local values of pressure, temperature, etc. The environmental conditions must either be brought to a standard known value, e.g. laboratory temperature control, or they must be measured and an appropriate correction applied.

Application errors occur when the introduction of the instrument alters the variable it is measuring, e.g. a thermometer conducting heat away from the measured fluid.

Partial immersion of a thermometer bulb in the measured fluid can cause significant measurement errors with liquid-filled thermometers. In vapour-pressure thermometers, this type of error occurs only if the liquid–vapour interface is not immersed.

Ambient temperature effects occur when the transmission and display elements of a thermometer are at the temperature of the surroundings, and not at the measured temperature. Ambient temperature changes can lead to measurement errors if the fluid and the capillary have different expansion coefficients. Compensation can sometimes be provided by installing a dummy capillary and Bourdon tube, or by using a bimetal compensating strip in the pointer linkage. This effect can be minimized by ensuring that the bulb volume is very much larger than that of the transmission system. However, this is an unsatisfactory arrangement for high-quality instruments because a large bulb increases the thermal capacity of the thermometer, and hence its thermal lag (see Sec. 5.6.2).

Head errors occur if the head of liquid in a vertical expansion-type thermometer exerts a fluid pressure on the bulb, causing it to expand and recording an apparent lowering of temperature. This will not occur if the axis is horizontal. The error is negligible in systems that are filled under high pressure, but it may be significant for vapour-pressure thermometers. Liquid-filled thermometers should be calibrated in their normal operating position to eliminate this type of error.

Ambient pressure changes can cause measurement errors because pressure springs usually measure gauge pressures. This error is negligible for systems filled at high pressures but may be significant for vapour-pressure thermometers.

Ageing is the slow contraction of glass after heating to high temperatures. Thermometers are annealed and slowly cooled during manufacture to reduce this effect. Therefore, thermometers should be used at high temperatures for short periods only.

5.6.2 Dynamic errors

Dynamic errors occur because of the thermal capacity of the sensing element (transducer) and heat transfer between the measured fluid and the thermometer itself. When a thermometer is suddenly placed in the fluid, a temperature step change occurs. Many practical situations experience fluctuating temperatures, e.g. ovens, and the temperature change is usually a ramp or sinusoidal function. *Thermal lag* is the time taken for a transducer to reach a new temperature before it can send the appropriate signal to the display. This time depends upon the thermal capacity of the instrument (which should be minimized) and the rate of heat transfer to the transducer (which should be maximized). Temperature differences between the transducer and the measured substance may be due to conduction or convection of heat away from the hot substance through the transducer to the environment, or by radiation to and from the transducer (radiation effects are predominant only above 600°C).

   A thermometer bulb in a stationary fluid has a lower rate of heat input than if it is immersed in a moving fluid; the fluid velocity can have a considerable effect. At very high fluid velocities, the fluid adjacent to the thermometer is brought to rest and its kinetic energy is converted into internal energy. This effect can also lead to measurement errors.

5.7 ELECTRICAL METHODS

A very basic knowledge of chemistry (related to the structure of atoms and molecules) and electricity is assumed. The following points are stated briefly, without detailed descriptions, to act as a refresher in these topics.

5.7.1 Basic principles

All matter is composed of very small particles called atoms. An atom consists of a nucleus made up of neutrons (having no electric charge) and protons (positive electric charge). Electrons are negatively charged particles moving in orbits around the nucleus. A proton and a neutron have approximately equal mass, which is 1840 times the mass of an electron. In its normal state an atom is electrically neutral, i.e. possessing equal numbers of protons and electrons. If an atom loses one or more of its electrons it becomes positively charged, i.e. ionized, and is known as a positive ion. Similarly, the gain of an electron produces a negative ion. Electrons revolve around the nucleus in a number of orbits or shells; the first, second and third orbits are complete when they contain 2, 8 and 8 (or 18) electrons respectively. The inert gases are particularly stable because their atoms have completely filled outer shells and it is very difficult to add or remove an electron. However, both copper (29 electrons) and silver (47 electrons) have only one electron in the outermost shell. If an electromotive force (e.m.f.) is applied to a wire made of these metals, the outer electron of each atom is driven from the negative pole (of the battery) to the positive pole. This flow of electrons constitutes an *electric current* and these metals are good *electrical conductors*.

Semiconductors have electrical conductivity values between those of a conductor and an insulator. Silicon (14 electrons) and germanium (32 electrons) are important semiconductors, each having 4 electrons in the outer shell. It would be expected that these elements act as good conductors, however the *pure* crystals are very poor conductors. This is because each atom shares its valence (outer) electrons with another atom, and strong covalent bonds exist between the atoms. These atoms act as if their outer shells were completely filled with electrons. In practice, germanium and silicon act as partial conductors because of the presence of impurities, and the release of electrons caused by thermal energy. The conductivity increases as the temperature rises.

The introduction of an impurity such as arsenic (having 5 outer electrons) into a crystal of pure germanium means that covalent bonds are formed between four electrons in each of the atoms of the elements, and one arsenic electron from each atom remains as a free conducting electron. The impurity atom 'donates' a free electron to the pure crystal and it becomes conducting. This type of semiconductor is known as an *n*-type because negatively charged particles are available. The introduction of a trivalent substance, e.g. indium having 3 outer electrons, into a pure germanium crystal also causes covalent bonds to form. However, each group is deficient in one electron (known as a 'hole') and current can pass through the crystal structure by the transfer of an electron from one atom deficiency to the next; this is called an 'acceptor' impurity. Germanium with a trivalent impurity is termed a *p*-type semiconductor.

In *transistors* the concentration of the impurity required is very small, e.g. 1 part in 10^8, but it can increase the conductivity by 16 times. A crystal of *n*-type or *p*-type germanium is a linear conductor that can transmit a current in either direction, i.e. the magnitude of the current is unaffected by reversing the applied voltage. However, for a crystal having an 'n-type/p-type' junction, the magnitude of the current depends upon the polarity of the e.m.f., i.e. a small current when the e.m.f. is in one direction and large when it is reversed. Such a crystal has applications as a rectifer and it is known as a 'junction diode'.

5.7.2 The thermo-electric effect

Two electrical conductors of dissimilar materials (e.g. copper and iron) are joined to form a circuit as shown in Fig. 5.7. When the junctions are at different temperatures (T_1 and T_2), small e.m.f.s (e_1 and e_2) are generated at the junctions. The algebraic sum of these e.m.f.s causes a current to circulate (from copper to iron) and this phenomenon is known as the *Seebeck effect* (discovered by Thomas Seebeck in 1822). Seebeck listed 35 metals in order of their relative thermo-electric e.m.f.s; a portion of this list is

$$\text{Ni—Pt—Cu—Ti—Pb—Sn—Cr—Au—Ag—Zn—Fe.}$$

A current flows across the hot junction from the earlier to the later metal in the series. The resultant e.m.f. is the same for any particular pair of metals with junctions at constant temperatures; it is not affected by the size of the conductors, the areas in contact, or the joining method. The e.m.f. does depend upon the choice of metals and the junction temperatures, and it may therefore be used as a basis for temperature measurement. A junction of metals used for this purpose is called a *thermo-junction*, and for most thermo-junctions the temperature–e.m.f. relationship is very nearly linear. *Thermocouple* is the term used for a single thermo-junction consisting of two wires joined at one end. The combination of two such junctions (Fig. 5.7) is called a *thermocouple circuit*.

Later work (after Seebeck) indicated that thermo-electric e.m.f.s are due to two separate effects. The *Peltier effect* (after Peltier, 1834) occurs when a current flows across the junction of two metals. Heat is absorbed at the junction when the current flows in one direction, and is liberated if the current is reversed. The heat flow is related to the quantity of electricity and should not be confused with the Joule heating effect, i.e. current^2 × resistance, which does not change to a cooling effect when the current is reversed.

The *Thomson effect* (after Professor W. Thomson, later Lord Kelvin) occurs in a

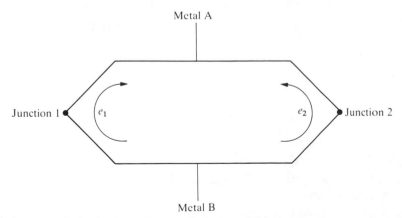

Figure 5.7 Basic thermocouple circuit. The e.m.f.'s (e_1 and e_2) are equal if the junction temperatures are the same. If metal A is iron and metal B is copper and junction 2 is at a higher temperature than junction 1, then current will flow in an anticlockwise direction, i.e. from junction 2 through metal A (iron) and from junction 1 through metal B (copper).

thermocouple circuit; it is observed as an e.m.f. (additional to the Peltier effect) occurring in each material of a thermocouple as a result of the temperature gradient between its ends.

Using some simplifying assumptions, it can be shown that the resultant e.m.f. (of these two effects) in a thermocouple is

$$e = a(T_1 - T_2) + b(T_1^2 - T_2^2)$$

where a and b are (almost) constants, and temperatures T_1 and T_2 are in kelvins. Temperature–e.m.f. values are tabulated in BS 1041: Part 4.

In order to measure the e.m.f. due to thermo-electric effects, it is necessary to introduce a measuring instrument into the circuit. Additional materials and junctions are introduced which may be at different temperatures; two laws are used to account for these practical effects.

The *law of intermediate metals* states that if one or both of the junctions of a thermocouple are opened and one or more metals are interposed, the resultant e.m.f. is unaltered provided that all the new junctions are at the same temperature as the original junction between which they are positioned. This law, when applied to the circuit shown in Fig. 5.8, means that if the reference junction (T_R) is replaced by the galvanometer, all the junctions (A, B, C, D) must also be at the temperature T_R, so that the resultant e.m.f. is unaltered.

The *law of intermediate temperatures* states that the e.m.f. of a thermocouple having junctions at temperatures of T_1 and T_3 is the algebraic sum of the e.m.f.s of two couples of the same materials having junctions at temperatures of T_1 and T_2, and T_2 and T_3 respectively. This law is illustrated in Fig. 5.9.

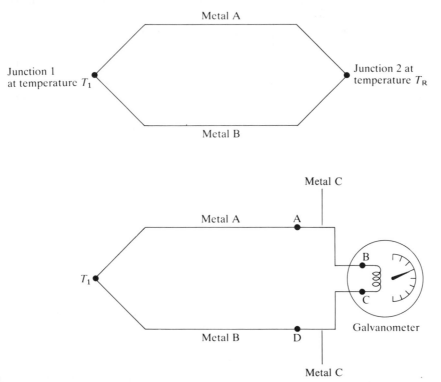

Figure 5.8 Illustration of the law of intermediate metals. New junctions A, B, C, D must all be at temperature T_R so that resultant e.m.f. is unchanged.

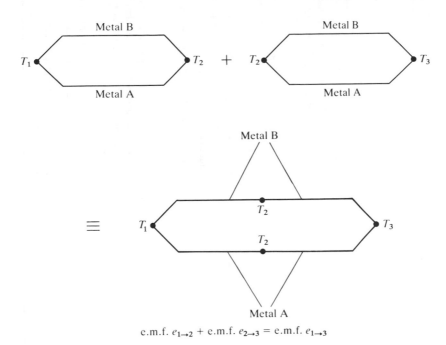

e.m.f. $e_{1\rightarrow2}$ + e.m.f. $e_{2\rightarrow3}$ = e.m.f. $e_{1\rightarrow3}$

Figure 5.9 Illustration of the law of intermediate temperatures.

Several combinations of metals are commonly used in the construction of thermocouples; these are classified as either base-metal thermocouples or rare-metal thermocouples. Base-metal thermocouples are preferred as they are more sensitive, cheaper and have nearly linear characteristics. However, rare-metal thermocouples are more suitable for high-temperature applications because of their higher melting points and resistance to oxidation. Common requirements in industrial situations are for a high output e.m.f., stability of e.m.f., sensitivity, resistance to chemical change caused by the working fluids or environment, mechanical strength (over the operating range) and cheapness. A comparison of some common thermocouples is presented in Table 5.1.

Thermocouples are usually made from wires of the two constituent metals. The e.m.f. generated is not dependent upon the size of the wires, therefore their size is limited only by the mechanical strength. Thicker wires have a longer life but they increase the rate at which heat is conducted away from the junction. It is important that the wires are not subjected to any strain as this reduces the e.m.f. and produces a low reading. A concentric thermocouple consists of a wire of one metal inside a tube of the other, and the junction at one end of the tube. The tube also acts as a protective sheath. Other metal sheaths with good thermal conductivities are also used and refractory sheaths are used for very high temperatures. Horizontal rather than vertical sheaths are used to avoid strain on the thermocouple. Impurity elements in the thermocouple wires affect the e.m.f. output and can also lead to mechanical failure due to embrittlement.

The reference junction must be maintained at a known constant temperature. Ice–water mixtures are often used in the laboratory, but temperature-controlled containers are preferred for industrial applications. For less precise work, ambient temperature may be used as the reference if the junction is shielded from temperature fluctuations. However, thermocouple tables giving e.m.f.–temperature values are usually quoted at one particular temperature value (normally 0°C) of the reference junction. At other reference temperatures, a calibration correction is required.

Table 5.1 Comparison of industrial thermocouples

Name (positive wire written first)	Composition	Temperature range (°C)	Maximum spot temperature (°C)	Sensitivity (mV/°C)	Characteristics
Base-metal thermocouples					
Copper/Constantan	Cu/approx. 40% Ni, 60% Cu	−250 to +400	500	0.05	High corrosion resistance in oxidizing and reducing atmospheres and with condensed moisture up to 350°C. Requires protection from acid fumes
Iron/Constantan	Fe/approx. 40% Ni, 60% Cu	−200 to +850	1100	0.05	Low cost, suitable for reducing atmospheres; corrosion in presence of moisture, oxygen and sulphur-bearing gases
Chromel/Alumel	90% Cr, 10% Ni/94% Ni, 2% Al, +Si and Mn	−200 to +1100	1300	0.04	Resistant to oxidizing (not reducing) atmospheres, attacked by carbon-bearing gases, sulphur and cyanide fumes
Chromel/Constantan	(each as above)	−200 to +850	1100	—	High e.m.f., suitable for oxidizing (not reducing) atmospheres
Rare-metal thermocouples					
Platinum–rhodium/platinum	90% Pt, 10% Rh/Pt	0 to +1400	1650	0.01	Low e.m.f., good resistance to oxidizing (not reducing) atmospheres; platinum is affected by metallic vapours and contact with metallic oxides
Tungsten–rhenium/tungsten–rhenium	95% W, 5% Re/72% W, 26% Re	0 to +2600	—	—	Non-oxidizing atmospheres only; 5% rhenium arm is brittle at room temperatures
Rhodium–iridium/iridium	90% Rh, 10% Ir/Ir	0 to +2100	—	—	Similar to platinum–rhodium/platinum thermocouple

Industrial thermocouples are often situated some distance from the display instrument. *Compensating leads* are installed from the head of the sensing element to the display in order to reduce the cost of long leads made from expensive materials. These leads have the same thermo-electric e.m.f. as the hot-junction metals, but possess inferior high-temperature properties. The intermediate junctions must be at the same temperature (even though this temperature may change) in order not to affect the reading (law of intermediate temperatures). The compensating leads do not really compensate in the normally understood sense of the word; more strictly they are extension leads. It is important that the correct cable is used with a particular type of thermocouple; the reference junction is then transferred to the display.

A *thermopile* is an arrangement of several thermocouples in series. The thermocouples are made of either fine silver and bismuth wires, or Chromel and Constantan wires. The e.m.f.s produced by the thermocouples are additive, and an instrument capable of measuring small e.m.f.s is used to provide a sensitive method of measuring radiation (see Sec. 5.8.2).

The output from a thermocouple circuit can be measured directly either by a moving-coil meter, an amplifying meter, or a potentiometer. A galvanometer-type instrument having a high resistance is often used to measure the small current circulating in a thermocouple circuit created by the resultant e.m.f. The galvanometer is usually more sensitive than a milliammeter or millivoltmeter. In principle, only a potentiometer can read the true e.m.f. since it takes no current; however, the error in the reading of a high-impedance amplifying meter is often negligibly small. Problems due to changes in the circuit resistance are virtually absent when amplifying meters or potentiometers are used; however, direct-deflection meters have the advantage of not requiring an auxiliary power source.

The following static errors may occur in thermocouple systems:

(a) Varying resistances because temperatures are different from the calibration value.
(b) Leakage currents, induced e.m.f.s due to close electric circuits, and electrolytic effects due to moisture at the junctions.
(c) The hot junction (above 600°C) reradiates a significant proportion of the heat it receives; a radiation shield may be used to overcome this problem.
(d) Condensation from gases and vapours on to the hot thermocouple junction. Probes are available which can separate the two phases and enable the true gas temperature to be measured.
(e) Conversion of the kinetic energy of a moving gas stream into internal energy, thus causing localized heating. The resultant measured temperature is the stagnation temperature; the true gas temperature can be calculated if the velocity and heat capacity (C_p) of the gas are known. This effect is significant for gas velocities above 50 m/s.

Errors also occur in thermocouple systems because of the dynamic response of the measuring junction and the e.m.f. or current-measuring instruments. The response of a galvanometer is usually faster than that of the measuring probe. However, the overall response of the system is slower than the item having the slowest response.

5.7.3 Electrical resistance

In the SI system of units, the ampere is the basic electrical unit and it is defined as follows.

The *ampere* is that constant current which if maintained in two straight parallel conductors of infinite length and negligible cross-section, and placed 1 m apart in vacuum, would produce a force equal to 2×10^{-7} newton per metre of length between the conductors.

The volt and the ohm are derived units and are defined as follows.

The *volt* is the potential difference between two points of a conducting wire carrying a constant current of 1 ampere, when the power dissipated between the points is equal to 1 watt.

The *ohm* is the resistance between two points of a conductor when a constant potential difference of one volt (applied between the two points) produces a current of 1 ampere in this conductor, and the conductor is not the source of any electromotive force (e.m.f.).

The definition of resistance is based on Ohm's law:

$$\text{potential difference} = \text{current} \times \text{resistance}$$

Units: volts   = amps ×   ohms

Using standard resistances and a standard cell, both ammeters and voltmeters may be calibrated using a potentiometer. A simple potentiometer consists of a length of uniform wire fixed along a length scale. The resistance of the wire per unit length is constant, and the voltage drop along the wire is uniform when a current flows. A standard cell is connected in series with a large resistance and a galvanometer, and a point is found on the potentiometer wire such that no current flows through the galvanometer. At this point, the potential drop along that length of the wire is equal to the e.m.f. of the standard cell. If another source of potential difference is connected to one end of the wire and to the galvanometer, then a new point can be found on the wire such that there is no deflection of the galvanometer.

The resistance of metallic conductors increases with temperature, i.e. they have positive temperature coefficients of resistance. The resistance of electrolytes, semiconductors and insulators decreases with increasing temperature. The metals normally used in practical electrical-resistance thermometry are platinum, copper and nickel. Platinum is the standard material (defining IPTS) because it can be manufactured in very pure form, and it possesses a stable and well-defined resistance–temperature relationship. The relationship between the resistance (R_T) of a platinum resistance element at a temperature T and its resistance (R_0) at 0°C, for the range 0°C to 630°C, is given by

$$T = \frac{1}{\alpha}\left(\frac{R_T}{R_0} - 1\right) + \delta\left(\frac{T}{100} - 1\right)\frac{T}{100}$$

where α and δ are constants. The relative values of α and δ are such that, at the lower end of the range, a platinum resistance element has a nearly linear characteristic as given by

$$R_T = R_0(1 + \alpha T)$$

For platinum, the value of α is 3.91×10^{-3} per °C, which provides an indication of the sensitivity of the material. The value of α depends upon the material, however other metals obey similar laws. The *fundamental interval* is the change in resistance corresponding to a temperature change from 0°C to 100°C. A *thermistor* is a semiconductor device with a negative temperature coefficient. A *resistance thermometer element* is any type of resistor (metallic or semiconductor) used for temperature measurement. The characteristics of a typical thermistor and a platinum-resistance thermometer are compared in Fig. 5.10. Thermistor characteristics are normally plotted on a logarithmic resistance scale as shown in Fig. 5.11. Relative resistance–temperature values for different metals are given in BS 1041: Part 3.

Resistance-type thermometer bulbs consist of wire or foil (or thin films deposited on insulating surfaces) which acts as the sensing element. A typical laboratory-type resistance thermometer is shown in Fig. 5.12, having a helical wire coil wound on a crossed mica former and enclosed in a Pyrex tube. The platinum is protected either by an inert gas or by vacuum in the tube. An industrial device typically consists of a grooved ceramic former with either a glass

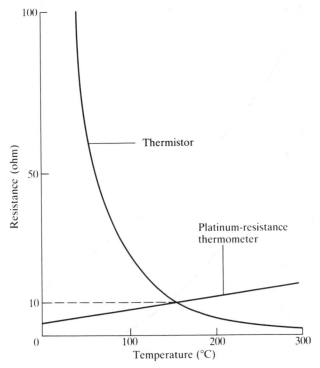

Figure 5.10 Comparison of the characteristics of a typical thermistor and a metallic resistance element.

coating or a stainless steel tube to protect the wire. Thin, etched grids of metal foil are also used as resistance elements; they may be bonded to a plastic support and attached to a surface. Foil-type sensors may be open-faced or coated; their response is much faster than bulb-type elements. Industrial platinum resistance thermometers are usually designed in accordance with BS 1904, although higher precision thermometers are also produced.

Thermistors have large negative temperature coefficients of resistance, typically (a negative value of) 10 times that of platinum. Their temperature–resistance relationship can be expressed in the form

$$R_T = R_0 a \exp(b/T)$$

where a and b are constants, and T is the absolute temperature (K). The elements are made of small pieces of sintered ceramic material, e.g. metallic oxides of manganese, cobalt, iron, etc., in the form of beads, discs and rods. Thermistor elements have a high sensitivity and fast response times.

Metallic resistance thermometers usually operate in the range $-220°C$ to $+600°C$, and thermistors from $-100°C$ to $+160°C$. The fast response of both thermistors and metallic resistance thermometers is due to the small thermal capacities of the active elements. The use of a protective sheath increases the thermal capacity, as does the former for a coil-type element, but it also reduces the response rate.

The standard method of using resistance thermometer elements for temperature measurement is by incorporation into a Wheatstone bridge. The bridge may be unbalanced, manually balanced or self-balancing. Compensation is necessary to account for the temperature gradient (not due to the measured temperature) of the lead wires from the sensing resistance to the

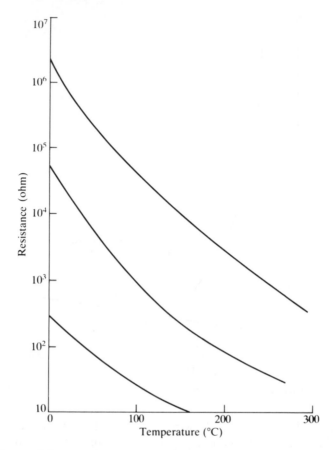

Figure 5.11 Typical characteristic curves for various thermistors.

measuring bridge. Several compensation methods are available, such as the three-wire system for a balanced bridge. The use of Wheatstone bridge-type circuits with thermistors produces a non-linear output which must be calibrated over the measuring range. An output voltage that is linear with temperature can be produced by using a half-bridge incorporating a sensing thermistor. Lead compensation is usually unnecessary because of the high sensitivity of thermistors.

5.8 RADIATION METHODS

Heat transfer occurs by conduction, convection and radiation, however only radiation is transmitted independently of matter, i.e. does not require contact between solid objects or fluids. All radiation consists of electromagnetic waves that are propagated through space with the same velocity; the extremes are long radio waves (wavelength approx. 2000 m) to X-rays (wavelength approx. 10^{-8} m). Small wavelengths are usually expressed either as microns (μm, 10^{-6} m) or Angstrom units (Å, 10^{-10} m). Visible light is only a small portion of the range, having wavelengths from 0.4 to 0.8 μm. Thermal radiation measurements are limited to the ultraviolet, visible and infrared radiations with wavelengths from 0.1 to 10 μm. The radiation travels in straight lines and can be reflected or refracted; the amount of radiation reaching a receiver of unit area is inversely proportional to the square of the distance of the receiver from the source.

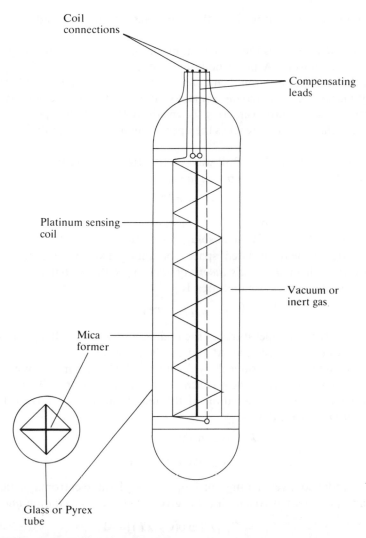

Figure 5.12 Typical laboratory-type resistance thermometer.

The International Practical Temperature Scale (IPTS) defines temperatures above 1064°C in terms of the radiation from a black body. However, radiation thermometers are used for temperature measurements from room temperature up to 5000°C if the conditions are corrosive or the temperature is too high for a thermocouple (maximum 2800°C for rare-metal couples). Temperature-measuring instruments using radiation are known as *radiation pyrometers*; they have a very fast response (a few microseconds) and do not require contact with the hot object. Pyrometers can be used to measure the temperature of a moving object, and for transient conditions.

5.8.1 Basic principles

The following theory is presented very briefly in order to provide an appreciation of the basis of operation of radiation pyrometers. The reader should consult a specialist heat transfer textbook for a more comprehensive treatment of the principles of thermal radiation exchange (see Bibliography).

A *black body* or *full radiator* is the *ideal* emitting surface; it emits more energy than any other body at the same temperature. A black body also absorbs all radiation falling upon it, at all temperatures, without any radiation being reflected or transmitted. A black body can be approximated by a large enclosure having a small opening, and the entire body is maintained at a constant temperature. Any radiation entering the enclosure through the opening is reflected many times on the inside surface before it reaches the opening again; the radiation is almost completely absorbed.

The total power or radiant flux of all wavelengths (Q_b) emitted by unit area of a perfectly black body is given by the *Stefan–Boltzmann law* as

$$Q_b = \sigma T^4$$

where σ is the Stefan–Boltzmann constant, having an accepted value of 5.6697×10^{-8} W/m^2 K^{-4}, and T is the absolute temperature (K) of the surface.

Consider a perfectly heat-insulated space containing two bodies; both bodies will be radiating and absorbing heat. *Prevost's theory of exchanges* states that the net radiant energy exchange (q) between the bodies per unit area is

$$q = \sigma(T_1^4 - T_2^4)$$

where T_1 and T_2 (K) are the temperatures of the bodies, and $T_1 > T_2$. If T_1 is much higher than T_2, then the T_2^4 term becomes insignificant and can be neglected.

The principle of operation of optical pyrometers is based upon the measurement of the spectral concentration of radiance at one wavelength emitted by a source. *Wien's displacement law* states that the wavelength of the radiation of the maximum intensity (λ_m) decreases as the absolute temperature (T) rises; that is

$$\lambda_m T = \text{constant}$$

$$= 2.898 \times 10^{-3} \text{ m K}$$

The energy is radiated over a range of frequencies of the electromagnetic spectrum; the distribution for any particular wavelength (λ) is given by *Planck's radiation law*:

$$Q_{b\lambda} = C_1 / \{\lambda^5 [\exp(C_2/\lambda T)] - 1\}$$

where $Q_{b\lambda}$ is the energy radiated at wavelength λ; C_1 and C_2 are constants. Energy-distribution curves calculated from this equation and showing the variation of spectral radiance with both the wavelength and temperature of a black body source can be found in most heat transfer textbooks (see Bibliography).

The *emissivity* ($\varepsilon < 1$) is defined as

$$\varepsilon = \frac{\text{total emissive power of a real surface}}{\text{total emissive power of an ideal black body}}$$

The emissivity of different types of surfaces varies at different wavelengths; values can be obtained from appropriate handbooks.

5.8.2 Total-radiation pyrometers

A *total-radiation pyrometer* (or *wide-band pyrometer*) receives almost all the radiation from a particular area of a hot object, and focuses it on to a sensitive temperature transducer. The

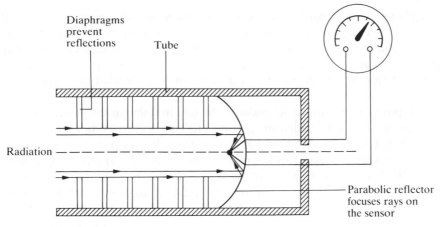

The sensor may be a resistance
element (conductor or thermistor type)
or a thermopile

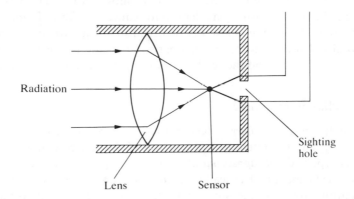

Figure 5.13 Total-radiation pyrometer.

pyrometer receives the maximum amount of radiant energy of all possible wavelengths, including both visible (light) and invisible (infrared) radiations. In general, the temperature-sensing element is capable of measuring radiation over a wide range of wavelengths, although the actual wavelengths received are usually restricted by the materials used in the optical system.

The radiation can be focused on the element by using either a lens system, a parabolic reflector, or a tube with diaphragms to prevent reflections. These devices are shown in Fig. 5.13. Ordinary glass lenses are unsatisfactory as they absorb radiation; the transmissibility of the glass determines the range of frequencies passing through. Mirrors are sometimes preferred to lenses for this reason. The radiation from the sighting hole of the furnace must cover the instrument, since radiation entering from another source will lead to incorrect readings. The cone of sight is given for each instrument such that larger areas must be viewed as the distance between the instrument and the source increases. Instruments having a fixed focus can only operate within a limited distance from the source.

A transducer (sensing element) can be a thermocouple or a resistance thermometer, but more often it is a thermopile consisting of several similar thermocouples connected in series. The rise in temperature of the element (caused by the radiation) is used as a measure of the source temperature. Therefore, the sensitive part of the transducer should have a small thermal capacity which produces a fast response (typically of the order of 1 second). Pyrometers incorporating thermocouple elements do not require compensating leads. The element measures its own temperature change (due to the radiation) relative to its surroundings, i.e. the instrument body. The reference junction is located within the instrument, but shielded from radiation from the source.

The fourth-power temperature dependence (Stefan–Boltzmann law) predicts the very non-linear characteristic of a total-radiation pyrometer. The sensitivity is also poor in the low-temperature range (0°C to 500°C), however most total-radiation pyrometers are now used for measurements in this range because optical pyrometers (see Sec. 5.8.3) give better results at higher temperatures. The low output signal is often of comparable magnitude to signals from other sources; this problem may be overcome by amplifying the transducer signal. The output signal is usually transmitted to either a moving-coil meter or a self-balancing potentiometer.

Conditions in a furnace are usually considered equivalent to a perfectly black body, however the presence of a flame, smoke or furnace gases may reduce the amount of radiation reaching the pyrometer. This effect can be partially overcome by using lenses made of Pyrex which do not transmit certain wavelengths. There will be a maximum allowable distance between the pyrometer and the source (e.g. 1.5 m when using glass lenses) to reduce the effect of radiation absorption by the cold atmosphere in the intervening space. A special spherical furnace has been developed for calibration of total-radiation pyrometers; comparison is made between the pyrometer reading obtained through a sight hole and the reading from a thermocouple situated inside the furnace.

5.8.3 Selective-radiation pyrometers

Selective-radiation pyrometers (or *narrow-band pyrometers*) are intended to receive the radiance of one wavelength only, and the intensity is a function of the source temperature. A widely used instrument of this type is the *disappearing-filament optical pyrometer* which is shown schematically in Fig. 5.14. The filament is electrically heated to a temperature such that its colour matches that of the radiation from the measured source; it then becomes invisible. If the filament temperature is too high, it appears bright against the source, and vice versa, as shown in Fig. 5.15.

In construction the pyrometer is similar to an ordinary telescope. The filter restricts the light passing through to the eye to a very narrow band (0.01 μm wide) at a wavelength of 0.65 μm. This value is particularly suitable because the intensity changes of bright metal surfaces with temperature are largest at this wavelength (i.e. the red portion of the visible spectrum). An absorption screen can be included for high-temperature work, allowing only a known proportion of the incident radiation to enter the instrument. The lens forms an image of the temperature source in the same plane as the filament. The filament and the superimposed image of the source are viewed through the eyepiece and the filter. An adjustable resistor is used to control the current through the filament (and hence its temperature), until the image of the filament 'disappears' into the image of the source. The filament current is read from an ammeter and using the known current–temperature relationship, the temperature of the filament is found and hence the temperature of the source. Usually the meter is calibrated to read the temperature directly.

Optical pyrometers are not suitable for recording or controlling temperatures, but they do provide an accurate method of measuring temperatures between 700°C and 3000°C. They are also used to check and calibrate total-radiation pyrometers. However, optical pyrometers can only be used at temperatures where visible radiation is emitted (i.e. above 700°C), also they require an operator and cannot provide a continuous readout.

Continuous optical pyrometers have been developed which enable a continuous readout to be

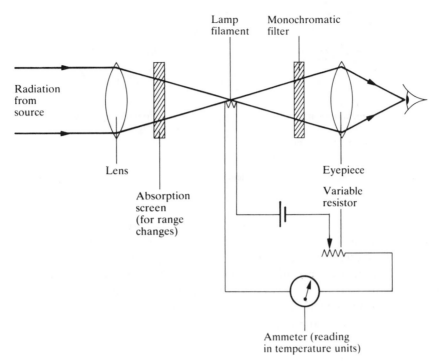

Figure 5.14 Disappearing-filament optical pyrometer.

obtained; they also operate on the principle of selective-radiation measurement. The receiving element is a photo-electric or photo-voltaic cell (solar cell) which produces an e.m.f. output dependent upon the intensity of the incident radiation. This type of pyrometer is usually calibrated by sighting on the filament of a tungsten-strip lamp whose temperature is known in terms of the lamp current. Other photo-sensitive cells may be used such as a photo-conductive cell, a photo-emissive cell, or a semiconductor-type element. These cells are responsive to various wavebands of radiation, which may be further narrowed using filters. The cells respond to infrared radiation and can be used to measure temperatures down to 400°C.

Continuous optical pyrometers have a very rapid response, of the order of a few milliseconds compared with several seconds for a total-radiation pyrometer. This improvement is because the sensing element responds directly to the radiation, rather than to a temperature change resulting from radiation absorption. The errors caused by the variation of the emissivity of metals due to oxidation are less at lower wavelengths, and this instrument also has a higher precision than the total-radiation pyrometer.

5.8.4 Sources of errors

Possible sources of error when using radiation pyrometers will now be considered:

(a) The radiation source may not be a perfectly black body and the emissivity of the actual surface should be determined. If the nature of the radiating surface changes, e.g. by surface oxidation, then its emissivity will change.

(b) Radiation may be absorbed by the atmosphere present between the source and the sensing element. Optical pyrometers respond to certain wavelengths and it is necessary to know

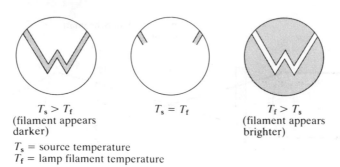

$$T_s > T_f$$
(filament appears
darker)

$$T_s = T_f$$

$$T_f > T_s$$
(filament appears
brighter)

T_s = source temperature
T_f = lamp filament temperature

Figure 5.15 Image viewed through a disappearing-filament pyrometer.

whether the atmosphere (or impurities such as smoke) will absorb that particular radiation.
(c) Fixed-focus pyrometers are correctly focused only when they are located at a specified distance from the source. If the source is too close to the pyrometer, some of the radiation will not be focused on the element although this can be taken into account during calibration. However, if the source is too far away, radiation from other sources can enter the pyrometer and fall on the sensing element. This situation cannot be allowed for during calibration and a larger element area is required to give a correct reading.

5.9 CALIBRATION OF TEMPERATURE-MEASURING INSTRUMENTS

Temperature-measuring instruments can be calibrated by measurement of the known primary and secondary fixed points as defined by *IPTS-68* (see Sec. 5.2). Boiling-point temperatures are particularly sensitive to changes in the pressure, whereas freezing-point temperatures are more susceptible to the effects of impurities in the materials. The fixed-point method is often used industrially, particularly for calibration of thermocouples. A common procedure is to immerse the temperature-sensing element in a molten metal and observe the 'arrest' point on the time–temperature graph as the metal is cooled. This point corresponds to a reduced cooling rate (the freezing point) owing to the latent heat of solidification.

If the temperatures to be measured are not close to the fixed points, it is preferable to calibrate the thermometer by comparison with another measuring instrument of known accuracy. This procedure may be quicker and more convenient than the use of fixed-point temperatures. A reference standard is required, e.g. a thermocouple or platinum-resistance thermometer; this device must have been accurately calibrated previously. Both the standard and the test instrument must be maintained at the same temperature. This is usually achieved by immersion in a thermostatically controlled bath containing either iso-pentane (for temperatures between $-150°C$ to $0°C$), water ($0°C$ to $100°C$), oils ($80°C$ to $300°C$), or a mixture of sodium and potassium nitrates ($200°C$ to $600°C$). Above $600°C$, a tubular-type furnace is used having an overall range from $100°C$ to $1800°C$. The accuracy is inferior to the liquid bath at lower temperatures because heat transfer then occurs mainly by conduction rather than radiation.

The instrument display may be calibrated independently if it can be separated from the transducer. Calibration is achieved by substituting electrical quantities, e.g. resistance or e.m.f., corresponding to particular temperatures. This calibration is common for resistance thermometers and thermocouple systems.

EXERCISES

5.1 Distinguish between temperature and heat.

5.2 How is the basic unit of temperature (kelvin; K) defined?

5.3 Describe the absolute thermodynamic temperature scale, and define the absolute zero of temperature.

5.4 List some of the primary and secondary fixed points (and their temperature values) on the International Practical Temperature Scale.

5.5 How are temperatures interpolated between (and above) the fixed points?

5.6 List the most common practical methods used for temperature measurement.

5.7 Describe the operation of:
  (a) a solid-rod thermostat;
  (b) a bimetal-strip thermometer;
  (c) the liquid-in-glass thermometer;
  (d) gas thermometers.

5.8 Describe some examples of chemical-change and physical-change methods of temperature measurement.

5.9 List with a brief discussion, the types of errors that may occur with temperature measurements.

5.10 Describe the principles of the thermo-electric and electrical resistance methods of temperature measurement.

5.11 Explain how a thermocouple operates. What is a thermopile?

5.12 What is a thermistor?

5.13 Define a black body and emissivity.

5.14 Write down the Stefan–Boltzmann law, Prevost's theory of exchanges, Wien's displacement law and Planck's radiation law.

5.15 Describe using schematic diagrams the operation of a total-radiation pyrometer, a disappearing-filament optical pyrometer and a continuous optical pyrometer.

5.16 How are temperature-measuring instruments calibrated?

5.17 Examine the temperature-measuring instruments that are used in your laboratory and identify their method of operation.

5.18 Obtain a range of catalogues and technical literature related to temperature measurement from several instrument manufacturers. Study this material and the descriptions given in instrument handbooks. Become aware of new developments in this field.

BIBLIOGRAPHY

Also see the Combined Bibliography to Chapters 3, 4 and 5 which follows this chapter.

British Standards

BS 1041: Temperature measurement. Part 2: Expansion thermometers, Section 2.1: Guide to selection and use of liquid-in-glass thermometers. *Part 3: Industrial resistance thermometry. Part 4: Thermocouples. Part 5: Radiation pyrometers. Part 7: Temperature/time indicators.*

BS 1794: Chart ranges for temperature recording instruments.

BS 1843: Colour code for twin compensating cables for thermocouples.

BS 1904: Specification for industrial platinum resistance thermometer sensors.

BS 2082: Code for disappearing-filament optical pyrometers.

BS 2765: *Dimensions of temperature detecting elements and corresponding pockets.*
BS 3166: *Thermographs (liquid-filled and vapour pressure types) for use within the temperature range* $-20°F$ *to* $220°F$
  ($-30°C$ *to* $105°C$).
BS 3231: *Thermographs (bimetallic type) for air temperatures within the range* $0°F$ *to* $140°F$ ($-20°C$ *to* $60°C$).
BS 4201: *Specification for thermostats for gas-burning appliances.*
BS 4892: *Guide to the measurement of thermal radiation by means of the thermopile radiometer.*
BS 4937: *International thermocouple reference tables. Parts 1 to 7 for different thermocouple pairs. Part 20: Specification
  for thermocouple tolerances.*

Standards of the American Society for Testing and Materials (ASTM)—*Volume 14.01: Temperature Measurement*

Liquid-in-glass thermometers (industrial)

E 1–83: *Specification for ASTM Thermometers.*
E 77–84: *Method for Verification and Calibration of Liquid-in-glass Thermometers.*

Thermometers (clinical)

E 667–81: *Specification for Clinical Thermometers (Maximum Self-registering, Mercury-in-glass).*
E 825–81: *Specification for Melting Point-type Disposable Fever Thermometer for Intermittent Human Temperature
  Determination.*
E 879–82: *Thermistor Sensors for Clinical Laboratory Temperature Measurements.*

Resistance thermometers (industrial)

E 644–78: *Test Method for Testing Industrial Resistance Thermometers.*

Reference baths

E 563–76: *Practice for Preparation and Use of Freezing Point Reference Baths.*

Nomenclature

E 344–84: *Terminology Relating to Thermometry and Hydrometry.*

Radiation

E 639–78: *Method of Measuring Total-Radiance Temperature of Heated Surfaces Using a Radiation Pyrometer.*

Thermal conductivity

C 177–76: *Test Method for Thermal Transmission Properties, Steady-State, by Means of the Guarded Hot Plate.*
C 236–80: *Test Method for Thermal Performance of Building Assemblies by Means of a Guarded Hot Box.*
C 518–76: *Test Method for Thermal Transmission Properties, Steady-State, by Means of the Heat Flow Meter.*
D 2717–83: *Test Method for Thermal Conductivity of Liquids.*

Thermal expansion

D 696–79: *Test Method for Coefficient of Linear Thermal Expansion of Plastics.*

Thermistors

E 879–82: *Specification for Thermistor Sensors for Clinical Laboratory Temperature Measurements.*

Thermocouples

E 207–83: *Test Method for Thermal EMF Test of Single Thermoelement Materials by Comparison with a Secondary
  Standard of Similar EMF–Temperature Properties.*
E 220–86: *Test Method for Calibration of Thermocouples by Comparison Techniques.*
E 230–83: *Temperature–Electromotive Force (EMF) Tables for Standardized Thermocouples.*
E 235–82: *Specification for Thermocouples, Sheathed, Type K, for Nuclear or for Other High-Reliability Applications.*

E 452–83: Test Method for Calibration of Refractory Metal Thermocouples Using an Optical Pyrometer.
E 574–76: Specification for Duplex, Base-Metal Thermocouple Wire with Glass Fiber or Silica Fiber Insulation (R 1982).
E 585–81: Specification for Sheathed Base-Metal Thermocouple Materials.
E 601–81: Test Method for Comparing EMF Stability of Single-Element Base-Metal Thermocouple Materials in Air.
E 608–84: Specification for Metal-Sheathed Base-Metal Thermocouples.
E 696–84: Specification for Thermocouple Wire, 97% Tungsten–3% Rhenium and 75% Tungsten–25% Rhenium.
E 710–79: Test Method for Comparing EMF Stabilities of Base-Metal Thermoelements in Air Using Dual, Simultaneous, Thermal-EMF Indicators.
E 780–81: Test Method for Measuring the Insulation Resistance of Sheathed Thermocouple Material at Room Temperature.
E 839–81: Test Method for Sheathed Thermocouples and Sheathed Thermocouple Material.
E 988–84: Tables for Temperature–Electromotive Force (EMF) for Tungsten–Rhenium Thermocouples.

Additional US Standards

Abbreviations used to designate the standards organizations are listed at the beginning of the book (page xiii).

Temperature measurement

API Manual of Petroleum Measurement Standards, 1985: Chapter 7—Temperature Determination; Section 2, Dynamic Temperature Determination; Section 3, Static Temperature Determination Using Portable Electronic Thermometers.
ASHRAE 41.1–74: Standard Measurements Guide—Section on Temperature Measurements.
ASME PTC 19.3–74, Part 3: Temperature Measuring Instruments and Apparatus—Performance Test Code (R 1985).
IEEE 119–74: Recommended Practice for General Principles of Temperature Measurement as Applied to Electrical Apparatus.
SAE ARP 175–48: Temperature Measurement, Well Insert Type.
SAE ARP 485–57: Temperature Measuring Devices Nomenclature.
SAE ARP 690 64: Standard Exposed Junction Thermocouple for Controlled Conduction Errors in Measurement of Air or Exhaust Gas Temperature.
SAE AS 793–66: Total Temperature Measuring Instruments (Turbine Powered Subsonic Aircraft).
SAE AS 8005–77: Minimum Performance Standard Temperature Instruments.
UL 873–79: Temperature-Indicating and Temperature-Regulating Equipment (19 February 1986).

Temperature transducers

SAE AS 196–48: Thermo-Sensitive Element (Iron–Constantan).
SAE AS 234–48: Thermo-Sensitive Element (Resistance Bulb Type).

Thermistors

EIA 275-A–71: Thermistor Definitions and Test Methods (R 1981).
EIA 309–65: General Specification for Thermistors, Insulated and Non-Insulated (R 1981).
EIA 337–67: General Specification for Glass Coated Thermistor Beads and Thermistor Beads in Glass Probes and Glass Rods; Negative Temperature Coefficient (R 1981).

Thermocouples

ISA MC96.1–82: Temperature Measurement Thermocouples.
SAE AIR 46–56: Preparation and Use of Chromel–Alumel Thermocouples for Turbojet Engines.
SAE ARP 464–58: Mount–Thermocouple.
SAE ARP 465A–71: Flange–Thermocouple.
SAE ARP 690–64: Standard Exposed Junction Thermocouple for Controlled Conduction Errors in Measurement of Air or Exhaust Gas Temperature.
SAE AS 196–48: Thermo-Sensitive Element (Iron–Constantan).

Other sources

Billing, B. F. and T. J. Quinn (eds), *Temperature Measurement*, Adam Hilger, Bristol, 1975.
Chapman, A. J., *Heat Transfer*, 4th edn, Macmillan, New York, 1984.
Holman, J. P., *Heat Transfer*, 5th edn, McGraw Hill, New York, 1981.

Kreith, F. and W. Z. Black, *Basic Heat Transfer*, Harper and Row, New York, 1980.

McAdams, W. A., *Heat Transmission*, 3rd edn, McGraw-Hill, New York, 1954.

Nicholas, J. V. and D. R. White, *Traceable Temperatures*, DSIR Science Information Division, New Zealand, 1982.

Quinn, T. J., *Temperature*, Academic Press, London, 1983.

Quinn, T. J., Temperature Standards, *Inst. Phys. Conf. Series*, No. 26, Chapter 1, 1975.

'The International Practical Temperature Scale of 1968', *Metrologia*, vol. 5, No. 2, April, 1969.

The International Temperature Scale of 1968, HMSO, London, 1969.

COMBINED BIBLIOGRAPHY—CHAPTERS 3, 4 AND 5

Benedict, R. P., *Fundamentals of Temperature, Pressure and Flow Measurement*, 3rd edn, Wiley, New York, 1984.

Considine, D. M. (ed.), *Process Instruments and Controls Handbook*, 3rd edn, McGraw-Hill, New York, 1985.

Considine, D. M. and S. D. Ross, *Handbook of Applied Instrumentation*, Robert E. Krieger Publishing Co., Melbourne, Fla, 1982.

Graham, A. R., *An Introduction to Engineering Measurements*, Prentice-Hall, Englewood Cliffs, New Jersey, 1975.

Handbook of Instruments and Instrumentation, Trade and Technical Press, London, 1977.

Jones, B. E. (ed.), *Instrument Science and Technology, Volume 1* (1982), *Volume 2* (1983), *Volume 3* (1985), Adam Hilger, Bristol.

Jones, E. B., *Instrument Technology. Volume 1: Measurement of Pressure, Level, Flow and Temperature*, 3rd edn (1974); *Volume 2: On-line Analysis Instruments*, 2nd edn. (1976); *Volume 3: Telemetering and Automatic Control* (1957); Newnes–Butterworths, London.

Miller, J. T. (ed.), *The Instrument Manual*, 5th edn, United Trade Press, London, 1975.

Norton, H. N. (ed.), *Sensor Selection Guide*, Elsevier Sequoia, Switzerland, 1984.

Pressure, Flow and Level Instrumentation Market (1984); *Process Analytical Instrumentation Market* (1984); *Process Control Equipment Market* (1983); Frost and Sullivan, New York.

Soisson, H. E., *Instrumentation in Industry*, Wiley, New York, 1975.

THREE

EXPERIMENTAL METHODS

SCOPE

Part Three describes the treatment of measured data including the use of numerical integration and differentiation (Chapter 6), calculation of statistical parameters and the use of statistical methods (Chapter 7), and the application of dimensional analysis to determine the relationship between related variables (Chapter 8).

In Chapter 7, the calculation of statistical parameters and the use of statistical methods for descriptive statistics and inferential statistics are described. The sections dealing with inferential statistics include sampling and the application of significance testing and confidence intervals. The Student-t distribution, chi-squared distribution and F-distribution are also described, and the treatment of errors. It is intended that readers will apply the statistical methods described to the analysis of their own experimental results. For this reason only basic principles are described and textbook problems are not included; there are many books available that include numerous examples and some of these are listed in the Bibliography at the end of Chapter 7.

TREATMENT OF MEASURED DATA

CHAPTER OBJECTIVES

> To encourage the reader to consider:
> 1. the treatment of measured data;
> 2. different types of equations that can be applied to measured data;
> 3. the use of numerical methods of integration;
> 4. the use of numerical methods of differentiation.

QUESTIONS

> - What are empirical equations?
> - How is numerical data presented accurately?
> - What methods are used to check numerical data and calculations?
> - When and how are the methods of numerical integration and numerical differentiation applied?

6.1 TREATMENT OF MEASURED DATA

Considerable effort is often required in order to obtain accurate data. It is essential that the accuracy and reliability are preserved during subsequent analysis, and later in its presentation. This section describes some elementary considerations when handling data, and also some basic methods that are available for interpreting the data.

6.1.1 Measured data

The 'result' or 'solution' of a problem is the important thing. If it is to be of any value, it must be free from inaccuracies. Not only must the method be correct, but the whole work from beginning to end must also be correct. Since everyone is likely to make mistakes, the only way to be reasonably sure of accuracy is to check results. Students are frequently encouraged to check their work (particularly in examinations) before it is finally submitted. This advice is often interpreted to mean reading through calculations, and performing the numerical calculations for a second time. However, although such an approach is useful, it should only be considered as a first stage for performing checks. Repeating calculations is not a particularly efficient way of finding mistakes—the brain appears to see what it wants to see, i.e. a previous answer as correct! Fortunately there are several techniques or methods that can be employed to help locate and eliminate errors.

Inaccurate or inconsistent results may be obtained owing to either mistakes or errors. *Mistakes* occur by misreading or misquoting figures (particularly the interchange of adjacent figures), and by performing incorrect sequences of operations. Mistakes may be found by repeating and rereading calculations, although this approach tends to be more efficient if performed by a third party. Mistakes may also be detected by their consequences upon the results, although this approach is less desirable and more time-consuming. *Errors* occur because of the use of an incorrect method or application, or the use of approximations. Although errors may be unavoidable, their magnitude can often be estimated. The main sources of numerical errors are *rounding off* and *truncation errors*. For example, the value of π is

$$3.1416 \text{ to 4 decimal places}$$

and
$$3.14 \quad \text{to 2 decimal places}$$

In lengthy calculations the resulting errors can accumulate and lead to considerable inaccuracy in the final answer. Many of the data with which engineers and scientists work are obtained experimentally and are subject to small errors.

Truncation errors occur when the terms of an infinite series are terminated at some point after which their magnitude (i.e. contribution to the final sum) is small. For example, the power series

(a) $e^x = 1 + x + \dfrac{x^2}{2!} + \dfrac{x^3}{3!} + \cdots + \dfrac{x^n}{n!} + \cdots \infty$

This series is convergent for all values of x.

(b) $\log(1 + x) = x - \dfrac{x^2}{2} + \dfrac{x^3}{3} - \dfrac{x^4}{4} + \cdots + (-1)^{n+1} \dfrac{x^n}{n} + \cdots$

This series is only convergent if $-1 < x \leqslant +1$.

(c) $\sin x = x - \dfrac{x^3}{3!} + \dfrac{x^5}{5!} - \dfrac{x^7}{7!} + \cdots$

where x is in radians.

Although the contribution of higher terms to the final solution may be negligible, the result will always be in error (or approximate).

The *reliability* of a number may be expressed in terms of either its precision or its accuracy. *Precision* is gauged by the position of the last reliable digit relative to the decimal point, whereas *accuracy* is measured by the number of significant figures. *Significant figures* are those known to be reliable and include any zeros not merely used to locate the decimal point. For example, the following measurements were made (using a micrometer) of the diameters of several ball bearings: 0.251, 0.197, 0.103, 0.097 mm. These diameters have all been measured to a precision of 0.001 mm, and to accuracies of three, three, three and two figures respectively.

The following statements apply to significant figures (SF):

(a) All non-zero digits are significant.
(b) Zero digits which lie between significant (non-zero) digits are significant, e.g. 2.3005 (5SF).
(c) Zero digits which lie to the right of both the decimal point and the last non-zero digit are significant, e.g. 19.00 (4SF), 1.20 (3SF), (0.901 (3SF), 0.006 10 (3SF).
(d) Zeros at the beginning of a decimal fraction are not significant, e.g. 0.003 41 (3SF) but 6.001 (4SF).
(e) Zeros at the end of a whole number are not significant if they are used only to locate the decimal point, e.g. 300 (1SF) but 300.0 (4SF). When one or more such zeros are known to be significant, this is indicated by the 'tilde' ($\tilde{\ }$) written over the last significant zero, e.g. 840$\tilde{0}$, 84$\tilde{0}$0.

The examples in Table 6.1 should help to clarify the definition of significant figures which may appear to be somewhat arbitrary from these statements.

The following rule is used when rounding off numbers:

– Figures are rejected if they represent less than half a unit in the last place to be retained.
– If they represent more than half a unit in the last place to be retained, the last remaining digit is increased by one.
– If the rejected figures represent exactly half a unit in the last place to be retained, then the last retained significant digit, if even, is left even or, if odd, is raised to the next even number.

The examples in Table 6.2 should help illustrate the rules that apply to rounding off.

Numbers are often rounded off either because the last digits are unreliable or because such a level of accuracy is not required for the computation.

For addition and subtraction, the answer is rounded off to correspond to the least precise of the numbers involved; for example, the addition

$$
\begin{array}{r}
119.25 \quad + \\
6.302\,9 + \\
*246.3 \quad + \\
18.007\,94 \\
\hline
389.860\,84
\end{array}
$$

* Least precise, first place of decimals, therefore the answer is given as 389.9.

Table 6.1 Examples of significant figures

Number	Significant figures	Number of significant figures
18.34	1, 8, 3, 4	4
8010	8, 0, 1	3
0.000 63	6, 3	2
0.000 450	4, 5, 0	3
6.500	6, 5, 0, 0	4
4.007 0	4, 0, 0, 7, 0	5
13 5$\tilde{0}$0	1, 3, 5, 0	4
13 50$\tilde{0}$	1, 3, 5, 0, 0	5
20 000	2	1

Table 6.2 Examples of rounding-off numbers

| Number | Rounded off to | | |
	four figures	three figures	two figures
6.5756	6.576	6.58	6.6
10.815	10.82	10.8	11
13.516	13.52	13.5	14
6.0813	6.081	6.08	6.1
346.25	346.2	346	350
346.55	346.6	347	350
345.45	345.4	345	350
345.85	345.8	346	350
40 039	40 040	40 0̃00	4̃0 000
40 651	40 650	40 700*	41 000

* 40 651 is closer to 40 700 than to 40 600.

For multiplication or division, the answer is rounded off to the number of significant figures equal to that of the least accurate quantity in the computation; for example

$$16.2914 \times 39.42 = 642.206\,988$$
$$\text{(6SF)} \qquad \text{(4SF)*}$$

*The product should be rounded off to four significant figures, i.e. 642.2.

The *standard index form* is a convenient way of writing a very large or very small number. The number is expressed as a number between 1 and 10 multiplied by an integral power of 10. For example

$$6\,785\,000 \equiv 6.785 \times 10^6$$

$$0.000\,123 \equiv 1.23 \times 10^{-4}$$

The accuracy of the number is indicated by the number of figures (including zeros) to the right of the decimal point, e.g. 13 000 can be written as 1.30×10^3 (3SF) or 1.300×10^3 (4SF). (This avoids the use of the tilde, 13 0̃00 and 13 0̃0̃0 respectively.)

The other advantages of using this 'scientific' notation are that the magnitude of the number can be seen directly from the exponent, and errors of writing or reading many zeros are avoided.

Before the widespread availability of calculators, the use of scientific notation was often of great benefit for evaluating computations involving several large and/or small numbers. For example

$$\frac{30\,000 \times 0.0065 \times 750\,000}{0.005 \times 250 \times 45\,000} = \frac{3 \times 10^4 \times 6.5 \times 10^{-3} \times 7.5 \times 10^5}{5 \times 10^{-3} \times 2.5 \times 10^2 \times 4.5 \times 10^4}$$

$$= \frac{3 \times 6.5 \times 7.5}{5 \times 2.5 \times 4.5} \times 10^{(4-3+5)-(-3+2+4)}$$

$$= 2.6 \times 10^3$$

Although this problem can be evaluated quite easily using a calculator, the use of the standard index form can provide a useful approximate answer (as a check against your accuracy when inserting numbers into a calculator). For example (again):

$$\frac{3 \times 10^4 \times 6.5 \times 10^{-3} \times 7.5 \times 10^5}{5 \times 10^{-3} \times 2.5 \times 10^2 \times 4.5 \times 10^4} = \frac{3 \times 1.3 \times 1}{5 \times 0.3} \times \frac{10^5}{10^2}$$

$$= \frac{1.3}{0.5} \times 10^3 = 2.6 \times 10^3$$

To conclude this section a few words about common sense—or what should be common sense! These comments are what I refer to as 'the curse of the calculator'. I have found that students nearly always want to present exactly what the calculator produces. As already discussed, the number of significant figures in the original data determines the accuracy of the final answer. The calculator will compute 59^2 as 3481, however the answer to be presented or used is 3500 (2SF). The situation is even worse when mathematical constants such as π and e are used as calculator functions. The circumference of a circle of diameter 2.4 cm can then be computed ($\pi \times$ diameter) as 7.539 822 369 cm. Again despite the ten-digit display, the answer is only accurate to two significant figures, i.e. 7.5 cm.

Finally the common sense. Students need to think about the accuracy that is required from practical considerations. Suppose all measurements are made or are available to an accuracy of five significant figures (is this necessary?) and a required tank diameter may be calculated (say) as 3.6789 m. However, it is unlikely in engineering situations that such a tank would be constructed (or be necessary) to the nearest 1/10 mm. The sensible answer would be:

tank diameter = 3.6789 m (5SF)

$\approx$ 3 m 68 cm (approx.)

Students appear reluctant to make approximations based on practical considerations, and yet when large amounts of data are being presented the inclusion of unnecessary digits merely obscures any comparisons that are required.

6.1.2 Equations

There are two types of equations that the student will encounter: these are analytical (or theoretical) equations and empirical equations.

Analytical equations are based upon the underlying assumptions that are made and the theoretical (mathematical) development that is used to arrive at the required final equation. In most branches of engineering the theory related to the fundamental areas of study is well established and can be found in many different textbooks. It is not usually necessary to derive an equation before it can be used to evaluate data. However, it should be remembered when intending to use analytical equations that certain initial assumptions, boundary conditions or limiting values apply to their derivation. It is essential to check that these assumptions apply to the situation and the data that are to be tested. Comparison of data with an inappropriate theory is not only invalid but also misleading.

Empirical equations are based upon observation and experience. Available data are analysed (often presented graphically) in order to establish the relationship between the main variables. Some data are presented and used in graphical form (in nomographs, see Sec. 9.2.1) although it is usually more convenient if a relationship can be expressed as an equation. The data that are used will have been obtained from particular experimental systems and it is important to establish that these systems are comparable or compatible with your own situation. It is often thought that the more data that are available, then the 'better' the empirical relationship that is obtained— quantity implying quality! However, if a large number of different experimental systems have been used to obtain the data, then some of these systems may not have been reliable or applicable to the particular situation. In such cases, the variation in the data may lead to empirical equations which are only generally applicable, or have large uncertainties in certain situations. It is also worth considering whether the range over which the data were measured is generally applicable, or whether several mechanisms/effects may have occurred in different ranges. In this case, the use of several empirical equations may be more appropriate.

Equations are used to check the applicability and reliability of physical data. Equations are also used to compare the results obtained from a physical experiment with the predicted results obtained from the (one hopes!) appropriate theory. If considerable effort has been made to ensure that accurate data are obtained, then this accuracy should be preserved during any subsequent analysis. Errors can occur from the use of equations and these errors should be minimized if the work is to be acceptable and worth while.

Having ascertained that the equations to be used are appropriate to the situation under investigation (or that the data may be used to test whether the theory is applicable in this particular situation), the first check which *should be* performed is for the correctness of the terms of the equation. It is often assumed that the equations written in books or journal articles are correct. However, typographic errors do occur and to proceed with an analysis based upon an incorrect equation will not yield any useful results. The equations to be used should be obtained from different and independent sources, not those sources quoted in the original book or article, and then compared. It is important to check the values of any constants and the exponents, and the definitions of the symbols used in the equations. Finally, it is necessary to check the units that should be used for each term in the equation, the units of the constants if these are not dimensionless, and the units in which the calculated value will be obtained. Consider the equation for the volume of a sphere. Is it

$$V = 4\pi r^3 \quad \text{or} \quad V = \tfrac{4}{3}\pi r^3?$$

Are the following equations correct?

$$s = \left(\frac{u - v}{2}\right)t \qquad \text{(laws of motion)}$$

$$V = I^2 R \qquad \text{(electricity)}$$

$$\Delta P = 32\,\frac{v\mu L}{D^2} \qquad \text{(pressure drop)}$$

$$\frac{\tau}{R} = \frac{T}{J} = \frac{G}{\theta L} \qquad \text{(torsion equation)}$$

$$Q = \sigma^4 \varepsilon T \qquad \text{(Stefan–Boltzmann radiation law)}$$

$$Q = \tfrac{2}{3}bH^{2/3} \qquad \text{(flow over a rectangular weir)}$$

Even though a particular subject or equation may be very familiar, mistakes are not always immediately recognizable.

Some errors in writing equations can be found by the use of dimensional checking. Every equation should be checked for dimensional homogeneity before it is used. A discussion of units and dimensions is presented in Chapter 1 and dimensional analysis and dimensionless groups are discussed in Chapter 8. If any constants present in an equation are not dimensionless, their appropriate values and dimensions should be ascertained and carefully checked. For mechanical systems, the fundamental quantities are mass (or force), length and time. In purely electrical systems, it is convenient to use the fundamental dimensions of volt, ampere and second. For thermal systems, three convenient units are quantity of heat, temperature and time. If only one type of physical system is being considered, three fundamental dimensions are sufficient to describe all quantities in that system. However, if another type of system is introduced into the equations, another fundamental dimension (chosen from the new system) is necessary to relate completely all quantities in the two combined systems. If force, length and time are retained as the fundamental mechanical dimensions, then temperature can be added for thermal equations and charge for electrical equations. Equations containing a combination of two or more types of quantities should not present problems of dimensional checking.

Note that a dimensional check will not check purely numerical factors, signs or missing terms. However, it will disclose many common errors and has the advantage of being easy to apply. The method can be used to check the original equations, the solution, or any intermediate step. Obviously the equations must be kept in purely literal form, because in a numerical equation the dimensions of the quantities are obscured.

Another method of checking is to work the problem through by an entirely different method. This may double the amount of work involved, but it provides an excellent check on the mathematical aspects.

Calculations, both literal and numerical, can be compared with the results obtained by other workers. If this method is to be at all reliable, then the work must be entirely independent. When examining the work of others, it is extremely easy to be led into the same mistakes that they made; always guard against this possibility.

Except for dimensional checking, the other methods described require considerable extra work and are not always suitable. Probably the most obvious method of checking is to examine the results in the light of experience and to be satisfied that they are really reasonable in magnitude. Surprisingly, this type of check is not often used.

The *limiting case method* is a quick and easy check that is particularly useful when combined with dimensional checking. The principle of the method is to simplify the equation by the removal of enough quantities (using zero or infinity as a limit) to allow the correctness of the remaining terms to be determined. Either the original equations can be rewritten and solved, or the equations can be interpreted by physical reasoning. The check should be applied several times with different quantities removed each time, as the quantities removed cannot be checked. This method will usually determine missing or extra terms; it is a good check on the form of the remaining terms, and it often indicates incorrect signs. Consider a very simple example, the equation for the total surface area of a cylinder:

$$A = 2\pi r^2 + 2\pi r L$$

When $L = 0$ (or is very small), the expression becomes

$$A = 2\pi r^2$$

This is the area presented by the two sides of a disc of negligible thickness. If $r = 0$ or ∞, then $A = 0$ or ∞ as would be expected. Letting $A = 0$, then:

$$r(r + L) = 0$$

i.e. $r = 0$ or $r = -L$, neither condition being a practical possibility.

Finally in this section, consider the following hypothetical equations:

$$\text{(a) } C = 3.82L^2 - 4t$$

$$\text{(b) } C = 4L^2 - 3.82t$$

$$\text{(c) } C = 2.38L^2 - 4t$$

It would be possible to use dimensional checking to determine whether the exponents of L and t were correct. However, it would not be possible to substitute values of L and t to obtain an order of magnitude check for C. The values of the constants are very similar, and differences between computed values of C would be small. However, if the equation was written as

$$C = 38.2L^2 - 4t$$

the value of C would be very much greater than predicted by cases (a), (b) and (c), for comparable values of L and t.

Note: Angles and exponents are dimensionless; for example

$$\tan \theta = \left(\frac{Ct - x}{Z} \right)$$

and

$$P = Q \exp(x/y)$$

(P and Q have the same units.)

6.2 NUMERICAL INTEGRATION AND DIFFERENTIATION

Problems can sometimes be formulated in terms of mathematical equations which then enable an analytical solution to be obtained. However, this closed-form solution may not be acceptable to an engineer who needs a numerical result. The time required to obtain a solution in closed form (if this is possible) may have been better spent in performing a numerical procedure.

If the equations to be solved contain derivatives, there are several possibilities:

1. If there is a single equation that can be put into the form $dy/dx = f(x)$, the solution can be expressed as $y = \int f(x)\,dx$.
    (a) If $f(x)$ can be expressed in equation form, the solution $y = \int f(x)\,dx$ may be evaluated by integral calculus.
    (b) If $f(x)$ is expressed empirically, then the solution $y = \int f(x)\,dx$ can be evaluated by numerical or graphical integration.
2. If the equations involve derivatives in such a manner that they cannot be expressed as $y = \int f(x)\,dx$, the methods of differential equations must be used. An analytic solution can be obtained if all the terms can be written explicitly in equation form. Alternatively, graphical or numerical methods of integration must be used. These methods produce a result in graphical form (rather than as an equation), and the procedure must be repeated for each problem. Therefore, mathematical methods are usually preferred because of their general applicability.

6.3 GRAPHICAL AND NUMERICAL INTEGRATION

The definite integral $y = \int_b^a f(x)\,dx$ can be evaluated by plotting $Z = f(x)$ against x, and determining the area under the curve between the limits $x = a$ and $x = b$. Several methods exist for determining this area; the four most widely used are:

(a) counting the squares;
(b) trapezoidal rule;
(c) Simpson's rule;
(d) graphical integration.

6.3.1 Counting the squares

Counting the squares enclosed under the curve, as shown in Fig. 6.1, can be a very accurate method if appropriate scales are chosen for the axes, and graph paper with small squares is used. For squares that are cut by the curve:

> If the cut square is greater than or equal to half a square, then add one square to the total; otherwise ignore the cut square.

The cut squares will tend to cancel each other and they usually represent only a small proportion of the total area.

6.3.2 Trapezoidal rule

The trapezoidal rule is used to evaluate an integral of the form

$$\int_{x_0}^{x_n} f(x)\,dx$$

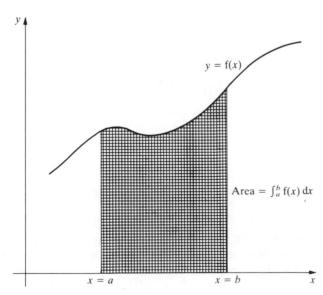

Figure 6.1 Evaluating an integral (finding the area under a curve) by counting the squares.

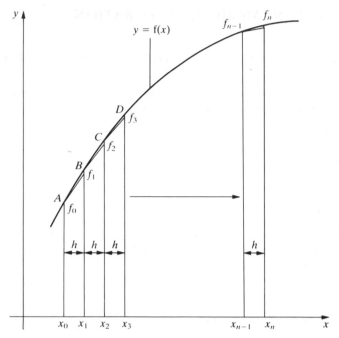

Figure 6.2 Evaluating an integral (finding the area under a curve) by the trapezoidal rule.

The interval or range from x_0 to x_n is divided into n equal strips of width h, where

$$h = \frac{x_n - x_0}{n}$$

Adjacent ordinates on the curve ($f_0, f_1, f_2, \ldots, f_n$, in Fig. 6.2) are joined by a series of straight lines AB, BC, etc. The sum of the areas of the resulting trapezoids is approximately equal to the area under the curve $y = f(x)$. The areas of the trapezoids in Fig. 6.2 are

$$A_1 = \left(\frac{f_0 + f_1}{2}\right) h$$

$$A_2 = \left(\frac{f_1 + f_2}{2}\right) h$$

$$\vdots$$

$$A_n = \left(\frac{f_{n-1} + f_n}{2}\right) h$$

The sum of the above areas (approximating to the area under the polygonal curve) is

$$\tfrac{1}{2}h(f_0 + f_1) + \tfrac{1}{2}h(f_1 + f_2) + \cdots + \tfrac{1}{2}h(f_{n-1} + f_n) = h[\tfrac{1}{2}(f_0 + f_n) + f_1 + f_2 + \cdots + f_{n-1}]$$

The error obtained when using this method depends upon the shape of the curve in the interval over which the approximation is made. Using the trapezoidal rule for the curve shown in Fig. 6.2 would provide an underestimate for the area.

If the integrand contains no points of inflection, then an overestimate (of the area) can be obtained by drawing the tangent at alternate points. This method assumes that n is even.

The area of the first strip (now of width $2h$) is approximately $2h \times f_1$ and the total area is

$$2h[f_1 + f_3 + f_5 + \cdots + f_{n-1}] \qquad (n \text{ even})$$

It is evident that smaller values of h will produce greater accuracy but will require more computation.

Problem Show that $\displaystyle\int_0^1 \sin x \, dx$ using ten intervals is

$$0.459\,33 \quad \text{for a first approximation (deficit)}$$

and

$$0.460\,47 \quad \text{for a second approximation using strips of width } 2h.$$

Also show that the correct answer (obtained analytically) to 5 decimal places is $0.459\,70$.

6.3.3 Simpson's rule

Unlike differentiation, integration is a 'smoothing' process; the errors that occur when a given curve is replaced by an approximating polynomial tend to cancel one another out. Whereas the trapezoidal rule consists of fitting a linear equation between each pair of adjacent ordinates, greater accuracy can be obtained by fitting a second-degree equation (a parabola) to three points. (Simpson's rule also gives exact results for a third-degree curve passing through the three equidistant points.)

The range of the abscissa (defined by the limits of the integrand) is divided into an *even* number $(2n)$ of strips. The portions of the curves ABC, CDE, etc., are replaced by arcs of parabolas as shown in Fig. 6.3. The area of the trapezoid is

$$A_{\mathrm{t}} = \left(\frac{f_0 + f_2}{2}\right) 2h$$

The height of the curve above the trapezoid (shown shaded in Fig. 6.3) is given by

$$y = d\left[1 - \frac{x^2}{h^2}\right]$$

The x-axis is shifted so that f_1 occurs at x_0. The shaded area (A_{p}) between the trapezoid and the curve is obtained by

$$A_{\mathrm{p}} = \int_{-h}^{+h} y \, dx$$

Substituting the above expression for y, the area can be shown to be

$$A_{\mathrm{p}} = (\tfrac{2}{3}d)2h$$

The centre height (d) is

$$d = f_1 - \left(\frac{f_0 + f_2}{2}\right)$$

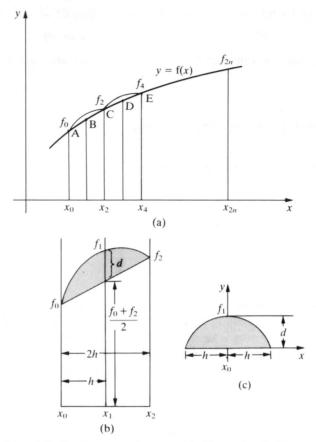

Figure 6.3 Evaluating an integral (finding the area under a curve) by Simpson's rule: (a) approximating a section of the curve by parabolas; (b) defining the strips; (c) shifting the 'origin' of a strip.

Therefore

$$A_1 = A_p + A_t$$

$$= \frac{2}{3}\left[f_1 - \left(\frac{f_0 + f_2}{2}\right)\right]2h + \left(\frac{f_0 + f_2}{2}\right)2h$$

$$= \frac{h}{3}(f_0 + 4f_1 + f_2)$$

Similarly

$$A_2 = \frac{h}{3}(f_2 + 4f_3 + f_4)$$

Because each area covers two increments (h), the number of increments must be even. The total area is given by

$$A = \int_{x_0}^{x_{2n}} f(x)\, dx \approx \frac{h}{3}\left[(f_0 + f_{2n}) + 2(f_2 + f_4 + f_6 + \cdots + f_{2n-2}) + 4(f_1 + f_3 + \cdots + f_{2n-1})\right]$$

This is Simpson's rule.

An alternative derivation is also presented as follows. Again using an even number of strips of width h and taken in pairs, e.g. $x_0 \rightarrow x_2$ of width $2h$. Replace the portions of the curves ABC, CDE, etc., by arcs of parabolas.

Shift the origin to x_0, therefore $x = X + x_0$, where X is a dummy variable. Let the equation of the parabola of ABC be

$$y = y_0 + bX + cX^2$$

The area of the two strips under ABC is

$$\int_0^{2h} (y_0 + bX + cX^2)\,dx = 2h(y_0 + bh + \tfrac{4}{3}ch^2)$$

$$= \tfrac{1}{3}h(y_0 + 4y_1 + y_2)$$

since $y_1 = y_0 + bh + ch^2$ and $y_2 = y_0 + 2bh + 4ch^2$.

Applying this result n times to the intervals $(x_0 \rightarrow x_2)$, $(x_2 \rightarrow x_4)$, etc., the total area is

$$A = \int_{x_0}^{x_{2n}} f(x)\,dx \approx \frac{h}{3}\left[(f_0 + 4f_1 + f_2) + (f_2 + 4f_3 + f_4) + \cdots\right.$$

$$\cdots + \left. (f_{2n-2} + 4f_{2n-1} + f_{2n})\right]$$

$$\approx \frac{h}{3}\left[(f_0 + f_{2n}) + 2(f_2 + f_4 + f_{2n-2}) + \cdots\right.$$

$$\cdots + \left. 4(f_1 + f_3 + \cdots + f_{2n-1})\right]$$

Stated in words:

The total area is equal to 1/3 of the common interval (h) multiplied by (the sum of the first and last ordinates + twice the sum of the remaining even ordinates + four times the sum of the odd ordinates)

An advantage of Simpson's rule is that larger increments can be chosen than with the trapezoidal method, but the same accuracy can still be obtained. As the size of h decreases, so the accuracy improves.

Repeat the previous problem for the trapezoidal rule, and show that by Simpson's rule:

$$\int_0^1 \sin x\,dx = 0.459\,70 \text{ (to 5 decimal places)}$$

6.3.4 Graphical integration

Graphical integration described here is essentially a graphical application of Simpson's rule. The method is simple and it has the advantage that the increment size can be varied, as the method is used, to suit particular portions of the curve, e.g. one large increment can be used where the curve is approximately linear.

The plot of the function to be integrated should be divided by vertical lines into sections that can be approximated by parabolic curves. Since a parabola has no point of inflection, lines of division should pass through all inflection points. Further subdivisions should be used as

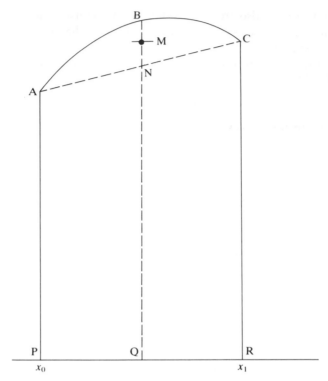

Figure 6.4 Graphical integration (nomenclature for the first strip).

necessary to make the sections approach parabolic curves. The areas of all sections can then be added together.

The basis of this method will now be described with reference to Fig. 6.4. Assume that the curve $y = f(x)$ can be approximated within the interval P to R by a second-degree equation, as for Simpson's rule. The average height of the trapezoid ACPR is QN, where Q is the mid-point of the line PR. The average height of the portion ABC above the trapezoid is approximately equal to two-thirds of NB, i.e. NM. Therefore, the mean ordinate of the curve ABC is the height QM.

To find the total area under the curve ABC, draw the straight line AC and the middle vertical line BQ. Locate the point M either by eye or measurement to be two-thirds of the distance between N and B. The area under the curve is equal to QM × PR, i.e. average height × base length.

A procedure will now be described that avoids the need to calculate the area of each individual section. The average height of each section is determined as previously described (with reference to Fig. 6.4), i.e. points M_1, M_2, etc., in Fig. 6.5. Refer to Fig. 6.5 and consider the first two increments bounded by PA to RC, and RC to TE. Draw the straight line $M_1 M_2$ and locate the point M_{12} above the mid-point (Z_1) of the base PT. This applies a weighting to the average heights M_1 and M_2 according to the width of their bases, and it can be shown by simple geometry that the area of the first two sections (i.e. total base × height) is

$$PT \times Z_1 M_{12} = (PR \times QM_1) + (RT \times SM_2)$$

Proceeding to the third section, a straight line is drawn between points M_{12} and M_3 (see Fig. 6.5) and the point M_{123} is located above the mid-point of the total base PV. The area of the three combined sections is $PV \times Z_2 M_{123}$ (total base length × height).

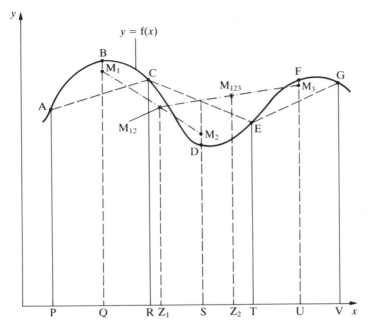

Figure 6.5 Procedure for graphical integration.

This procedure can be continued indefinitely and only one (final) calculation is required:

total base length of interval × final average height

= total area enclosed under the curve

Problem The following tabulation details the experimentally determined velocity of a fluid flowing in a 12 cm diameter cylindrical pipe, at various distances from the pipe centre.

Radius (r, cm) i.e., distance from pipe centre	0	1.2	2.4	3.6	4.8	5.4	5.7	6.0
Velocity (v, m/s)	0.390	0.383	0.363	0.331	0.279	0.234	0.200	0

Calculate the total flow rate in m³/hour.
Determine the relationship between flow rate and $\int v \, d(r^2)$.

6.4 NUMERICAL DIFFERENTIATION

Most graphical methods of obtaining a derivative, e.g. finding the local value of the slope, are inaccurate. Those methods that do provide a satisfactory answer are time-consuming because of the care that is required in the calculations. Numerical differentiations, and tabular methods in particular, are therefore usually preferred.

6.4.1 Finite difference method

Assume that numerical data are available for the curve $y = f(x)$, and that h is a small increment in x. Two estimates can be made of the value of $f'(x)$, ie dy/dx, at the point x, these are:

$$f'(x) = \frac{f(x + h) - f(x)}{h}$$

and

$$f'(x) = \frac{f(x) - f(x - h)}{h}$$

These are the *forward and backward differences*, respectively.

The *central difference* often provides a better estimate; this is

$$f'(x) = \frac{f(x + h) - f(x - h)}{2h}$$

Example 6.1 Consider the following 'smoothed' data for the velocity (v) of a car as a function of time (t):

Time (t, s)	0	2.5	5.0	7.5	10.0	12.5	15.0	17.5	20.0	22.5
Velocity (v, km/h)	90	70	54	42	33	26	20	16	12	9

Time (t, s)	25.0	27.5	30.0	32.5	35.0	37.5	40.0
Velocity (v, km/h)	7	5	4	3	2	1	0

Use these data to find the first derivative, i.e. the acceleration dv/dt. The time intervals are wide where the velocity changes rapidly, and the central difference equation should then be used. The forward and backward differences are used to calculate the initial and final accelerations. From these data and calculations, it can be shown that the central difference value is the average of the accelerations calculated using the forward and backward difference equations. The values obtained are:

t (s)	0	2.5	5.0	7.5	10.0	12.5	15.0	17.5	20.0	22.5
dv/dt (km/h per s)	-80	-72	-56	-42	-32	-26	-20	-16	-14	-10

t (s)	25.0	27.5	30.0	32.5	35.0	37.5	40.0
dv/dt (km/h per s)	-8	-6	-4	-4	-4	-4	-4

By plotting these data on suitable semi-logarithmic graph paper, the following empirical equation is obtained:

$$dv/dt = -90 \exp(-0.101t)$$

The acceleration is negative, i.e. a retardation, and the units are km/h per s. The empirical equation above can be multiplied by an appropriate conversion factor to yield the acceleration in m/s²:

$$dv/dt = -90 \times \frac{1000}{3600} \exp(-0.101t) \quad \text{m/s}^2$$

$$= -25 \exp(-0.101t) \quad \text{m/s}^2$$

Example 6.2

x	0.350	0.399	0.401	0.450
$f(x)$ (4SF)	0.3429	0.3885	0.3903	0.4350

For the smaller (middle) interval:

$$f'(x) = \frac{f(0.401) - f(0.399)}{0.401 - 0.399} = \frac{0.0018}{0.002} = 0.9 \text{ (1SF)}$$

For the larger (outer) interval:

$$f'(x) = \frac{f(0.450) - f(0.350)}{0.450 - 0.350} = \frac{0.0921}{0.10} = 0.921 = 0.92 \text{ (2SF)}$$

The larger interval provides a better estimate, as only one significant figure is available for the value from the smaller interval.

In this example, $f(x) = \sin x$, therefore $f'(x) = \cos x$. Hence at $x = 0.4$:

$$f'(x) = \cos x = 0.9211 \text{ (4DP)}$$

6.4.2 Step-by-step integration method

The step-by-step integration method can be used to evaluate the solution of y vs x, if the differential equation can be expressed as $dy/dx = f(x, y)$.

Let the initial condition be $y = y_0$ at $x = 0$. Choose a small increment Δx such that the variable y will not change significantly (say <10 per cent), depending upon the required accuracy.

(a) Calculate dy/dx at $x = 0$ by substituting the initial condition into the differential equation $dy/dx = f(x, y)$.
(b) Calculate an approximate value of y at $x = 0 + \Delta x$, by assuming that the slope remains constant over the increment:

$$y_1 = y_0 + \Delta x \left.\frac{dy}{dx}\right|_0$$

(c) Calculate the slope $\left.\dfrac{dy}{dx}\right|_1$ at y_1 by substituting $y = y_1$ and $x = \Delta x$ into the differential equation.
(d) The value of y_1 obtained in (b) is in error because the slope is not constant. A more accurate value is

$$\left.\frac{dy}{dx}\right|_{av} = \frac{1}{2}\left(\left.\frac{dy}{dx}\right|_0 + \left.\frac{dy}{dx}\right|_1\right)$$

Recalculate y_1 using this average slope:

$$y_1 = y_0 + \Delta x \left.\frac{dy}{dx}\right|_{av}$$

(e) Repeat steps (c) and (d) until successive values of y_1 are within the required accuracy.
(f) Repeat the procedure across the next increment until successive values of y_2 converge, then the third increment, etc.

Note: If the increment Δx is chosen small enough, only a few iterations are required to produce a constant value of y. The number of iterations will also be considerably reduced if step (b) is omitted in favour of steps (c) and (d).

Example 6.3 A load of 218 g is to be brought to rest by a resistance (including friction) as given in the following tabulation:

Velocity (v, m/s)	12	10	8	6	4	2	0
Resistance (F, N)	9.4	7.3	5.6	4.2	3.1	2.4	2.1

Therefore, the resistance is a function of velocity, i.e. $F(v)$. The initial condition of the load is $v = 12.0$ m/s at $t = 0$. The load is also subjected to a forward thrust equivalent to $4(1 - t)$ newtons.

Hence

$$dv/dt = f(v, t)$$

It can be shown that

$$\frac{dv}{dt} = 0.218 \times 9.81[4(1 - t) - F(v)]$$

that is

$$\frac{dv}{dt} = 2.14[4(1 - t) - F(v)]$$

Choose $\Delta t = 0.1$ s.

(a) $v = 12$ m/s at $t = 0$; $F(v) = 9.4$ N. Therefore

$$\left.\frac{dv}{dt}\right|_0 = -11.6 \text{ m/s}^2$$

(b) At $t = 0.1$ s:

$$v_1 = v_0 + \Delta t \left.\frac{dv}{dt}\right|_0$$

$$v_1 = 12 + (0.1)(-11.6) = 10.84 \text{ m/s}$$

(c) From the tabulation (i.e. plot the values), for $v_1 = 10.84$ m/s then $F = 8.15$ N. Therefore

$$\left.\frac{dv}{dt}\right|_1 = 2.14[4(1 - 0.1) - 8.15]$$

$$= -9.73 \text{ m/s}^2$$

(d) $\left.\dfrac{dv}{dt}\right|_{av} = \dfrac{1}{2}(-11.6 + -9.73) = -10.67 \text{ m/s}^2$

$$v_1 = v_0 + \Delta t \left.\frac{dv}{dt}\right|_{av}$$

$$= 12 + (0.1)(-10.67)$$

$$= 10.93 \text{ m/s}$$

Therefore determine a new value of F from the table, and a new value of

$$\left.\frac{dv}{dt}\right|_1 = -9.95 \text{ m/s}^2$$

Hence

$$\left.\frac{dv}{dt}\right|_{av} = \frac{1}{2}(-11.6 - 9.95) = -10.78 \text{ m/s}^2$$

$$v_1 = 12 + (0.1)(-10.78) = 10.92 \text{ m/s}$$

This is sufficiently close to the previous trial and the calculations can proceed to the next increment ($t = 0.3$ s). Results of the calculations are tabulated as follows:

t	v	dv/dt	$(dv/dt)_{av}$
0	12.00	−11.6	
0.1	10.84	−9.73	−10.67
	10.93	−9.95	−10.78
	10.92		
0.2	9.92	−8.66	−9.31
	9.99	−8.78	−9.37
	9.98		
0.3	9.10	−7.92	−8.35
	9.14	−8.02	−8.40
	9.14		

Note: The step-by-step integration method can be extended to differential equations other than the simple type:

$$\frac{dy}{dx} = f(x, y)$$

The original differential equation is reduced to an equation containing only a single first derivative by substituting new variables in differential form. As an example:

$$\frac{d^3y}{dx^3} + A\frac{d^2y}{dx^2} + B\frac{dy}{dx} + Cy = f(x)$$

where the coefficients A, B and C can be functions of x and y.

Let $$m = \frac{dy}{dx}, \quad \text{then} \quad \frac{d^2m}{dx^2} + A\frac{dm}{dx} + Bm + Cy = f(x)$$

Let $$n = \frac{dm}{dx}, \quad \text{then} \quad \frac{dn}{dx} = f(x) - An - Bm - Cy$$

Alternatively

$$\frac{d^2y}{dx^2} + A\frac{dy}{dx} + B = 0$$

where A and B are not functions of x.

Let
$$\frac{dy}{dx} = Z, \quad \text{then} \quad \frac{dZ}{dx} = -(AZ + B)$$

Therefore

$$\frac{dZ}{dx} \bigg/ \frac{dy}{dx} = \frac{dZ}{dy}$$

$$= -\frac{(AZ + B)}{Z}$$

If A and B are functions of y but not of x, a solution can be obtained using step-by-step integration. Determine $Z = f(y)$; hence

$$x = \int \left(\frac{1}{Z}\right) dy \quad \text{or} \quad x = -\int \frac{1}{AZ + B} dZ$$

Numerical or graphical integration may be required.

EXERCISES

Numerical problems are deliberately not included in these exercises; readers should apply the methods and ideas described in this chapter to the treatment of their own experimental results. This will provide a more useful exercise than merely working through a series of arbitrarily defined problems.

6.1 Assess some of your written work, especially laboratory reports, with respect to the accuracy and precision of the numerical data. Determine whether the numerical data are expressed in an appropriate form, i.e. whether they use the standard index form of notation, number of significant figures, rounding-off numbers, etc.

6.2 What methods do you normally use to check your numerical data and the calculations that are performed?

6.3 Explain the difference between analytical and empirical equations. What are the advantages and limitations of using each type of equation?

6.4 Describe what is meant by dimensional checking.

6.5 Explain the use of the limiting case method for checking equations.

6.6 What types of errors are not revealed by using the methods given in Exercises 6.4 and 6.5.

6.7 Apply the methods of numerical integration and numerical differentiation described in Secs 6.3 and 6.4 to the analysis of data obtained in your laboratory classes.

If additional practice of problem solving (similar to the worked examples given in this chapter) is required, then the reader can substitute other functions having known analytical solutions in the problems, e.g. numerical integration or differentiation of $\tan x$ between appropriate limits.

BIBLIOGRAPHY

Adey, R. and C. Brebbia, *Basic Computational Techniques for Engineers*, Wiley, New York, 1982.
Austin, J. and M. Isern, *Technical Mathematics*, 3rd edn, Holt, Rinehart and Winston, New York, 1983.

Bender, C. M. and S. A. Orszag, *Advanced Mathematical Models for Scientists and Engineers*, McGraw-Hill, New York, 1978.

Calter, P., *Schaum's Outline of Technical Mathematics*, McGraw-Hill, New York, 1979.

Colton, D. L. and R. P. Gilbert (eds), *Constructive and Computational Methods for Differential and Integral Equations*, Springer-Verlag, Berlin, 1974.

Ferziger, J. H., *Numerical Methods for Engineering Applications*, Wiley, New York, 1981.

Fried, I., *Numerical Solutions of Differential Equations*, Academic Press, Orlando, Fla, 1979.

Hall, G. and J. M. Watt (eds), *Modern Numerical Methods for Ordinary Differential Equations*, Oxford University Press, 1976.

Jeffrey, A., *Basic Mathematics for Engineers and Technologists*, Thomas Nelson, London, 1974.

Johnston, R. L., *Numerical Methods: A Software Approach*, Wiley, New York, 1982.

Lambert, J. D., *Computational Methods in Ordinary Differential Equations*, Wiley, New York, 1973.

Ortega, J. M. and W. G. Poole, *An Introduction to Numerical Methods for Differential Equations*, Pitman, Marshfield, MA, 1981.

Reddy, J. N., *Applied Functional Analysis and Variational Methods in Engineering*, McGraw-Hill, New York, 1986.

Stephenson, G., *Worked Examples in Mathematics for Scientists and Engineers*, Longman, London, 1985.

Tuma, J. J., *Engineering Mathematics Handbook*, 2nd edn, McGraw-Hill, New York, 1979.

Wells, G. L. and P. M. Robson, *Computation for Process Engineers*, International Ideas, Philadelphia, 1974.

STATISTICS AND TREATMENT OF ERRORS

CHAPTER OBJECTIVES

To encourage the reader to consider:
1. the applications of descriptive statistical methods;
2. the applications of inferential statistical methods;
3. the ways of dealing with small samples;
4. the use of non-parametric statistics;
5. the use of regression and correlation techniques for the analysis of data;
6. interpolation and extrapolation, and the graphical interpretation of data;
7. appropriate methods for error analysis.

QUESTIONS

- Which statistical parameters are usually calculated in order to describe numerical data?
- Which statistical methods are applied to the analysis of experimental results?
- What is a representative sample? How is it obtained?
- Why are samples and sampling methods important?
- How are correlation and regression analysis applied?

7.1 INTRODUCTION

This chapter does not provide detailed coverage of the basic introductory topics of statistics, e.g. mean, standard deviation, probability, etc. These are usually referred to as *descriptive statistics*. It has been assumed (although perhaps this is incorrect) that the student has already studied this material. For those who have not, or who require some revision work, there are *very* many adequate textbooks concerned with the fundamentals of statistics. A bibliography of suitable texts is not included here; it will be a useful exercise for the student to survey the books that are actually available in the library (probably several shelves), and to select one or two that suit the individual. The skills of information retrieval and qualitative assessment required in this exercise will be valuable not only for experimental work but also for many aspects of the course, especially project work.

This chapter includes in Sec. 7.2 a summary of statistical terms for descriptive statistics, which includes the relevant formulae and brief explanatory notes. This is followed by selected topics, in Sec. 7.3, which are often referred to as *inferential statistics*. These topics include data sampling, descriptions of some specialized statistical techniques, and finally consideration of correlation, curve fitting and error analysis.

As this chapter is intended to provide only an introduction (or a refresher) to the subject, and to act as a reference/revision source, detailed problems are not included.

If 'standard' problems were included, they would be inadequate for the needs of most students when analyzing their experimental data. It is hoped that students wishing to apply statistical techniques to their data will be able to refer to several textbooks in order to find problems/examples relevant to their particular work.

7.2 SUMMARY OF STATISTICAL TERMS AND FORMULAE— DESCRIPTIVE STATISTICS

As already stated, this section provides only a summary of terms and formulae commonly used in statistics. It is intended to provide a reference source for the student concerned with engineering experimentation. Students should be capable of investigating for themselves such topics as probability, sampling and correlation. At the same time, students should be asking what applications these techniques have in experimentation, discussing some of their own ideas of typical situations, and applying these methods to the treatment of their own experimental results.

7.2.1 Frequency distributions

Data are often presented in frequency distributions and some examples are shown in Fig. 7.1.

7.2.2 Measures of average and dispersion

The *mode (modal value, modal class interval)* is the most popular measured value, i.e., the value occurring most often or with highest frequency. The modal class interval and modal value are indicated on the histogram and the frequency polygon (Fig. 7.1).

The *median* is the $[(n + 1)/2]$ observation or measurement when measured values are arranged in order of magnitude. The median value corresponds to the *50th percentile value*.

The range of measurements below the *25th percentile* is known as the *lower quartile*, that above the *75th percentile* is known as the *upper quartile*. Between the 25th and 75th percentiles is the *inter-quartile range*.

For a set of *n* values, $x_1, x_2, \ldots, x_n$, the *arithmetic mean* ($\bar{x}$) is given by:

$$\bar{x} = \frac{\sum x}{n}$$

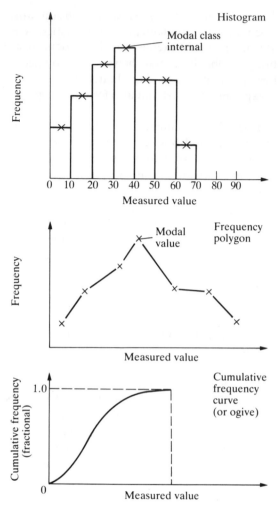

Figure 7.1 Presentation of data–frequency distributions (from Ray, *Elements of Engineering Design*, Prentice-Hall International (UK) Ltd, 1985).

The *variance* (σ^2) is

$$\sigma^2 = \frac{\sum x^2}{n} - \left(\frac{\sum x}{n}\right)^2 = \frac{\sum x^2}{n} - (\bar{x})^2$$

The *standard deviation* (σ) is

$$\sigma = \sqrt{\left(\frac{\sum x^2}{n} - (\bar{x})^2\right)}$$

The *coefficient of variance* provides some indication of the relative importance of the standard deviation referred to the mean. It indicates the spread of values about the mean, although it is less useful when the mean is very small in absolute size.

$$\text{Coefficient of variance} = \frac{\text{standard deviation}}{\text{mean}}$$

The *geometric mean* is

$$\sqrt[n]{(x_1 \times x_2 \times \cdots \times x_n)}$$

The values of the arithmetic mean, the median and the mode for a set of measurements are usually close together. The value of the arithmetic mean is greater than that of the geometric mean.

Weighted averages The arithmetic mean of n quantities $x_1, x_2, x_3, \ldots, x_n$ is given by

$$\frac{x_1 + x_2 + \cdots + x_n}{n} = \frac{\sum x}{n}$$

If each quantity is weighted by amounts $w_1, w_2, \ldots, w_n$ respectively, then the *weighted arithmetic mean* is

$$\frac{w_1 x_1 + w_2 x_2 + \cdots + w_n x_n}{w_1 + w_2 + \cdots + w_n} = \frac{\sum_{i=1}^{n} w_i x_i}{\sum_{i=1}^{n} w_i}$$

The weights (w_i) are used to provide more (large value of w) or less (small value of w) importance to the effect of $x_1, \ldots, x_n$ in calculating the final weighted average.

For a frequency distribution, the mean ($\sum fx / \sum f$) is a weighted mean of the values of x with the frequencies acting as the weights. The weighted arithmetic mean is usually calculated, although the idea can be applied to other averages. For example

$$\text{weighted geometric mean} = \sqrt[N]{[(x_1)^{w_1} + (x_2)^{w_2} + \cdots + (x_n)^{w_n}]}$$

where $N = w_1 + w_2 + \cdots + w_n$.

Taking logarithms:

$$\log(\text{weighted geometric mean}) = \frac{w_1 \log (x_1) + w_2 \log (x_2) + \cdots w_n \log (x_n)}{N}$$

$$= \frac{\sum_{i=1}^{n} w_i \log (x_i)}{\sum_{i=1}^{n} w_i}$$

Therefore, the logarithm of the weighted geometric mean of $x_1, \ldots, x_n$ is the weighted arithmetic mean of their logarithms.

7.2.3 Index numbers

One method of comparing different sets of data is to reduce the data to purely relative numbers by comparison with a fixed (base) value. These relative numbers are the simplest index numbers. The purpose for which the data are required often indicates the base to be used. The base can refer to an average of several sets of data taken over a period of time, but it should represent a recent 'normal' or average situation.

Assume that data are available for the price of commodity A over a period of years. Taking year 1 (say) as the base year, the *price relative* (index) for commodity A is

$$\text{Year } x: \text{Price relative} = \frac{\text{price in year } x}{\text{price in year 1}} \times 100 \text{ per cent}$$

The *cost of living index* is well known; however it is a very difficult matter to decide which items are to be included and the weighting to be applied to each item.

7.2.4 Crude and standardized death rates

The *crude death rate* for a given location is the number of deaths per 1000 population.

$$\text{Crude death rate} = \frac{\text{total number of deaths}}{\text{total population}} \times 1000$$

This value is unsuitable for comparisons between different locations as it does not take into account the distribution of the population between different age groups.

If the distribution by age and sex is available for the entire population, then a *standard population* can be calculated for each age group. The crude death rates are calculated for each group under consideration and the total deaths per 1000 of the standard population are found (crude death rate × standard population). The sum of these values is known as the *standardized death rate*.

7.2.5 Time series

A *time series* is a set of observations taken at different times, usually at equal intervals. The observations can be plotted against time. An example is the daily average temperature.

Variations in the set of observations can be attributed to long-term (secular) trends, seasonal variations (short cyclical variations) and residual variations (long-term cyclical movements and erratic variations). The analysis of a time series consists of attempting to separate any variation into the three components of trend, seasonal variation and residual.

7.2.6 Moving averages

The *moving average of order n* for a given set of numbers $x_1, x_2, \ldots, x_{n+1}, x_{n+2}, \ldots$ is given by the following sequence of arithmetic means:

$$\frac{x_1 + x_2 + \cdots + x_n}{n}; \qquad \frac{x_2 + x_3 + \cdots x_{n+1}}{n}; \qquad \frac{x_3 + x_4 + \cdots x_{n+2}}{n}; \qquad \text{etc.}$$

If the data are obtained each month, then a 3 month, 6 month or 12 month moving average could be calculated (as appropriate). The moving average is centred at the middle of the period to which it refers. A moving average can be used to 'average out' seasonal variations if its period is known or can be deduced. The choice of the correct order of a moving average is important, so that most of the cyclical variation is eliminated.

7.2.7 Probability

Probability is based upon relative frequency, i.e. the proportion of times that an event has previously occurred. The value of a probability can therefore change with time. If one event occurs r_1 times and a second event occurs r_2 times, each from n trials, then the two events are *equally likely* if

$$\lim_{n \to \infty} \left(\frac{r_1}{n} \right) = \lim_{n \to \infty} \left(\frac{r_2}{n} \right)$$

The probability of a *desirable event* (p) is

$$p = \frac{r}{n}$$

where r is the number of desirable events which occurred out of a total of n equally likely events.
The probability of an *undesirable event* (q) is

$$q = \frac{n - r}{n} = 1 - \frac{r}{n} = 1 - p$$

Therefore, $p + q = 1$.

$p = 0$ indicates an impossible event.

$p = 1$ indicates an event that *must* occur.

If two events A and B can *never* occur simultaneously, i.e. $P(A \text{ and } B) = 0$, then A and B are *mutually exclusive events*.

If two events A and B *can* occur simultaneously and they are not mutually exclusive, then

$$P(\text{either A or B}) = P(A) + P(B) - P(\text{both A and B})$$

Note: 'either A or B' means one or both events, i.e. 'inclusive or'.

Independent events are those in which the occurrence or non-occurrence of one event has no effect upon the other event. Therefore, if events A and B are independent, then

$$P(A \text{ and } B) = P(A) \times P(B)$$

Dependent events are those in which the occurrence or non-occurrence of one of the events has a direct effect on the probability of occurrence of the other.

The probability that A will occur, given that event B has occurred (or will occur), is written as $P(A \mid B)$, i.e. probability of A given B. This is a *conditional probability*. For mutually exclusive events:

$$P(A \mid B) = P(B \mid A) = 0$$

For independent events:

$$P(A \mid B) = P(A)$$

because the occurrence of B has no effect on the probability of A.

$$\text{Also, } P(\text{both A and B}) = P(A) \times P(B \mid A)$$

$$= P(B) \times P(A \mid B)$$

For *repeated trials*, if the conditions remain unchanged, then the trials are independent and the above multiplication rule is still valid.

Repeated trials are independent if the 'data' are replaced after each trial so that the original sample is still available.

7.2.8 Permutations and combinations

The number of *permutations* of n items selected in groups of r is

$$^{n}P_{r} = n(n - 1)(n - 2)\ldots(n - r + 1)$$

$$= \frac{n!}{(n - r)!}$$

Note: $n! = {}^{n}P_{n}$

For permutations, the group ABC is considered a separate entity from BAC, etc.

The number of *combinations* of n items selected in groups of r, where the order of the items is immaterial (e.g. ABC is equivalent to BCA, etc.), is

$$^{n}C_{r} = \frac{n!}{r!(n - r)!}$$

7.2.9 Probability distributions

A *Bernoulli trial* is an experiment that has exactly two mutually exclusive outcomes. The probability of r successes in a sequence of n independent Bernoulli trials, where $P(\text{success}) = p$ and $P(\text{failure}) = 1 - p = q$, is

$$\frac{n!}{r!(n - r)!} p^{r}q^{n-r}$$

The binomial probability distribution
The probability of an event occurring at a single trial is p, and the probability of the same event not occurring is q (where $p + q = 1$). Then the probabilities $P(0), P(1), \ldots, P(n)$ of $0, 1, \ldots, n$ occurrences of the event in n independent trials are given by the terms of the *binomial expansion*:

$$(q + p)^{n} = q^{n} + nq^{n-1}p + \frac{n(n - 1)}{2!} q^{n-2}p^{2} + \cdots + p^{n}$$

$$= P(0) + P(1) + P(2) + \cdots P(n)$$

This is known as the *binomial probability distribution*, with mean value $= np$ and variance $= npq$.

The Poisson distribution
In a particular situation, if the average of a number of events is μ, then the probabilities of $0, 1, 2, \ldots, n$ events occurring are given by the consecutive terms of the *Poisson distribution*:

$$e^{-\mu} + \mu e^{-\mu} + \frac{\mu^{2} e^{-\mu}}{2!} + \frac{\mu^{3} e^{-\mu}}{3!} + \cdots + \frac{\mu^{n} e^{-\mu}}{n!} = 1$$

i.e. P (zero events) $= e^{-\mu}$

The mean $(\mu) = np$, as for the binomial distribution, and the variance $= np$.

The binomial and the Poisson distributions are both *discrete* probability distributions, and they can be represented by histograms. If the number of trials is large and the probability of success (p) is not too small, so that the mean $(np) > 5$, then the binomial distribution approximates to

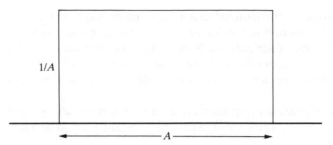

Figure 7.2 The rectangular distribution.

the continuous normal distribution (see Sec. 7.2.10). If the mean of the Poisson distribution is >25, then it also approximates to the continuous normal distribution.

The *rectangular distribution* is a simple example of a continuous distribution defined by a *probability density function*. The rectangular distribution is defined by

$$h(x) = 1/A \qquad 0 \leqslant x \leqslant A$$

where A is a constant. The range of values of x can be taken over any range of width A. The distribution is shown in Fig. 7.2.

The mean and median are both equal to $A/2$, and the distribution has no mode. The variance is equal to $A^2/12$.

7.2.10 The normal distribution

When n is large, the *normal (probability) distribution*, or error curve, or Gaussian distribution, is indistinguishable from the binomial distribution. The characteristic shape of the normal distribution curve is shown in Fig. 7.3. The normal distribution has been widely studied in statistics and has the advantage that it is easy to use.

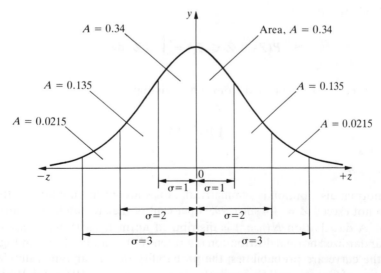

Figure 7.3 The normal distribution curve (mean = 0; variance = 1) (from Ray, *Elements of Engineering Design*, Prentice-Hall International (UK) Ltd, 1985).

Many distributions are encountered that are approximately normal, e.g. most sets of random errors. The normal distribution is also important as a 'limiting distribution', i.e. it can be used as an approximation to other distributions. Non-normal distributions can often be approximated by assuming that they are normally distributed. Also the means of samples of size n taken from any population are themselves approximately normally distributed (this approximation improves as n increases).

The normal distribution is a continuous distribution; it is bell-shaped and symmetrical about the mean as shown in Fig. 7.3 (for mean = 0). The general equation describing any symmetrical normal distribution is

$$y = f(x) = \frac{1}{\sigma \sqrt{(2\pi)}} \exp[-(x - \mu)^2/2\sigma^2]$$

where x can have all values from $-\infty$ to $+\infty$, and the parameters μ and σ^2 represent the mean and variance of the distribution respectively. The spread of the curve about the mean (μ) depends upon the standard deviation (σ). The area under the curve between $-\infty \leqslant x \leqslant +\infty$ is equal to 1. The probability that x lies between two particular values a and b is given by

$$P(a \leqslant x \leqslant b) = \int_a^b f(x) \, dx$$

This is a difficult integral to evaluate and tables of areas are more appropriate. Tables are available for the areas under the *standard* normal distribution, related to the *standardized normal variate* (Z). Part of these tables is included at the end of this chapter (see Table 7.8). The following equation is used to convert the x values for any distribution (mean μ, standard deviation σ) into standardized form:

$$Z = \frac{x_i - \mu}{\sigma}$$

Note: $Z = 0$ when $x_i - \mu = 0$, or $x_i = \mu$.

$Z = 1$ when $x_i - \mu = \sigma$.

$$P(Z_2 \leqslant Z \leqslant Z_1) = \int_{Z_2}^{Z_1} f(Z) \, dZ$$

Integrals of this type can be evaluated from tables giving:

$$\int_{-\infty}^{Z} f(Z) \, dZ$$

for $-\infty \leqslant Z \leqslant +\infty$.

Since the normal distribution is symmetrical, tables are given only for positive values of Z. Most tables do not exceed $Z = +4$ since 99.99 per cent of the area under the curve lies between $-4 \leqslant Z \leqslant +4$. A distribution $N(\mu, \sigma^2)$ is distributed normally with mean μ and variance σ^2. Therefore, a standardized normal distribution is written $N(0, 1)$ and is shown in Fig. 7.3. The areas shown under the curve are probabilities; the probability that a random value falls within ± 1 standard deviation of the mean is 0.68. Similarly, the probabilities are 0.95 and 0.993 that a random value falls within ± 2 or ± 3 standard deviations respectively.

The probability that $Z \leqslant 1.0$ is

$$P(Z \leqslant 1.0) = \int_{-\infty}^{1} f(Z) \, dZ$$

$$= \int_{-\infty}^{0} f(Z) \, dZ + \int_{0}^{1} f(Z) \, dZ$$

$$= 0.5 + 0.34 \text{ (from tables or Fig. 7.3)}$$

$$= 0.84$$

The probability that $Z \geqslant 2.0$ is

$$1 - P(Z \leqslant 2.00) = 1 - [0.5 + 0.475]$$

$$= 0.025$$

Any rounding-off errors in the variate x must be taken into account when calculating the standardized variate Z. For example, if the range of x is between 10 and 20, the range of values of Z is given by

$$Z_2 = \frac{20.5 - \mu}{\sigma} \quad \text{and} \quad Z_1 = \frac{9.5 - \mu}{\sigma}$$

The normal distribution can be used as an approximation to the binomial distribution if $np \geqslant 5$ (see Sec. 7.2.9). The mean and variance of the binomial distribution (np, npq) are used for the values of μ and σ^2 in the corresponding normal distribution.

The Poisson distribution can also be approximated by the normal distribution if the mean (np) > 25. The mean and variance in the normal distribution can both be approximated by np from the Poisson distribution.

7.2.11 Skewness and kurtosis

With actual data, effects may occur that change the shape of the normal distribution as shown in Fig. 7.4. *Skewness* is the tendency of a frequency distribution to be unbalanced in respect to its centre. Examples are the skewed distributions in Fig. 7.4. The *coefficient of skewness* (α_3) can be calculated from the following formulae:

$$\alpha_3 = \frac{\sum (x_i - \bar{x})^3}{N\sigma^3} \quad \text{or} \quad \alpha_3 = \frac{\sum (x_i - \bar{x})^3 \times f_i}{N\sigma^3}$$

A positive value of α_3 means that the distribution is skewed to the right. The absolute value of α_3 indicates the degree of skewness, e.g. $\alpha_3 = +2.0$ indicates that the peak is skewed to the right by two standard deviations. If $\alpha_3 = 0$, the distribution is perfectly symmetric. An alternative formula is

$$\alpha_3 = 3(\text{mean} - \text{median})/\text{standard deviation}$$

Kurtosis is a measure of the 'peakedness' of a symmetric distribution, e.g. a sharp or flat peak. The *coefficient of kurtosis* (α_4) is calculated from

$$\alpha_4 = \frac{\sum (x_i - \bar{x})^4}{N\sigma^4}$$

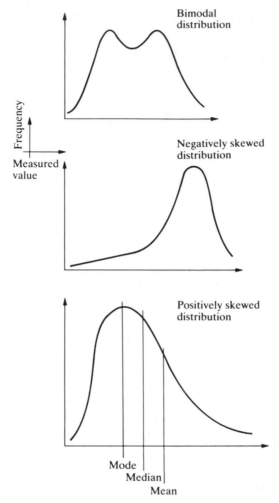

Figure 7.4 Effects on the normal distribution (from Ray, *Elements of Engineering Design*, Prentice-Hall International (UK) Ltd, 1985).

For a flat distribution, $\alpha_4 \approx 2.0$; for a standard normal distribution, $\alpha_4 \approx 3.0$; and for a peaked distribution, $\alpha_4 \approx 4.0$.

7.3 INFERENTIAL STATISTICS

7.3.1 Sampling

It is often necessary to take samples from a population and to use the sample data as a basis for decision making. It may be that the population is indefinable or infinite, or the testing procedure is destructive or, as is most common, that the whole population is too large to be tested economically.

Measurements obtained from a sample (e.g. mean, standard deviation, etc.) are known as *statistics* (singular *statistic*); the corresponding population measurements are called *parameters*.

The notation used here is defined in Table 7.1. *Statistical inference* is the technique of drawing conclusions about a population as a whole by examining or testing a sample only.

A sample should be fully representative of the population from which it is taken. However, this is very difficult to achieve in practice and some bias is usually inevitable. It is important to be able to control the extent of this bias, as far as possible. *Sampling errors* are differences in calculated figures which occur either as a result of bias, or because of an inherent loss of information when a sample is taken. The 'error' here is the difference between the observed figure and its value if the full data were available. Personal bias is usually avoidable but cyclic trends need to be avoided or observed.

A *simple sample* is obtained when the selection process provides an equal chance of every item in the population being selected. A *random sample* is selected in an unbiased manner, without replacement of the items for possible reselection. *Quota sampling* is used to obtain a representative sample, taking into account the distribution of categories within a population.

7.3.2 The sampling distribution

The sample mean can be used to provide a point estimate of the population mean. This should be an *unbiased estimate*; it is necessary to know how close the estimate is to the true value, i.e. the standard deviation of its probability distribution. The probability distribution of a sample statistic is known as a *sampling distribution*. A sample of given size from a defined population will vary in its content. A statistic of the sample will also vary and its distribution can be found.

Table 7.1 Statistical measures of populations and samples

Measure	Notation	
	Population	Sample
Number of items	N	n
Mean	μ	$\bar{x}$
Standard deviation	σ	s

The standard deviation of the sample mean ($\bar{x}$) is given by $\sigma/\sqrt{n}$ (this measures the precision of the unbiased estimator $\bar{x}$ of the parameter μ). If σ is unknown, then the precision can be obtained from the standard error of the sample mean (σ_n), where $\sigma_n = s/\sqrt{n}$. The standard error is based on the sample alone. The variance of the sample mean (var($\bar{x}$)) is equal to σ^2/n, whereas the expected value (i.e. mean) of the sample variance ($E(s^2)$) is given by: $E(s^2) = \sigma^2$.

The standard deviation of a sampling distribution is known (historically) as a standard error (this avoids confusion between the standard deviation of the sample itself and the standard errors of its statistics).

The larger the observed sample mean, the better the estimate of the population mean. The population should be much larger than the sample, by at least a factor of 20. It not, the following formula should be used:

$$\sigma_n^2 = \frac{\sigma^2}{n} \times \frac{N-n}{N-1}$$

where N is the population size.

The distribution of the sample mean is approximated by the normal distribution, whatever the population distribution (unless n is very small or the population is very far from normal). The standard deviation of the population must be established independently of the sample. If the sample itself is the only data for estimating it, then the t-distribution is used (see Sec. 7.3.5) which is a modified form of the normal distribution.

The conclusions made about the sampling distribution can be expressed either by using a significance test or as a confidence interval. These techniques are described in Secs 7.3.3 and 7.3.4.

7.3.3 Significance testing

Assume that the estimates of the parameters are normally distributed, and standardize each parameter so that its Z value follows the standard normal distribution $N(0, 1)$ (or the special situation of the t-distribution described in Sec. 7.3.5). If this Z value then occurs in the tail of the normal distribution (see Fig. 7.3), it may be due to one of two reasons. Either a not very probable event has occurred, or the original assumption (*null hypothesis*) about the parameters was incorrect. In statistical testing it is concluded that the second possibility is the correct one—but the probability of this conclusion being wrong is equal to the probability of the rate event mentioned in the first possibility. This probability defines the limits of the tail(s) of the distribution, and the critical values of Z outside which a *significant result* is obtained. A value of Z inside these critical values gives a *non-significant result*.

There are two possibilities for the range of values of Z where a significant result is obtained; these are shown in Fig. 7.5. A *two-tailed test* encloses 95 per cent of the curve between values of $Z = +1.96$, whereas a *one-tailed test* has a critical Z value of $+1.64$ (check these values in statistical tables of the standard normal distribution—see Table 7.8 at the end of this chapter). The values in Table 7.2 indicate particular common significance levels at critical Z values for one-tailed and two-tailed tests.

The choice of a one-tailed or two-tailed test depends upon the conditions of the problem; a single-tailed test is performed when only changes in a given direction (increase or decrease) are of interest. Two-tailed tests are more common. If a result is obtained outside the critical range of Z (i.e. inside the shaded area in Fig. 7.5), then it is stated that *the result is significant at the 5 per cent level*. The *confidence limits* are the two x values that have a 95 per cent chance (say) of enclosing the true value of x. The confidence interval is the distance between these limits. The following formula can be used if the population standard deviation is known:

$$\text{confidence interval} = \bar{x} \pm (\sigma Z)$$

The choice of the significance level to be used is a crucial and difficult decision. A 5 per cent level (or between 1 and 5 per cent) is usually called 'significant' even though it will be in error one

Table 7.2 Significance points for the standardized normal variate (Z)

	Two tails	One tail
5 per cent level	1.96	1.64
2 per cent level	2.33	2.06
1 per cent level	2.58	2.33

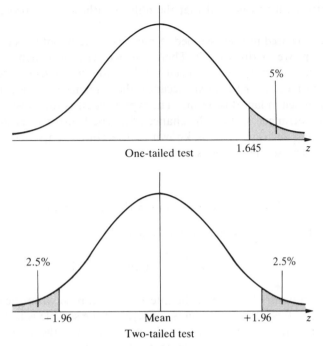

Figure 7.5 Significance testing using the normal distribution.

time in twenty. The 1 per cent level (or between 0.1 and 1 per cent) is usually called 'highly significant' even though this will also be wrong one time in every hundred.

The confidence level is not (although it is sometimes wrongly taken to be) the probability that the unknown parameter lies within a stated interval. The parameter is not a random variable; it has a fixed value although this value is unknown. Instead the outer values of the limits are the random variables and, hence, determine the probability that they contain the fixed but unknown parameter. Therefore, there is a 95 per cent probability that the limits (± 1.96) will enclose the true value of Z.

Tests of this type using the Z variate should be used either for entire populations, or for samples larger than 30. For smaller samples, the Student-t test (see Sec. 7.3.5) should be used.

It may be necessary to test hypotheses about the parameters of a population. A *statistical hypothesis* is an assumption made about some parameter. This assumption could be completely verified if the entire population were examined. Often only estimates of the parameters obtained from random samples are available, and the assumptions must be tested using these estimates. The assumption about the parameters is called the *null hypothesis* denoted by H_0; the *alternative hypothesis* is denoted by H_1.

A null hypothesis is proposed either on the basis of symmetry or by assuming that an observed effect which appears to be systematic is actually due to chance. For example, assume a new drug has no effect and an apparent increase in the recovery rate of patients is no more than an ordinary variation due to chance. The probability is calculated that a result as far from the postulated one as that of the data, could occur in a random experiment. If this probability is over 5 per cent (say), no conclusions can be made. The evidence tends to support the null hypothesis rather than disprove it, but the hypothesis is not proven. If the probability is less than 5 per cent, the data are significantly different from the hypothesis, and therefore the hypothesis is unlikely to be correct. Alternatively, it can be said that the null hypothesis is 'disproved' at the 5 per cent level.

The null hypothesis need not be assumed. A more positive hypothesis could be assumed and then a test made to prove or disprove it. There are two ways in which a significance test can produce misleading results. A *type I error* occurs if the null hypothesis is determined as incorrect when it is actually correct. A *type II error* occurs if the evidence does not indicate that the null hypothesis is wrong when in fact it is wrong. The type II error is the most common and the test conditions should be arranged so that the chance of a type II error is as small as possible. The probability of avoiding such an error is known as the *power* of the test; the obvious way of achieving this is to increase the sample size.

7.3.4 Confidence interval for a population mean (large samples)

A confidence interval for the population mean is given by

$$\mu = \bar{x} \pm Z \frac{\sigma}{\sqrt{n}} \qquad \text{(see Table 7.1)}$$

An approximation occurs because of the interrelationship between the mean and standard deviation of the sample; this approximation is negligible if the sample is large. The sample mean ($\bar{x}$) is an unbiased estimate of μ, but s consistently underestimates the value of σ and it is not an unbiased estimator. However, an unbiased estimator for the population variance (σ^2) is given by the sample variance, and is defined as

$$s^2 = \frac{1}{(n-1)} \sum (x_i - \bar{x})^2$$

The $(n-1)$ term makes this an unbiased estimator.
(*Note:* the 'hat' over the $\hat{\sigma}$ is used to denote an estimated value.)
The 95 per cent confidence interval for the mean is given by

$$\mu = \bar{x} \pm 1.96 \frac{s}{\sqrt{n}} \sqrt{\left(\frac{n}{n-1}\right)}$$

$$\text{i.e.} \quad \mu = \bar{x} \pm 1.96 \frac{s}{\sqrt{(n-1)}}$$

In general terms:

$$\mu = \bar{x} \pm Z \frac{\hat{\sigma}}{\sqrt{n}}$$

where Z is the appropriate normal variate.

7.3.5 Student-*t* distribution (small samples)

Small samples are usually considered to be those where $n < 30$. In these cases the normal distribution and the normal variate Z are not used, but rather the Student-*t* distribution and Student-*t* statistic. These were developed by W. S. Gosset in 1908 who wrote under the pen name of 'Student'.

The Student-t distribution resembles the normal curve, and its equation is

$$y = c\left(1 + \frac{t^2}{v}\right) - \frac{(v + 1)}{2}$$

where c is a constant, dependent upon v—the *degrees of freedom*. The value of v used here is $(n-1)$. It is these degrees of freedom that complicate the use of the t-distribution; as the areas under the curve are dependent upon v it would take many tables (one for each value of v) to show them all (unlike the single table for the standard normal distribution). Tables that are used give the areas under the curve at certain critical points for all $v < 30$ (for $v \geqslant 30$, large sample methods can be used instead of the t-distribution). The critical points are 0.10, 0.05, 0.025, 0.010 and 0.005 representing those points where it is 90 per cent, 95 per cent, 97.5 per cent, 99 per cent and 99.5 per cent certain of the conclusion, respectively. These tables are satisfactory for nearly every case encountered, and an example is shown in Table 7.9 at the end of this chapter. The symbol $t_{0.05}$ is the value of t to the right of which 5 per cent of the area under the t-curve will be found with the given number of degrees of freedom.

The t-statistic is used in a similar manner to the Z test for the normal distribution, i.e. measuring the difference between the sample mean and the population mean to determine if it is significant. In this case if the population standard deviation (σ) is known, the value $[\sigma/\sqrt{(n - 1)}]$ is used. If σ is unknown, the value to be used is $[s\sqrt{n}/\sqrt{(n - 1)}]$. Hence

$$t = \frac{\bar{x} - \mu}{\sigma/\sqrt{(n - 1)}}$$

or

$$t = \frac{\bar{x} - \mu}{s\sqrt{n}/\sqrt{(n - 1)}}$$

Comparing Z and t, $Z = 1.96$ is significant at the 5 per cent level while t must be 2.13 with $v = 15$ to be equally significant. At the 1 per cent level, the difference is between values of Z and t of 2.58 and 2.95. Because these numbers represent standard deviations, they are directly comparable. Table 7.3 gives the significance points for t at 15 degrees of freedom (compare these with Table 7.2).

The confidence interval for a population mean using a small sample is given by

$$\mu = \bar{x} \pm t(\hat{\sigma}/\sqrt{n})$$

7.3.6 Significance testing—special cases

(a) *Difference of two population means*
Two populations have unknown means μ_1 and μ_2 and known variances σ_1^2 and σ_2^2. The means $\bar{x}_1$ and $\bar{x}_2$ are for independent random samples of size n_1 and n_2 from each population. ($\bar{x}_1 - \bar{x}_2$) is

Table 7.3 Significance points for the t-statistic

($v = 15$)	Two-tail test	One-tail test
5 per cent level	2.13	1.75
2 per cent level	2.60	2.29
1 per cent level	2.95	2.60

distributed approximately normally with a population mean of $(\mu_1 - \mu_2)$ and variance

$$\frac{\sigma_1^2}{n_1} + \frac{\sigma_2^2}{n_2}$$

To test any hypothesis about the difference $(\mu_1 - \mu_2)$, e.g. $\mu_1 = \mu_2$, obtain the standardized variable

$$Z = \frac{\bar{x}_1 - \bar{x}_2 - (\mu_1 - \mu_2)}{\sqrt{\left(\dfrac{\sigma_1^2}{n_1} + \dfrac{\sigma_2^2}{n_2}\right)}}$$

and test it against a value from the standard normal distribution tables.
If $\sigma_1^2 = \sigma_2^2 = \sigma^2$ (known), then

$$Z = \frac{\bar{x}_1 - \bar{x}_2 - (\mu_1 - \mu_2)}{\sigma\sqrt{\left(\dfrac{1}{n_1} + \dfrac{1}{n_2}\right)}}$$

If $\mu_1 = \mu_2$, then

$$Z = \frac{\bar{x}_1 - \bar{x}_2}{\sigma\sqrt{\left(\dfrac{1}{n_1} + \dfrac{1}{n_2}\right)}}$$

(b) *Difference between two population means, given two large samples (population variances unknown)*
Often the population variance is unknown and σ^2 has to be estimated by the sample variance:

$$s^2 = \frac{\sum (x_i - \bar{x})^2}{(n - 1)}$$

Take two large samples, one from each population (denoted by suffices 1 and 2). The variances of the two populations are assumed equal: $\sigma_1^2 = \sigma_2^2 = \sigma^2$. Test if the population of all such differences $(x_1 - x_2)$ has a mean $(\mu_1 - \mu_2)$; first find s^2 as an estimate of σ^2:

$$s^2 = \frac{(n_1 - 1)s^2 + (n_2 - 1)s^2}{n_1 + n_2 - 2} = \frac{\sum (x_i - \bar{x}_1)^2 + \sum (x_i - \bar{x}_2)^2}{n_1 + n_2 - 2}$$

Obtain:

$$Z = \frac{\bar{x}_1 - \bar{x}_2 - (\mu_1 - \mu_2)}{s\sqrt{\left(\dfrac{1}{n_1} + \dfrac{1}{n_2}\right)}}$$

and test using the normal distribution. If $\sigma_1^2 \neq \sigma_2^2$, use the statistic:

$$Z = \frac{\bar{x}_1 - \bar{x}_2 - (\mu_1 - \mu_2)}{\sqrt{\left(\dfrac{s_1^2}{n_1} + \dfrac{s_2^2}{n_2}\right)}}$$

(c) *Difference between two population means, given two small samples (population variances unknown)*

Assume $\sigma_1^2 = \sigma_2^2 = \sigma^2$, and use s^2 as an estimate of σ^2 from

$$s^2 = \frac{(n_1 - 1)s_1^2 + (n_2 - 1)s_2^2}{n_1 + n_2 - 2}$$

Then

$$t = \frac{\bar{x}_1 - \bar{x}_2 - (\mu_1 - \mu_2)}{s\sqrt{\left(\dfrac{1}{n_1} + \dfrac{1}{n_2}\right)}}$$

This statistic is tested against $t_{\alpha\%}$ for $v = n_1 + n_2 - 2$. If $\sigma_1^2 = \sigma_2^2$, then

$$t = \frac{\bar{x}_1 - \bar{x}_2 - (\mu_1 - \mu_2)}{\sqrt{\left(\dfrac{s_1^2}{n_1} + \dfrac{s_2^2}{n_2}\right)}}$$

This is known as the Fisher–Behrens problem; it requires special tables to obtain a solution because the t-statistic no longer follows the t-distribution. An approximate test has been given by Welch and Aspin using the t-statistic defined above; this follows the t-distribution approximately if v is given by

$$\frac{1}{v} = \frac{1}{(n_1 - 1)} \left\{ \frac{s_1^2/n_1}{s_1^2/n_1 + s_2^2/n_2} \right\}^2 + \frac{1}{(n_2 - 1)} \left\{ \frac{s_2^2/n_2}{s_1^2/n_1 + s_2^2/n_2} \right\}^2$$

It provides a satisfactory answer to most practical problems.

7.3.7 Chi-squared distribution

The t-statistic used for small samples can be defined as:

$$t = \frac{\bar{x} - \mu}{(s/\sqrt{n})}$$

$$= \frac{\bar{x} - \mu}{(\sigma/\sqrt{n})} \bigg/ \sqrt{\left(\frac{s^2}{\sigma^2}\right)}$$

$$= \frac{\bar{x} - \mu}{(\sigma/\sqrt{n})} \bigg/ \sqrt{\left(\frac{\chi^2}{v}\right)}$$

Therefore

$$\chi^2 = \frac{vs^2}{\sigma^2}$$

where χ^2 *follows the chi-squared distribution*, and v is the number of degrees of freedom associated with s^2 (v is also the number of independent variables which are used to calculate χ^2).

The chi-squared distribution is a probability distribution, since

$$\int_0^\infty f(\chi^2)\, d\chi^2 = 1$$

The mean is v and the variance is $2v$. The maximum value of $f(\chi^2)$ occurs when $\chi^2 = v - 2$ for $v \geqslant 2$. The shape of the distribution for various values of v is shown in Fig. 7.6.

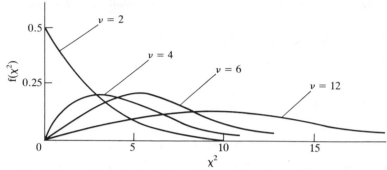

Figure 7.6 Shape of the chi-squared (χ^2) distribution for different values of the degrees of freedom (v).

Percentage points (χ_p^2) are given in tables for various values of v. χ_p^2 and p are defined as:

either

$$\frac{p}{100} = \int_{\chi_p^2}^{\infty} f(\chi^2)\, d\chi^2$$

or

$$\frac{p}{100} = \int_{0}^{\chi_p^2} f(\chi^2)\, d\chi^2$$

These situations are shown in Fig. 7.7. The first equation above defines the probability that a variable $\chi^2 > \chi_p^2$; and the second equation that $\chi_p^2 > \chi^2$. χ_p^2 is called the critical value of χ^2 for a given v and p. Tables of the distribution usually only cover v from 1 to 30 (in steps of 1) and from 30 to 100 (in steps of 10); linear interpolation is satisfactory for the higher values of v. An example of the χ^2 distribution is given in Table 7.10 at the end of this chapter. For values of $v > 100$, $\sqrt{(2\chi^2)}$ is approximately normally distributed with mean $= \sqrt{(2v - 1)}$ and unit variance (*note:* $\sqrt{(2\chi^2)}$ converges more rapidly than χ^2).

The standardized variable for the χ^2 distribution is

$$\frac{\chi^2 - v}{\sqrt{(2v)}}$$

For the $\sqrt{(2\chi^2)}$ distribution, the standardized variable is

$$\sqrt{(2\chi^2)} - \sqrt{(2v - 1)}$$

A comparison of the values of the standardized variables for the two distributions (χ^2 and $\sqrt{(2\chi^2)}$) is given in Table 7.4 for the 5 per cent level of p (for the case $\chi^2 > \chi_p^2$).

Table 7.4 Standardized variables for χ^2 and $\sqrt{(2\chi^2)}$ distributions

v	$\chi_{5\%}^2$	$\dfrac{\chi^2 - v}{\sqrt{(2v)}}$	$\sqrt{(2\chi^2)} - \sqrt{(2v - 1)}$
1	3.84	2.01	1.77
25	37.7	1.80	1.69
100	124.3	1.72	1.66

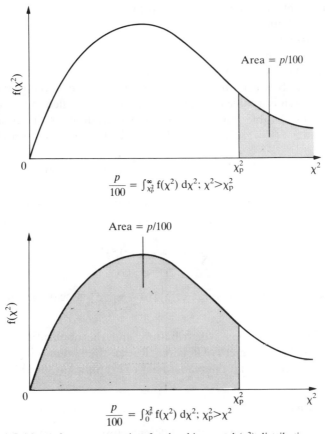

$$\frac{p}{100} = \int_{\chi_p^2}^{\infty} f(\chi^2) \, d\chi^2; \ \chi^2 > \chi_p^2$$

$$\frac{p}{100} = \int_0^{\chi_p^2} f(\chi^2) \, d\chi^2; \ \chi_p^2 > \chi^2$$

Figure 7.7 Alternative definitions of percentage points for the chi-squared (χ^2) distribution.

The chi-squared distribution has many uses, although to describe them and present suitable examples is beyond the scope of this chapter. For application of this distribution in particular situations, reference should be made to one of the many textbooks on statistics. Some of the applications of the chi-squared distribution will now be mentioned briefly:

(a) Testing for the variance (σ^2) of a population.
(b) Extending the number of degrees of freedom by using the additive property: $\chi^2 = \chi_1^2 + \chi_2^2$, and combining the data from different samples. The values of χ^2 become more reliable as the number of degrees of freedom increases.
(c) Setting up confidence intervals for the variance (σ^2) of a normal population.
(d) Measuring the discrepancies that exist between observed frequencies (o_i) and the expected (theoretical) frequencies (e_i) as given by the statistic

$$\chi^2 = \sum_{i=1}^{k} \frac{(o_i - e_i)^2}{e_i}$$

where the total frequency $= \sum_{i=1}^{k} o_i = \sum_{i=1}^{k} e_i.$

The value of χ^2 obtained is compared with the critical value $\chi^2_{\alpha\%}(\nu)$ from tables, and used to determine whether the difference is significant.

(e) Testing whether sample data were obtained from:
   (i) a binomial distribution;
   (ii) a Poisson distribution;
   (iii) a normal distribution.
   These tests employ the method of part (d) above.
(f) Finding the difference between observed frequencies in two or more groups. This test uses a *contingency table* which is a table of data that can be sub-totalled in two or more directions. For a two-dimensional table having h rows and k columns with observed and expected frequencies $o_{i,j}$ and $e_{i,j}$ ($i = 1, 2, \ldots, h; j = 1, 2, \ldots, k$), the test can be performed by using

$$\chi^2 = \sum_{i=1}^{h} \sum_{j=1}^{k} \frac{(o_{i,j} - e_{i,j})^2}{e_{ij}} \quad \text{against} \quad \chi^2_{\alpha\%}(v)$$

(g) Determining Yates correction, which is an approximation of $\sum [(o - e)^2/e]$ for small values of the number of degrees of freedom. If there is only one degree of freedom, then χ^2 is better approximated by

$$\sum_{i=1}^{k} \frac{(|o_i - e_i| - 0.5)^2}{e_i}$$

This will only be necessary if the value of $\sum [(o - e)^2/e]$ is near to, and above, the critical value $\chi^2_{\alpha\%}(1)$.

Note: The main difficulty with the χ^2 distribution, and the applications listed above, is the determination of the number of degrees of freedom. Rather than attempting to cover all problems or situations here, it is suggested that the reader refer to examples in suitable statistics textbooks.

7.3.8 The *F*-distribution

It is often necessary to determine the effect of different samples on some common property. One method that can be used is called *analysis of variance*. The method analyses the variance of samples into useful components in order to measure the statistical differences between samples, and hence to determine which effect was produced by each sample.

The *F-distribution* (named after R. A. Fisher) is an extension of the Student-t distribution. However, in the F-distribution a ratio of variances is formed in which the degrees of freedom may differ in the numerator *and* in the denominator of the ratio. Therefore, determination of the critical values of F requires two degrees of freedom, as well as a level of significance.

Suppose there are two independent sample variances s_1^2 and s_2^2, which are based on v_1 and v_2 degrees of freedom respectively, and are estimates of the variances σ_1^2 and σ_2^2 of two normally distributed populations. Therefore, there are two independent chi-squared distributions:

$$\chi_1^2 = \frac{v_1 s_1^2}{\sigma_1^2} \quad \text{and} \quad \chi_2^2 = \frac{v_2 s_2^2}{\sigma_2^2}$$

The ratio

$$F = \frac{v_2 \chi_1^2}{v_1 \chi_2^2} = \frac{s_1^2 \sigma_2^2}{s_2^2 \sigma_1^2}$$

follows the F-distribution. The equation of the curve representing this distribution is

$$y = K \frac{F^{(v_1/2) - 1}}{(v_2 + v_1 F)^{(v_1 - v_2)/2}} \quad (0 < F < \infty)$$

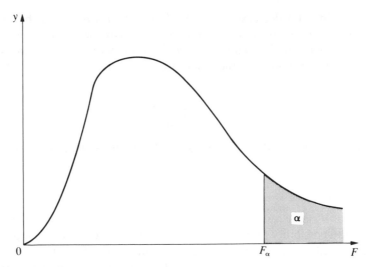

Figure 7.8 Typical F-distribution curve ($v_1 > 2$).

where K is a constant depending upon the values of v_1 and v_2. The curve for $v_1 > 2$ is shown in Fig. 7.8; this curve reaches a maximum at

$$F = \frac{(v_1 - 2)}{v_1} \frac{v_2}{(v_2 + 2)} < 1$$

This value approaches 1 as v_1 and v_2 become larger. Values of $F_{\alpha\%}$ are presented in tables for various combinations of α, v_1 and v_2, where ($\alpha/100$) is the probability of obtaining a value of $F > F_{\alpha\%}$ (an example is given in Table 7.11 at the end of this chapter).

The F-distribution can be used to test (at a given level of significance) whether there is any difference between the variances of two populations. The F-statistic can also be used to test the null hypothesis that the ratio of the population variances is (σ_1^2/σ_2^2). This requires the confidence interval for the ratio to be determined.

The analysis of variance has been developed to treat a wide variety of problems, but only the F-statistic have been mentioned here. In many problems and situations, the data must be adjusted and altered before being used to compute an F-value. These methods are described in more advanced textbooks.

7.4 NON-PARAMETRIC STATISTICS

Special tests have been devised for situations where the population is not closely approximated by the normal curve. The parameters of the population are not assumed and the tests are called *non-parametric tests* (sometimes referred to as *distribution-free methods*). The mean and standard deviation still have to be assumed, but they do not relate to the parameters of the population itself. Non-parametric tests are also used when it is impossible to obtain the data required to perform a Z-test or a t-test, but where it is possible to rank the data or categorize them as $+$ or $-$.

7.4.1 The sign test

If data are available as paired variates, it is possible to count the number of increasing and decreasing pairs. The expected 'mean' is zero, and the probability of an increase is assumed to be equal to the probability of a decrease ($p = 0.5$). A Z-test or a t-test can be used to determine whether an increase (or decrease) is significant, as previously described in Sec. 7.3.3.

Assume a sample of n paired variates where n_i show an increase (or decrease). For a binomial distribution, the mean ($\bar{n}$) is np and the standard deviation (σ) is $\sqrt{(npq)}$. A Z-value can be calculated:

$$Z = \frac{n_i - \bar{n}}{\sigma}$$

and compared with the critical Z-value (e.g., $Z = 1.96$ at the 5 per cent level).

7.4.2 The U-test

The U-test or *Mann–Whitney test* can sometimes be used when data are not paired. It is used to confirm (or deny) the possibility that two unlike samples belong to the same population. The U-test is often used to reduce computation time; the data are merely ranked in order of size and these ranking values used instead of actual values. The ranked values may give results as good as those obtained from using actual data (often a Z-test or a t-test could have been performed).

The method works as follows. Two samples n_1 and n_2 of events A and B respectively are combined to form one set of data and the items are ranked $1, 2, \ldots$, etc. (For tied rankings, an average rank is used for these values.) The rankings for each event (A and B) are separated and the average rank of each event is obtained. If R_1 is the sum of the ranks of sample size n_1 (and similarly R_2 and n_2), then R_1/n_1 and R_2/n_2 are compared. It is required to test if their difference is significant.

The test value of each event is determined:

$$U_A = n_1 n_2 + \frac{n_1(n_1 + 1)}{2} - R_1$$

and

$$U_B = n_1 n_2 + \frac{n_2(n_2 + 1)}{2} - R_2$$

The expected ranking ($\bar{R}$) for the population and its standard deviation (σ) are obtained:

$$\bar{R} = \frac{n_1 n_2}{2}$$

and

$$\sigma^2 = \frac{n_1 n_2 (n_1 + n_2 + 1)}{12}$$

A Z-test can now be used for each event, where

$$Z_A = \frac{U_A - \bar{R}}{\sigma}$$

and

$$Z_B = \frac{U_B - \bar{R}}{\sigma}$$

These Z values can be compared with the critical Z value to determine whether the differences are significant.

Note: Rank correlation methods are also non-parametric tests; the Spearman and Kendall methods of rank correlation are described in Sec. 7.5.2.

7.5 REGRESSION AND CORRELATION

We are often interested in the relationship (if one exists!) between two sets of data that are thought to vary together. Such a set of paired observations is known as a *bivariate distribution*, and the measure of the degree of relationship between two sets of data is called *correlation*. The method used to determine the best-fitting straight line for two sets of data is called *linear regression*.

These methods are used to predict what will happen with respect to sets of data in the future; for example, the probable success of a student in a degree course based upon school grades. Alternatively, they are used to predict one variable based upon the measurement of another, e.g. estimation of the strength of a material based upon measurement of its hardness (an easier property to measure experimentally).

However, a word of caution! The experimenter who uses these methods to try and 'make sense' of experimental data should consider carefully the cause and effect that has occurred. For example, suppose there is a high correlation between smoking and lung cancer. The statistics might just as easily be used to 'prove' that incipient lung cancer induces people to start smoking as that smoking causes lung cancer! Be careful about the claims that you make based upon statistics. There is always someone who is prepared to argue against you, and to use your data to prove their point, e.g. the Freedom to Smoke Society. It is a situation such as this that gives rise to the common phrases: 'statistics can be used to prove anything' or 'there are lies, damn lies and statistics'.

7.5.1 Linear regression

Suppose that two sets of data are obtained as paired items, and they can be presented graphically. The graph would appear as one of the three alternatives shown in Fig. 7.9, or as a graph somewhere between the three. This type of graph is known as a *scatter diagram*. The notation and axes are chosen so that x is the variable that will be used to estimate a value of y (i.e. x is often referred to as 'independent', although this term is inappropriate in statistics). This is an important point as the method is not directly reversible and the regression line of y on x is usually not equivalent to the regression line of x on y. Careful consideration of the assignment of x and y values to the paired data is necessary. A scatter diagram *must* be used in conjunction with linear regression analysis as this method is only suitable where the correlation is very close to linear. The data shown on the scatter diagram in Fig. 7.10 would be unsuitable for linear regression analysis, unless there was a very strong theoretical basis for assuming a linear relationship. In this situation (Fig. 7.10), additional data points would be required.

The data presented on the scatter diagram is to be used to construct the straight line (if this is appropriate) that best expresses the relationship between the sets of data. This is the *linear regression equation of y on x*. Only linear regression will be considered here, although it is possible to make estimates based upon a curve. However, a high correlation is still required and a curve is much more difficult to analyse mathematically. In the case of a non-linear relationship, it may be

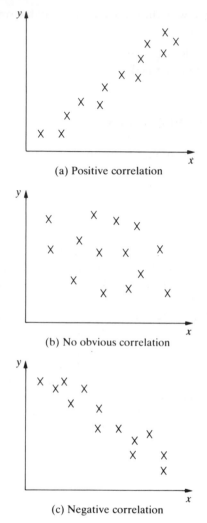

Figure 7.9 Examples of scatter diagrams.

possible to use logarithms (or other functions) of the data to obtain a straight line fit. Some possibilities are included in Sec. 7.8, and are shown in Table 7.5.

The paired data are plotted as shown in Fig. 7.11, and the regression line of y on x is also shown which is of the form:

$$y = mx + b$$

The slope (m) and 'y intercept' (i.e. the value of b) of the line are called the *parameters of the line*. The position of the regression line is determined such that $\sum d^2$ is a minimum, where d is the deviation of a point from the line as measured in the y-direction and shown in Fig. 7.11. The line also passes through ($\bar{x}, \bar{y}$) which is known as the *mean centre* of the bivariate distribution (note that $\sum d = 0$).

There are several ways of determining the linear regression equation; their derivations and discussion can be found in various statistics textbooks. The most common process is called the

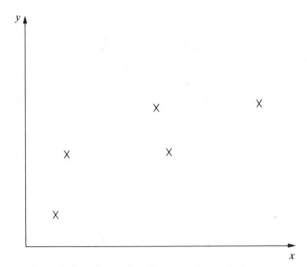

Figure 7.10 Scatter diagram with insufficient data points for regression analysis.

method of least squares but, as it is described in most books on statistics, only the final equations will be presented here.

The general regression line of y on x is given by

$$y = mx + b$$

Then

$$m = \frac{n \sum xy - \sum x \sum y}{n \sum x^2 - \left(\sum x\right)^2}$$

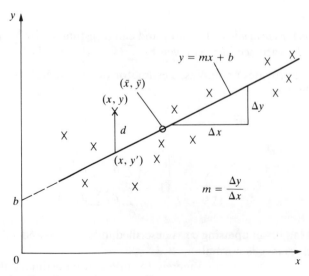

Figure 7.11 Linear regression line of y on x ($y = mx + b$)

and

$$b = \frac{\sum x^2 \sum y - \sum x \sum xy}{n \sum x^2 - \left(\sum x\right)^2}$$

However, once m has been determined from the above formula, the value of b can be obtained from the equation

$$\bar{y} = m\bar{x} + b$$

because the regression line passes through the point $(\bar{x}, \bar{y})$.

Only two points are needed to fix the position of the regression line. Usually the point $(x = 0, y = b)$ is chosen and also a convenient value of x (say 10 or 100) for which y can be determined.

As an alternative, the following pair of (simpler) equations can be used to determine the values of m and b:

$$\sum y = m \sum x + nb$$

and

$$\sum xy = m \sum x^2 + b \sum x$$

Known values from the given data are substituted into the two equations which are solved simultaneously for m and b.

Similar expressions for the equation of the regression line of y on x are given by

(i) $\quad y - \bar{y} = \dfrac{\sum d_x d_y}{\sum d_x^2} (x - \bar{x})$

where $d_x = (x_i - \bar{x})$ and $d_y = (y_i - \bar{y})$, and the gradient of the line is known as the *regression coefficient* of y on x.

(ii) $\quad y - \bar{y} = \dfrac{S_{x,y}}{S_x^2} (x - \bar{x})$

where $S_{x,y}$ is known as the *covariance* of x and y and can be estimated by $\sum (y - \bar{y})(x - \bar{x})/(n - 1)$. S_x^2 is the estimate of the variance of x as given by $\sum (x - \bar{x})^2/(n - 1)$.

The regression line provides an unbiased estimator of y on x, and the standard error of the estimate σ_e can be obtained from

$$\sigma_e^2 = \frac{S}{n - 1}$$

$$= \frac{\sum d_y^2}{(n - 1)} \left\{ 1 - \frac{\left(\sum d_x \sum d_y\right)^2}{\sum d_x^2 \sum d_y^2} \right\}$$

A confidence interval can be set up using σ_e as described in Secs 7.3.3 and 7.3.4, assuming that the distribution of the errors in y is normal.

A regression line should not be used merely to state a point estimate, it is also necessary to include either a scatter diagram or the correlation coefficient (see Sec. 7.5.2) to indicate the magnitude of the error to which the estimate is subject. Also, a regression line cannot be used reliably for extrapolation (i.e. outside the limits of the original data), at least not without some theoretical justification.

Note: The equations used to calculate the regression line of x on y are similar to those given above; they are:

$$x = m'y + b'$$

where

$$m' = \frac{n \sum xy - \sum x \sum y}{n \sum y^2 - \left(\sum y\right)^2}$$

and

$$b' = \frac{\sum y^2 \sum x - \sum y \sum xy}{n \sum y^2 - \left(\sum y\right)^2}$$

This line also passes through the point $(\bar{x}, \bar{y})$.

7.5.2 Correlation

The linear regression line is a straight line that provides the 'best' possible fit to the data. However, the method of least squares will still result in a straight line even if the data do not lie on a straight line! Therefore, what is also required is a single numerical measure (a statistic) that will indicate how closely a regression line matches given data. The *coefficient of correlation* (r) is the measure to be used. For the scatter diagrams shown in Fig. 7.9 we say that graph (a) shows a positive correlation, graph (b) shows zero correlation and graph (c) shows negative correlation.

Suppose that data have been obtained for the number of cars registered and the number of car accidents in several successive years, and the data show a high degree of positive correlation (e.g. Fig. 7.9a). There are five possible conclusions that can be made.

1. The sample was biased.
2. The correlation could reasonably have been due to chance, and no conclusions can be drawn.
3. As more cars use the roads, more accidents occur.
4. Because more accidents are occurring, more cars are required (reversing the cause and effect in 3).
5. The correlation is due to some other untested factor.

Statistics are used to ensure that the sample is free of bias, and to decide whether the degree of correlation is sufficient to exclude (within reasonable limits) the possibility that it might have been due to chance. Considerations of cause and effect are usually wider issues that require careful research and analysis.

Correlation measures the dispersion of paired data from a straight line. As the data points approach a straight line, the two regression lines of y on x and x on y approach each other. When their slopes are equal, the coefficient of correlation becomes 1. As the data become more dispersed, the two regression lines approach right angles to each other and eventually the coefficient becomes zero. The coefficient has values given by $-1 < r < +1$; the negative sign means that high values of x tend to be associated with low values of y, or vice versa (see Fig. 7.9(c)).

An equation was developed by Karl Pearson for determining the coefficient of correlation; it is known as the *product–moment method* (deviations from a mean are sometimes known as moments). The method compares the deviations of the actual data from the line of best fit, with the deviations of the same data from the common mean of these data. The *product–moment correlation coefficient* (r) or, more simply, the *correlation coefficient* is calculated from

$$r = \frac{n \sum xy - \sum x \sum y}{\sqrt{\left(n \sum x^2 - \left(\sum x\right)^2\right)\left(n \sum y^2 - \left(\sum y\right)^2\right)}}$$

Alternatively:

$$r = \frac{\sum d_x d_y}{\sqrt{\left(\sum d_x^2 \sum d_y^2\right)}}$$

The correlation coefficient is often misinterpreted as a percentage degree of correlation because it has a fractional value. A correlation of $r = 0.8$ is *not* twice as good as $r = 0.4$. It can be shown that $100r^2$ actually represents the percentages of variation in the dependent values that may be attributed to variations in the independent set of values. The 0.8 correlation coefficient represents a 64 per cent relationship, i.e. only 64 per cent of the variations have been explained and the remaining 36 per cent of the variations in the dependent values are probably due to causes other than the variations in the independent set of values. The 0.4 correlation coefficient represents only a 16 per cent relationship.

The significance of the product–moment coefficient, when the population value is assumed zero, is tested by calculating the t-statistic:

$$t = r \sqrt{\left(\frac{n - 2}{1 - r^2}\right)}$$

using tables of the t-distribution, where $(n - 2)$ is the degrees of freedom. There is also a 'standard deviation' of the coefficient that tests its degree of variation for different levels of confidence. It is not often used but it is described in more advanced statistics textbooks.

Rank correlation can be used to provide a quick approximation to the product–moment correlation coefficient, or when only ranking or objective values can be obtained. Positions in order of size are known as *ranks* and are obtained by allocating the integers 1 to n in place of the values x_i. For the smallest value of x_i substitute 1, and so on, and substitute n for the largest (or vice versa). Similar substitutions are made for y_i values.

Suppose that the pair of values (x_i, y_i) have been given the ranks (q, s), then *Spearman's rank correlation coefficient* (r_s or R) is given by

$$r_s = 1 - \frac{6 \sum D^2}{n(n^2 - 1)}$$

where $D = q - s$.

The calculation is extremely simple compared with the calculation of r. If tied ranks occur, each value is given the mean rank. This is actually in error, but it is only significant if there are many tied ranks. When n is greater than 30, the significance of r_s is tested by calculating the standard error:

$$\sigma_r = \frac{1}{\sqrt{(n-1)}}$$

and using tables of the normal distribution, or tables of percentage points for the significance of r_s.

Kendall's rank correlation coefficient (r_k or τ) provides an alternative procedure. It is slightly more complicated, but data can be included after the initial calculation and it has advantages in testing the significance. The two sets of values (x_i, y_i) are replaced by ranks as previously described. The paired ranks are arranged so that one set is in order: 1, 2, ..., n; the number of inversions (Q) are then counted. This will be easier to visualize from an example:

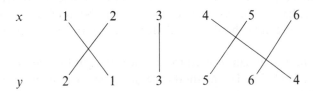

Corresponding ranks are joined by straight lines and Q is equal to the number of intersections of these lines. In this example, $Q = 3$. Kendall's coefficient is defined as

$$r_k = 1 - \frac{4Q}{n(n-1)}$$

The distribution of r_k has a mean of zero and variance of $[2(2n+5)/(9n(n-1))]$ and it tends rapidly to normality. Tables of the distribution function are available. Tied ranks are unimportant if there are only a few of them.

Both rank coefficients (r_s and r_k) vary between ± 1 but, except at the limits, the values are not the same. Some information is lost in the ranking process and the product–moment coefficient is more accurate. If the correlation is not linear, ranking methods may give a better measure of the true degree of correlation. However, transformation of one or both sets of data (e.g. as logarithmic functions) may give the 'best' results.

7.6 INTERPOLATION AND EXTRAPOLATION

Interpolation is the use of data *between* known points, and *extrapolation* is a method of obtaining data *outside* the range of known points. If either technique *must* be used, then it should be done only after careful consideration and the results regarded with caution. This is especially true of extrapolation, the results of which are more uncertain.

A straight line obtained by the method of least squares can be used to interpolate readings, or to extrapolate beyond the measured data points. Although the dangers of extrapolating beyond the range of the measured data are obvious, problems can also occur with interpolation. If there is a change in the physical phenomenon being studied, e.g. a change from laminar to turbulent fluid flow, or heat transfer occurring mainly by radiation rather than convection, then a large number of measurements should be made in the region where these changes have significant effects. The true effect of these changes on the variables can then be established.

7.7 GRAPHICAL ANALYSIS

Having obtained experimental data, it is usually necessary to present these data graphically and to compare the plot with theoretical predictions. However, before commencing curve-plotting, the experimenter should give considerable thought to the type of information that it is hoped to obtain. It will be necessary to understand the physical phenomena involved in an experiment; only then can the full range of possible information be obtained from the graphical displays.

What is usually required from the data is an analytical expression relating the variables that were measured. The analytical relation is easy to obtain when the data can be approximated by a straight line. However, the functional form of a curve is difficult to identify and the task becomes easier if the data can be plotted so as to obtain a straight line for other types of functional relationships. Knowledge of the associated theory can be used to estimate the particular functional form that represents the data, and then the appropriate type of plot can be easily selected.

The plotting methods that can be used to produce straight lines for several different types of functions are given in Table 7.5. The method of least squares can be used in each case to obtain the best straight line to fit the data.

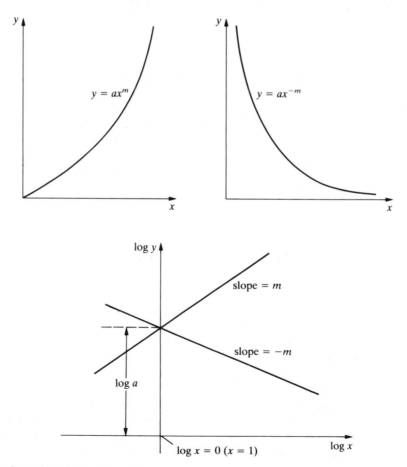

Figure 7.12 Transformation of power law curves.

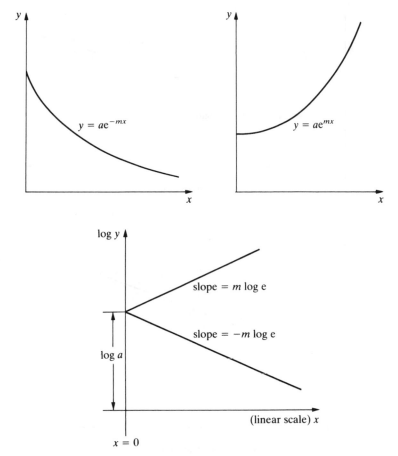

Figure 7.13 Transformation of exponential curves.

A process known as *rectification* is used to transform the equation of a curve into a straight line by changing the coordinates. Some transformations for rectifying equations are also given in Table 7.5. These transformations can often be used to obtain an empirical equation by trial and error. If the form of the empirical equation is known, an appropriate transformation can be chosen initially. Two common transformations are shown in Figs 7.12 and 7.13.

If a straight line ($y = ax + b$) is obtained, then a is the slope and b is the value of y when $x = 0$. It is tempting to plot the data so that b can be found from the intercept of the straight line with $x = 0$, even though the results may be unsuitable for this type of plot, as shown in Fig. 7.14(a). This type of plot (extending the range of the data) results in poor accuracy in the coefficients obtained. If the origin is far from the lower point, the origin should be suppressed as shown in Fig. 7.14(b). An accurate value of the slope can be obtained, and b determined by calculation.

Sometimes the curve cannot be transformed simply and other methods must be used. It is possible to substitute each of the n data pairs $(x_1, y_1) \ldots (x_n, y_n)$ into the equation of the polynomial which passes through all the (n) points:

$$y = a_0 + a_1x_1 + a_2x_2 + \cdots + a_{n-1}x^{n-1}$$

This will yield n simultaneous equations which can be solved to give the n coefficients of the

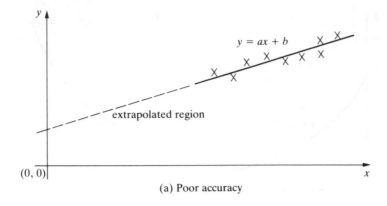

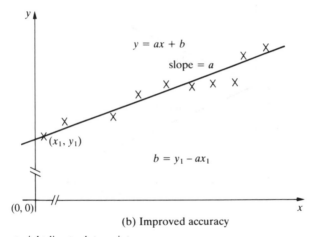

Figure 7.14 Fitting a straight line to data points.

polynomial. Often an equation of the fourth order is sufficiently accurate, as the higher terms are small. Finite difference methods (discussed in Sec. 6.4.1 for numerical differentiation) can be used to find the order of a suitable fitting equation, if this cannot be deduced from the shape of the curve.

7.8 ERROR ANALYSIS

Errors and mistakes were considered in Sec. 6.1.1. In this section we will look more closely at the occurrence of errors, their analysis and estimation. However, before discussing errors it will be useful to make some general observations about data analysis:

(a) The data should be consistent, and any 'obviously' incorrect data points should be discarded. This decision should not be subject to personal bias and whims, but should be based upon the use of consistent statistical analysis. If too many data points need to be discarded, the entire experimental procedure should be investigated.
(b) A statistical analysis of data is only appropriate when measurements are repeated several times.

Table 7.5 Methods of plotting various functions to obtain straight lines

Original function	Transformed function	Functions to be plotted	Parameters Slope	Parameters Intercept
$y = ax + b$	—	y vs x (linear paper)	a	b
$y = ax^b$	$\log y = \log a + b \log x$	$\log y$ vs $\log x$ (log–log paper)	b	$\log a$
$y = a\,e^{bx}$	$\log y = \log a + bx \log e$	$\log y$ vs x (log–linear paper)	$b \log e$	$\log a$
$y = \dfrac{x}{a + bx}$	$\dfrac{1}{y} = \dfrac{a}{x} + b$	$\dfrac{1}{y}$ vs $\dfrac{1}{x}$ (linear paper)	a	b (extrapolated)
$y = a + bx + cx^2$	$\dfrac{y - y_1}{x - x_1} = b + c(x + x_1)$	$\dfrac{y - y_1}{x - x_1}$ vs x (linear paper)	c	$b + cx_1$
$y = \dfrac{x}{a + bx} + c$	$\dfrac{x - x_1}{y - y_1} = (a + bx_1) + x\left(b + \dfrac{b^2}{a}x_1\right)$	$\dfrac{x - x_1}{y - y_1}$ vs x (linear paper)	$b + \dfrac{b^2}{a}x_1$	$a + bx_1$
$y = a\,e^{bx + cx^2}$	$\dfrac{\log(y/y_1)}{(x - x_1)} = b \log e + c(x + x_1) \log e$	$\log\left[\left(\dfrac{y}{y_1}\right)^{1/(x - x_1)}\right]$ vs x (log–linear paper)	$c \log e$	$b \log e + cx_1 \log e$

Note: (x_1, y_1) is any pair of values from the data.

(c) Estimate the uncertainties or errors in the results, as discussed in this section.

(d) Study the theory related to the experimental work and try to estimate the expected trends.

(e) Compare the results with what is already known and available.

An *error* can be defined as the difference between the true value and the measured value. In most situations, the true value is never known. Therefore, the value of an error is unknown, otherwise we could correct our data and eliminate the error! In certain artificial cases, the magnitude of an error may be known. For example, the value of π is known and, therefore, calculation of its value from measurements of the diameter and circumference of a circle can be used to find the magnitude of the error.

An error is a mistake or inaccuracy which occurs *every time* we try to measure something; it is a result of the fallibility of either the experimenter or the measuring equipment. Errors occur in *every* measurement and they are impossible to eliminate. The aim in experimental work should be to minimize the number and size of the errors that occur, and whenever possible to estimate their magnitude. One question that is sometimes asked is: Can we measure or estimate the magnitude of an error and, therefore, adjust our measurements accordingly? The answer is that it is impossible to measure something without using the phenomenon itself to influence the measuring instrument. This implies that the measurement changes the phenomenon. If the equipment was adjusted to eliminate the error, this adjustment would itself produce an error. It is impossible to remove totally the influence of variables that affect the process but which are not wanted in the measurement.

For example, when using a thermometer to measure the temperature of a liquid, some heat from the liquid will be used to heat up the thermometer itself if it was not initially at the liquid temperature. Even when an equilibrium situation is achieved, there will be some heat loss from the portion of the thermometer not immersed in the liquid. The measurement system can be made progressively more sophisticated to try and minimize the errors, but some error will always occur. What is important is to reduce the magnitude of the error so that the desired (or required) level of accuracy is obtained from the experimental work.

When errors occur, incorrect measurements are obtained and, hence, incorrect calculations are performed. This may lead to wrong decisions being made, unsatisfactory performance, accidents and/or failure. Ultimately the experimenter may be held responsible for a whole chain of events by having allowed an unacceptable or unrecognizable error to occur.

Errors may be categorized into two groups: these are accountable errors and unaccountable errors. *Accountable errors* (sometimes called 'systematic errors' or 'fixed errors') are usually reproducible and occur because of deficiencies in instrument response (see Secs 2.4 to 2.8). These errors can often be detected and measured in supplementary tests. They may occur for the following reasons:

(a) *Scale errors*, e.g. zero errors, scale divisions, incorrect scale mounting.

(b) *Static response errors*, i.e. non-uniformity of action, may be due to design, manufacturing defects, wear, incorrect maintenance, temperature, etc.; also hysteresis effects that are caused by friction.

(c) *Dynamic response errors* occur because the difference between the instrument reading and the measured quantity varies because of changes in the actual quantity over time.

(d) *Interference* from, or rather the presence of, the actual measuring device can produce changes in the quantity to be measured; for example a flow-measuring device restricting the flow rate. In this case it may be possible to use remote-sensing instruments to avoid or minimize this problem, e.g. an optical flow-measuring system using a dye flow tracer, or an optical pyrometer for temperature measurement.

(e) *Personal error* must be systematic if it is to be detected and corrected, e.g. parallax error, number prejudice, incorrect interpolation between scale divisions.

Unaccountable errors are *random* errors that occur in all measurements. If allowance is made for all accountable errors, and repeated measurements are made of a supposedly 'invariant' quantity, the data will still exhibit some *scatter*. This may be due to a lack of precision in allowing for accountable errors, but it is also due to many minor independent influences. These influences, such as vibrations, draughts, atmospheric pressure changes, are often considered to be individually unimportant, of approximately equal magnitude, and acting to increase or decrease the reading in a random manner. Mistakes by the experimenter *should* be infrequent; they are probably random and unlikely to be amenable to analysis, e.g. incorrect scale readings, transposition of figures. Statistical methods described in Sec. 7.3.3 can be used to test whether these random errors are likely to be significant.

The treatment and propagation of errors will now be considered. It is necessary to determine the uncertainty of a particular measurement, and to be able to specify (consistently) this uncertainty in analytical form. An experiment often requires several measurements to be made; these are then manipulated arithmetically to obtain the calculated result. It is necessary to consider the effect of a combination of errors, and to obtain the uncertainty in the final result. This can be performed in several ways.

One method is to assume that the error in the final result is equal to the maximum error in any parameter used to calculate the result. Alternatively, all the errors could be combined to provide the maximum (worst) error in the result. For example, consider the calculation of electric power (in watts) from the product of the measurements of voltage and current. Suppose these values are 250 V $\pm$ 2 V and 10 A $\pm$ 0.4 A, then the nominal power is 250 $\times$ 10 = 2500 W. The maximum and minimum variations in power are:

$$\text{maximum power} = 252 \times 10.4 = 2620.8 \text{ W}$$

$$\text{and minimum power} = 248 \times 9.6 = 2380.8 \text{ W}$$

As these results are only accurate to three and two significant figures respectively, i.e. 2620 W to 2400 W, the uncertainty in the power varies between +4.8 per cent and −4.0 per cent. This method produces an uncertainty that is too severe, and although this may be considered prudent as a safety factor it may also lead to acceptable results being rejected. There is often no good reason to assume that the maximum or minimum readings should coincide. This method should only be used to provide an initial approximate inspection of data and to provide an indication of whether the experimental system appears satisfactory.

A more precise method of calculating the propagation of errors and uncertainties is available. Suppose that the calculated result (P) is a *given* function of the independent variable (x), i.e. $P = f(x)$. Differentiation of this equation yields

$$\frac{dP}{dx} = \frac{df(x)}{dx}$$

or

$$\Delta P = \frac{df(x)}{dx} \Delta x$$

It follows that

$$U_p^2 = \left[\frac{df(x)}{dx} \right]^2 U_x^2$$

where U denotes the uncertainty. To manipulate the errors as percentages, the equation can be written:

$$\frac{U_p}{P} = \left| \frac{1}{f(x)} \frac{df(x)}{dx} \right| U_x$$

This form of the equation is especially useful in dealing with functions of the form $P = kx^n$.

When more than two independent variables are involved, the chain rule states:

$$\Delta P = \frac{\partial f(x)}{\partial x_1} \Delta x_1 + \frac{\partial f(x)}{\partial x_2} \Delta x_2 + \cdots$$

The worst case is given by:

$$\Delta P = \left| \frac{\partial f(x)}{\partial x_1} \Delta x_1 \right| + \left| \frac{\partial f(x)}{\partial x_2} \Delta x_2 \right| + \cdots$$

assuming that the increments are equivalent to the errors or uncertainties in P and x_i.

The combined uncertainty or error is better estimated by the equations (although not exact):

$$\Delta P = \left[\left(\frac{\partial f(x)}{\partial x_1} \Delta x_1 \right)^2 + \left(\frac{\partial f(x)}{\partial x_2} \Delta x_2 \right)^2 + \cdots \right]^{1/2}$$

or

$$U_p^2 = \left(\frac{\partial f(x)}{\partial x_1} \right)^2 U_{x_1}^2 + \left(\frac{\partial f(x)}{\partial x_2} \right)^2 U_{x_2}^2 + \cdots$$

This equation gives approximately the same significance level for both ΔP and the Δx_i values. The uncertainties U_{xi} should all be defined consistently, i.e. to the same confidence limit:

$U_i = s_i$ is equivalent to a 68 per cent confidence limit

$U_i = 2s_i$ is equivalent to a 95 per cent confidence limit

Confidence limits are often not quoted and it may be necessary to establish appropriate values.

The uncertainty in the calculated result (U_p) depends the squares of the uncertainties in the independent variables (U_{xi}). Therefore, the smaller uncertainties (by a factor of 5 or 10) can generally be neglected and the largest uncertainties predominate. This should be considered when designing the experiment and identifying the experimental technique.

Note: If the calculated result (P) is the difference of two independent variables, e.g. $P = x - y$, the percentage uncertainty can become very large if these variables (x and y) are approximately equal quantities. This problem may not be apparent when a computer is used to perform the calculations.

EXERCISES

Numerical problems are deliberately not included in these exercises; readers should apply the methods and ideas described in this chapter to the treatment of their own experimental results. This will provide a more useful exercise than merely working through a series of arbitrarily defined problems.

7.1 Apply the statistical methods described in this chapter to your own data, i.e. to the results obtained from laboratory studies.

7.2 Explain the difference between descriptive and inferential statistics.

7.3 Calculate some common measures of average and dispersion. These parameters are frequently required as a preliminary analysis of measured data.

7.4 Calculate a weighted average, an index number and a moving average.

7.5 Probability theory is used for the assessment of equipment reliability. Determine the probability of equipment failure for a particular (or proposed) experiment and hence determine the need/advantage of installing duplicate (back-up) items of equipment. Reference should be made to a suitable introductory text describing reliability engineering (see Bibliography).

7.6 What is a probability distribution?
What is the difference between discrete and continuous probability distributions?

7.7 Identify some practical situations that are described by the binomial, Poisson and normal distributions.

7.8 Describe some practical problems associated with obtaining 'representative' samples.

7.9 Explain the meaning of the terms:
(a) simple sample;
(b) random sample;
(c) unbiased estimate;
(d) sampling distribution.

7.10 What is the standard error of the sample mean?

7.11 Describe how to estimate the population mean and variance using sampling data.

7.12 Use significance testing and confidence intervals (Sec. 7.3.3) to analyse some experimental results.

7.13 What is a 'small' sample?
What problems occur when analysing the results obtained using small samples?
Explain the use of the Student-t statistic.

7.14 Explain the use of the chi-squared distribution, and (if possible) apply the distribution to the analysis of your results.

7.15 Explain the use of the F-distribution.

7.16 Apply some non-parametric statistical tests (Sec. 7.4) to your numerical data; assess the accuracy of the results compared with an 'exact' numerical analysis.

7.17 Draw a scatter diagram.
Identify the independent and dependent variables.
Obtain the appropriate regression equation.
Calculate the coefficient of correlation.
Interpret and discuss the data based upon the results obtained using these statistical methods.

7.18 Discuss the problems associated with interpolation and extrapolation.

7.19 Obtain data for two (non-linearly) related variables and obtain a mathematical expression for the relation between the variables (see Sec. 7.7).

7.20 Identify and discuss the errors that may influence particular experimental studies. Calculate the uncertainties in the measured values of particular variables.

EXAMPLES OF STATISTICAL TABLES FOR PARTICULAR DISTRIBUTIONS

Table 7.6 Binomial probabilities
Table 7.7 Poisson probabilities
Table 7.8 The normal distribution function
Table 7.9 Two-tail percentage points of the t-distribution
Table 7.10 Percentage points of the chi-squared distribution
Table 7.11 Percentage points of the F-distribution

Table 7.6 Binomial probabilities for the expansion of $(q + p)^n$

The table gives (in columns) the value of $^nC_x(1 - p)^{n-x}p^x$, for values of p from 0.1 to 0.9 (increments of 0.1), of $n = 3$, 4 and 5 and of x from 0 to n (integer values)

x	$p = 0.1$	$p = 0.2$	$p = 0.3$	$p = 0.4$	$n = 3$ $p = 0.5$	$p = 0.6$	$p = 0.7$	$p = 0.8$	$p = 0.9$
0	0.7290	0.5120	0.3430	0.2160	0.1250	0.0640	0.0270	0.0080	0.0010
1	0.2430	0.3840	0.4410	0.4320	0.3750	0.2880	0.1890	0.0960	0.0270
2	0.0270	0.0960	0.1890	0.2880	0.3750	0.4320	0.4410	0.3840	0.2430
3	0.0010	0.0080	0.0270	0.0640	0.1250	0.2160	0.3430	0.5120	0.7290

x	$p = 0.1$	$p = 0.2$	$p = 0.3$	$p = 0.4$	$n = 4$ $p = 0.5$	$p = 0.6$	$p = 0.7$	$p = 0.8$	$p = 0.9$
0	0.6561	0.4096	0.2401	0.1296	0.0625	0.0256	0.0081	0.0016	0.0001
1	0.2916	0.4096	0.4116	0.3456	0.2500	0.1536	0.0756	0.0256	0.0036
2	0.0486	0.1536	0.2646	0.3456	0.3750	0.3456	0.2646	0.1536	0.0486
3	0.0036	0.0256	0.0756	0.1536	0.2500	0.3456	0.4116	0.4096	0.2916
4	0.0001	0.0016	0.0081	0.0256	0.0625	0.1296	0.2401	0.4096	0.6561

x	$p = 0.1$	$p = 0.2$	$p = 0.3$	$p = 0.4$	$n = 5$ $p = 0.5$	$0 = 0.6$	$p = 0.7$	$p = 0.8$	$p = 0.9$
0	0.5905	0.3277	0.1681	0.0778	0.0313	0.0102	0.0024	0.0003	0.0000
1	0.3281	0.4096	0.3602	0.2592	0.1563	0.0768	0.0284	0.0064	0.0004
2	0.0729	0.2048	0.3087	0.3456	0.3125	0.2304	0.1323	0.0512	0.0081
3	0.0081	0.0512	0.1323	0.2304	0.3125	0.3456	0.3087	0.2048	0.0729
4	0.0004	0.0064	0.0283	0.0768	0.1563	0.2592	0.3602	0.4096	0.3280
5	0.0000	0.0003	0.0024	0.0102	0.0313	0.0778	0.1681	0.3277	0.5905

Table 7.7 Poisson probabilities ($e^{-\mu}\mu^x/x!$) for values of x from 0 to 8 (integer values) and values of the mean μ from 0.1 to 2.5 (increments of 0.1)

Mean μ	x								
	0	1	2	3	4	5	6	7	8
0.1	0.9048	0.0905	0.0045	0.0002	0.0000	0.0000	0.0000	0.0000	0.0000
0.2	0.8187	0.1637	0.0164	0.0011	0.0001	0.0000	0.0000	0.0000	0.0000
0.3	0.7408	0.2222	0.0333	0.0033	0.0003	0.0000	0.0000	0.0000	0.0000
0.4	0.6703	0.2681	0.0536	0.0072	0.0007	0.0001	0.0000	0.0000	0.0000
0.5	0.6065	0.3033	0.0758	0.0126	0.0016	0.0002	0.0000	0.0000	0.0000
0.6	0.5488	0.3293	0.0988	0.0198	0.0030	0.0004	0.0000	0.0000	0.0000
0.7	0.4966	0.3476	0.1217	0.0284	0.0050	0.0007	0.0001	0.0000	0.0000
0.8	0.4493	0.3595	0.1438	0.0383	0.0077	0.0012	0.0002	0.0000	0.0000
0.9	0.4066	0.3659	0.1647	0.0494	0.0111	0.0020	0.0003	0.0000	0.0000
1.0	0.3679	0.3679	0.1839	0.0613	0.0153	0.0031	0.0005	0.0001	0.0000
1.1	0.3329	0.3662	0.2014	0.0738	0.0203	0.0045	0.0008	0.0001	0.0000
1.2	0.3012	0.3614	0.2169	0.0867	0.0260	0.0062	0.0012	0.0002	0.0000
1.3	0.2725	0.3543	0.2303	0.0998	0.0324	0.0084	0.0018	0.0003	0.0001
1.4	0.2466	0.3452	0.2417	0.1128	0.0395	0.0111	0.0026	0.0005	0.0001
1.5	0.2231	0.3347	0.2510	0.1255	0.0471	0.0141	0.0035	0.0008	0.0001
1.6	0.2019	0.3230	0.2584	0.1378	0.0551	0.0176	0.0047	0.0011	0.0002
1.7	0.1827	0.3106	0.2640	0.1496	0.0636	0.0216	0.0061	0.0015	0.0003
1.8	0.1653	0.2975	0.2678	0.1607	0.0723	0.0260	0.0078	0.0020	0.0005
1.9	0.1496	0.2842	0.2700	0.1710	0.0812	0.0309	0.0098	0.0027	0.0006
2.0	0.1353	0.2707	0.2707	0.1804	0.0902	0.0361	0.0120	0.0034	0.0009
2.1	0.1225	0.2572	0.2700	0.1890	0.0992	0.0417	0.0146	0.0044	0.0011
2.2	0.1108	0.2438	0.2681	0.1966	0.1082	0.0476	0.0174	0.0055	0.0015
2.3	0.1003	0.2306	0.2652	0.2033	0.1169	0.0538	0.0206	0.0068	0.0019
2.4	0.0907	0.2177	0.2613	0.2090	0.1254	0.0602	0.0241	0.0083	0.0025
2.5	0.0821	0.2052	0.2565	0.2138	0.1336	0.0668	0.0278	0.0099	0.0031

Table 7.8 The normal distribution function*

Z	0.00	0.01	0.02	0.03	0.04	0.05	0.06	0.07	0.08	0.09
0.0	0.5000	5040	5080	5120	5160	5199	5239	5279	5319	5359
0.1	0.5398	5438	5478	5517	5557	5596	5636	5675	5714	5753
0.2	0.5793	5832	5871	5910	5948	5987	6026	6064	6103	6141
0.3	0.6179	6217	6255	6293	6331	6368	6406	6443	6480	6517
0.4	0.6554	6591	6628	6664	6700	6736	6772	6808	6844	6879
0.5	0.6915	6950	6985	7019	7054	7088	7123	7157	7190	7224
0.6	0.7257	7291	7324	7357	7389	7422	7454	7486	7517	7549
0.7	0.7580	7611	7642	7673	7704	7734	7764	7794	7823	7852
0.8	0.7881	7910	7939	7967	7995	8023	8051	8078	8106	8133
0.9	0.8159	8186	8212	8238	8264	8289	8315	8340	8365	8389
1.0	0.8413	8438	8461	8485	8508	8531	8554	8577	8599	8621
1.1	0.8643	8665	8686	8708	8729	8749	8770	8790	8810	8830
1.2	0.8849	8869	8888	8907	8925	8944	8962	8980	8997	9015
1.3	0.9032	9049	9066	9082	9099	9115	9131	9147	9162	9177
1.4	0.9192	9207	9222	9236	9251	9265	9279	9292	9306	9319
1.5	0.9332	9345	9357	9370	9382	9394	9406	9418	9429	9441
1.6	0.9452	9463	9474	9484	9495	9505	9515	9525	9535	9545
1.7	0.9554	9564	9573	9582	9591	9599	9608	9616	9625	9633
1.8	0.9641	9649	9656	9664	9671	9678	9686	9693	9699	9706
1.9	0.9713	9719	9726	9732	9738	9744	9750	9756	9761	9767
2.0	0.9772	9778	9783	9788	9793	9798	9803	9808	9812	9817
2.1	0.9821	9826	9830	9834	9838	9842	9846	9850	9854	9857
2.2	0.9861	9864	9868	9871	9875	9878	9881	9884	9887	9890
2.3	0.9893	9896	9898	9901	9904	9906	9909	9911	9913	9916
2.4	0.9918	9920	9922	9925	9927	9929	9931	9932	9934	9936

* The figures in the table are the areas under the standard normal curve from minus infinity to the given variate Z, i.e. the probability that a random variate will be less than Z.

To find the area up to a Z value of 0.35, look down the left-hand column to $Z = 0.3$ and then across the top of the table to 0.05. The point of intersection in the table gives an area of 0.6368.

The table given above can be continued and when $Z = 4.00$, the area is 0.999 97.

Table 7.9 Two-tail percentage points of the *t*-distribution*

v	2P						$120/v$
	20%	10%	5%	2%	1%	0.1%	
1	3.078	6.314	12.71	31.82	63.66	636.6	
2	1.886	2.920	4.303	6.965	9.925	31.60	
3	1.638	2.353	3.182	4.541	5.841	12.92	
4	1.533	2.132	2.776	3.747	4.604	8.610	
5	1.476	2.015	2.571	3.365	4.032	6.869	
6	1.440	1.943	2.447	3.143	3.707	5.959	
7	1.415	1.895	2.365	2.998	3.499	5.408	
8	1.397	1.860	2.306	2.896	3.355	5.041	
9	1.383	1.833	2.262	2.821	3.250	4.781	
10	1.372	1.812	2.228	2.764	3.169	4.587	12
12	1.356	1.782	2.179	2.681	3.055	4.318	10
20	1.325	1.725	2.086	2.528	2.845	3.850	6
30	1.310	1.697	2.042	2.457	2.750	3.646	4
40	1.303	1.684	2.021	2.423	2.704	3.551	3
60	1.296	1.671	2.000	2.390	2.660	3.460	2
120	1.289	1.658	1.980	2.358	2.617	3.373	1
∞	1.282	1.645	1.960	2.326	2.576	3.291	0

* The probabilities are that the tabulated value will be numerically exceeded, and therefore apply to a two-tailed test.
For a one-tailed test, the percentages are divided by two (P rather than $2P$).
Interpolation v-wise should be linear in $120/v$.

Table 7.10 Percentage points of the chi-squared distribution*

v	P						
	99%	95%	10%	5%	2.5%	1%	0.1%
1	0.000 16	0.003 9	2.706	3.841	5.024	6.635	10.83
2	0.020 1	0.103	4.605	5.991	7.378	9.210	13.82
3	0.115	0.352	6.252	7.816	9.351	11.35	16.27
4	0.297	0.711	7.780	9.488	11.14	13.28	18.47
5	0.554	1.15	9.236	11.07	12.83	15.08	20.51
6	0.872	1.64	10.64	12.59	14.45	16.81	22.46
7	1.24	2.17	12.02	14.07	16.02	18.49	24.36
8	1.65	2.73	13.36	15.51	17.53	20.09	26.13
9	2.09	3.33	14.68	16.92	19.02	21.67	27.89
10	2.56	3.94	15.99	18.31	20.48	23.21	29.59
12	3.57	5.23	18.55	21.03	23.34	26.22	32.91
15	5.23	7.26	22.31	25.00	27.49	30.58	37.70
20	8.26	10.85	28.41	31.41	34.17	37.57	45.32
25	11.52	14.61	34.38	37.65	40.65	44.31	52.62
30	14.95	18.49	40.26	43.77	46.98	50.89	59.70
40	22.16	26.51	51.81	55.76	59.34	63.69	73.40
50	29.71	34.76	63.17	67.50	71.42	76.15	86.66
60	37.48	43.19	74.40	79.08	83.30	88.38	99.61
70	45.4	51.7	85.5	90.5	95.0	100.4	112.3
80	53.5	60.4	96.6	101.9	106.6	112.3	124.8
90	61.7	69.1	107.6	113.1	118.1	124.1	137.2
100	70.1	77.9	118.5	124.3	129.6	135.8	149.5

* The probabilities are that a given value will be exceeded, i.e. $x \geqslant \chi_p^2$. For $v \leqslant 100$, linear interpolation in v is satisfactory.

Table 7.11 Percentage points (5 per cent and 1 per cent) of the *F*-distribution*

v_2	v_1												
	1	2	3	4	5	6	8	10	12	15	20	24	∞
1	161	199	216	225	230	234	239	242	244	246	248	249	254
	4050	5000	5400	5620	5760	5860	5980	6060	6110	6160	6210	6230	6370
2	18.5	19.0	19.2	19.3	19.3	19.3	19.4	19.4	19.4	19.4	19.5	19.5	19.5
	98.5	99.0	99.2	99.2	99.3	99.3	99.4	99.4	99.4	99.4	99.5	99.5	99.5
3	10.1	9.55	9.28	9.12	9.01	8.94	8.85	8.79	8.74	8.70	8.66	8.64	8.53
	34.1	30.8	29.5	28.7	28.2	27.9	27.5	27.2	27.1	26.9	26.7	26.6	26.1
4	7.71	6.94	6.59	6.39	6.26	6.16	6.04	5.96	5.91	5.86	5.80	5.77	5.63
	21.2	18.0	16.7	16.0	15.5	15.2	14.8	14.6	14.4	14.2	14.0	13.9	13.5
5	6.61	5.79	5.41	5.19	5.05	4.95	4.82	4.74	4.68	4.62	4.56	4.53	4.36
	16.3	13.3	12.1	11.4	11.0	10.7	10.3	10.1	9.89	9.72	9.55	9.47	9.02
6	5.99	5.14	4.76	4.53	4.39	4.28	4.15	4.06	4.00	3.94	3.87	3.84	3.67
	13.7	10.9	9.78	9.15	8.75	8.47	8.10	7.87	7.72	7.56	7.40	7.31	6.88
8	5.32	4.46	4.07	3.84	3.69	3.58	3.44	3.35	3.28	3.22	3.15	3.12	2.93
	11.3	8.65	7.59	7.01	6.63	6.37	6.03	5.81	5.67	5.52	5.36	5.28	4.86
10	4.96	4.10	3.71	3.48	3.33	3.22	3.07	2.98	2.91	2.85	2.77	2.74	2.54
	10.0	7.56	6.55	5.99	5.64	5.39	5.06	4.85	4.71	4.56	4.41	4.33	3.91
12	4.75	3.89	3.49	3.26	3.11	3.00	2.85	2.75	2.69	2.62	2.54	2.51	2.30
	9.33	6.93	5.95	5.41	5.06	4.82	4.50	4.30	4.16	4.01	3.86	3.78	3.36
15	4.54	3.68	3.29	3.06	2.90	2.79	2.64	2.54	2.48	2.40	2.33	2.29	2.07
	8.68	6.36	5.42	4.89	4.56	4.32	4.00	3.80	3.67	3.52	3.37	3.29	2.87
20	4.35	3.49	3.10	2.87	2.71	2.60	2.45	2.35	2.28	2.20	2.12	2.08	1.84
	8.10	5.85	4.94	4.43	4.10	3.87	3.56	3.37	3.23	3.09	2.94	2.86	2.42
24	4.26	3.40	3.01	2.78	2.62	2.51	2.36	2.25	2.18	2.11	2.03	1.98	1.73
	7.82	5.61	4.72	4.22	3.90	3.67	3.36	3.17	3.03	2.89	2.74	2.66	2.21
∞	3.84	3.00	2.60	2.37	2.21	2.10	1.94	1.83	1.75	1.67	1.57	1.52	1.00
	6.63	4.61	3.78	3.32	3.02	2.80	2.51	2.32	2.18	2.04	1.88	1.79	1.00

* For each pair of values given in the table, the upper figure will be exceeded with probability 5 per cent, and the bottom one with probability 1 per cent.

v_1 must correspond with the sample having the greater mean square deviation.

BIBLIOGRAPHY

Book, S. A., *Statistics: Basic Techniques for Solving Applied Problems*, McGraw-Hill, New York, 1977.

Breipohl, A. M., *Probabilistic System Analysis: An Introduction to Probabilistic Models, Decisions and Applications of Random Processes*, Wiley, New York, 1970.

Brook, R. J. and G. C. Arnold, *Applied Regression Analysis and Experimental Design*, Marcel Dekker, New York, 1985.

Chatfield, C., *Statistics for Technology: A Course in Applied Statistics*, 2nd edn, Chapman & Hall, London, 1978.

Choi, S. C., *Introductory Applied Statistics in Science*, Prentice-Hall, Englewood Cliffs, New Jersey, 1978.

Clarke, G. M. and D. Cooke, *A Basic Course in Statistics*, Edward Arnold, London, 1978.

Cox, D. R. and E. J. Snell, *Applied Statistics: Principles and Examples*, Chapman & Hall, London, 1981.

Daniel, W. D., *Applied Nonparametric Statistics*, Houghton Mifflin, Boston, MA, 1978.

Eadie, W. T., D. Drijard, F. E. James, M. Roos and B. Sadoulet, *Statistical Methods in Experimental Physics*, North-Holland, Amsterdam, 1971.

Ellis, R. B., *Statistical Inference: Basic Concepts*, Prentice-Hall, Englewood Cliffs, New Jersey, 1975.

Erickson, B. H. and T. A. Nosanchuk, *Understanding Data*, McGraw-Hill, New York, 1977.

Fischer, F. E., *Fundamental Statistical Concepts*, Canfield Press, California, 1973.

Guttman, I., *Introductory Engineering Statistics*, 3rd edn, Wiley, New York, 1982.

Huntsberger, D. V. and P. Billingsley, *Elements of Statistical Inference*, 5th edn, Allyn and Bacon, Newton, MA, 1981.

Kapur, K. C. and L. R. Lamberson, *Reliability in Engineering Design*, Wiley, New York, 1977.

Lipson, C. and N. J. Sheth, *Statistical Design and Analysis of Engineering Experiments*, McGraw-Hill, New York, 1972.

McCuen, R. H., *Statistical Methods for Engineers*, Prentice-Hall, Englewood Cliffs, New Jersey, 1985.

Meddis, R., *Statistical Handbook for Non-Statisticians*, McGraw-Hill, New York, 1975.

Mendenhall, W. and T. Sincich, *Statistics for Engineering and Computer Science*, Dellen Publishing Co., Santa Clara, CA, 1984.

Miller, I. and J. Freund, *Probability and Statistics for Engineers*, 3rd edn, Prentice-Hall, Englewood Cliffs, New Jersey. 1985.

Milton, J. S. and J. C. Arnold, *Probability and Statistics in the Engineering and Computing Sciences*, McGraw-Hill, New York, 1985.

O'Connor, P. D. T., *Practical Reliability Engineering*, Heyden and Son, London, 1981.

Polak, P., *Systematic Errors in Engineering Experiments*, Gulf Publishing Co., Houston, 1979.

Robertson, A. G., *Quality Control and Reliability*, Thomas Nelson, London, 1971.

Runyon, R. P., *Nonparametric Statistics: A Contemporary Approach*, Addison-Wesley, Reading, MA, 1977.

Siegel, S., *Nonparameteric Statistics for the Behavioral Sciences*, McGraw-Hill, New York, 1976.

Smith, C. O., *Introduction to Reliability in Design*, McGraw-Hill, New York, 1976.

Wassell, H. J. H., *Reliability of Engineering Products*, Engineering Design Guide Number 38, Oxford University Press, 1980.

American Society for Testing and Materials (ASTM)—Volume 14.02: Standard Practices for Statistical Methods

E 141–69: *Acceptance of Evidence Based on the Results of Probability Sampling*

E 122–72: *Choice of Sample Size to Estimate the Average Quality of a Lot or Process.*

E 691–79: *Conducting an Interlaboratory Test Program to Determine the Precision of Test Methods.*

E 178–80: *Dealing with Outlying Observations.*

E 29–67: *Indicating which Places of Figures are to be Considered Significant in Specified Limiting Values.*

E 105–58: *Probability Sampling of Materials.*

E 177–71: *Use of the Terms Precision and Accuracy as Applied to Measurement of a Property of a Material.*

E 456–83: *Definitions of Terms Related to Statistical Methods.*

British Standards

BS 2846: *Guide to statistical interpretation of data.*

    Part 1: *Routine analysis of quantitative data.*

    Part 2: *Estimation of the mean: confidence interval.*

    Part 3: *Determination of a statistical tolerance interval.*

    Part 4: *Techniques of estimation and tests relating to means and variances.*

Part 5: Power of tests relating to means and variances.
Part 6: Comparison of two means in the case of paired observations.
Part 7: Tests of departure from normality.

BS 5532: Statistical terminology.
Part 1: Glossary of terms relating to probability and general terms relating to statistics.
Part 3: Glossary of terms relating to the design of experiments.

EIGHT

DIMENSIONAL ANALYSIS

CHAPTER OBJECTIVES

To encourage the reader to consider the dimensional consistency of equations, and to use the methods of dimensional analysis for the interpretation of experimental results.

QUESTIONS

- How is the principle of dimensional homogeneity used to check the validity of equations?
- How are the methods of dimensional analysis applied to the interpretation of experimental results?

8.1 TYPES OF EQUATIONS

Before studying the method of *dimensional analysis*, it is important to read Chapter 1 concerning units and dimensions and to appreciate the difference between these terms. Two types of equations are encountered: those that are *dimensionally homogeneous* (or universally valid) and *dimensionally inhomogeneous* equations.

Consider the equation relating the surface area (A) of a sphere to its radius (r):

$$A = 4\pi r^2 \tag{8.1}$$

This equation is dimensionally homogeneous; the constant (4) is a *true constant* and it is dimensionless. The constant has the same value whatever system of *consistent* units is used. (The dimensionless quantity π is the ratio of circumference to radius of a circle.)

In engineering heat transfer, the heat transfer coefficient (h; SI units: $W/m^2 K$) is an important variable. Its value is often determined from *empirical equations* containing dimension-less groups (considered later in this chapter). Air is a common heat transfer medium for natural convection, and by substituting the properties of air (under particular and appropriate conditions) into the appropriate empirical equation, a simplified form of the equation can be obtained. An example is given by

$$h = 1.24 \, \Delta T^{1/3} \tag{8.2}$$

for natural thermal convection from a cylindrical surface into air at 10^5 Pa absolute pressure, where T is measured in kelvins.

Equation (8.2) is an example of a dimensionally inhomogeneous equation; the constant (1.24) is known as a *suppressed constant*. By rearranging Eq. (8.2), thus:

$$\frac{h}{\Delta T^{1/3}} = \text{constant} \tag{8.3}$$

it can be seen that the dimensions of the constant are ($\text{W/m}^2 \, \text{K}^{4/3}$).

The value of the constant in Eq. (8.2) changes if the units in which h and ΔT are measured are changed. For example:

$$h = 0.18 \, \Delta T^{1/3} \tag{8.4}$$

where h is measured in $\text{Btu/h ft}^2 \, {}^\circ\text{F}$ and T is measured in ${}^\circ\text{F}$.

The reader could verify Eqs (8.2) and (8.4) by substituting the appropriate properties into the equation for natural convection heat transfer. This equation can be found in most engineering heat transfer textbooks, e.g. Kreith and Black.[1]

The *principle of homogeneity of equations* states that each term of a universally valid equation must have the same dimensions. Consider the equation relating the total surface area (A) of a solid cylinder to its radius (r) and length (l):

$$A = 2\pi r^2 + 2\pi r l \tag{8.5}$$

Dimensions: $\qquad\qquad\qquad\quad L^2 \quad L^2 \qquad L \times L$

The terms of Eq.(8.5) are homogeneous and the same answer is obtained whatever system of units is used. (*Note:* the symbols in an equation represents numbers only, i.e. the *measures* of the quantities concerned.) Universally valid equations can always be converted into a dimensionless form by isolating the constant. For example, dividing Eq. (8.5) by (πr^2) and re-arranging:

$$\frac{A}{\pi r^2} - 2\frac{l}{r} = 2$$

Dimensions: $\qquad\qquad\qquad\quad \dfrac{L^2}{L^2} \quad \dfrac{L}{L} \qquad \text{none}$

8.1.1 Normalization of equations

The way in which an equation is written affects how easy it is for that equation to describe a particular situation. It also influences the level of understanding of the phenomenon character-ized by the equation. The units in which the quantities are measured do not necessarily produce a clear understanding of the situation; often what is required is a comparison between different values.

Consider as an example, the radioactive decay of an element. The mass (m) remaining after a certain time (t) is related to the initial mass (m_0) by the expression

$$m = m_0 \exp(-0.693t/t_0)$$

where t_0 is the half life of the element. The half life is the time required for 50 per cent of the radioactive mass to decay, i.e. 50 per cent of the radiation released. The equation can also be written in terms of the numbers (n; n_0) of radioactive nuclei, thus:

$$n = n_0 \exp(-0.693t/t_0)$$

The mass (or number of nuclei) and the half life of a particular element can be obtained from a handbook of chemical data (e.g. *Lange's Handbook of Chemistry*[2]). It is therefore possible to draw a graph of m (or n) against t for any element. However, it is preferable to plot (m/m_0) against (t/t_0) as shown in Fig. 8.1. This curve is universally valid for all radioactive elements. The equations can be written:

$$M = \exp(-0.693T)$$

or

$$N = \exp(-0.693T)$$

where

$$M = m/m_0$$

$$N = n/n_0$$

$$T = t/t_0$$

The process of transforming an equation into units such as M, N and T is known as *normalization*. The normalized variable usually has a limited range, e.g. $1 \geqslant M \geqslant 0$, although in this case an infinite time is required for all nuclei to decay ($T \to \infty$). If a graph of m against t is plotted, it has the same shape as Fig. 8.1.

Now consider the laminar flow of water in a circular pipe of radius R. The velocity (v) as a

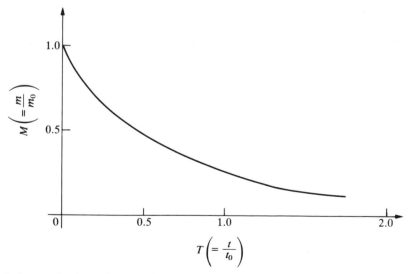

Figure 8.1 Typical curve showing radioactive decay of an element.

function of distance (y) across the pipe is given by the equation

$$v = v_{max}\left[1 - \left(\frac{y}{R}\right)^2\right]$$

where v_{max} is the maximum velocity which occurs at the centre of the pipe, and the velocity is zero at the pipe walls. Typical results may be plotted as shown in Fig. 8.2(a). Similar graphs would be obtained for other pipe sizes with different (but laminar) flowrates. However, the equation can be normalized thus:

$$\frac{v}{v_{max}} = 1 - \left(\frac{y}{R}\right)^2$$

i.e.
$$V = 1 - Y^2$$

where
$$V = v/v_{max}$$

and
$$Y = y/R$$

The normalized velocity distribution can now be drawn as shown n Fig. 8.2(b). The velocity at any point in a pipe of known radius can be determined if the maximum velocity (at the centre) is measured. Figure 8.2(b) is only universal for low flow rates. The velocity can be expressed as a function of distance from the lower surface of the pipe. The equation is

$$V = 4Y - 4Y^2$$

This situation is shown in Fig. 8.2(c).

Another common situation in which normalized equations are easier to use is the change in position or magnitude of a force due to a sinusoidal fluctuation, for example, a mass oscillating on a spring or the vibration of the free end of a cantilever beam. The use of a normalized Z variable for analysis of the normal distribution curve in statistics enables calculations to be performed more easily, and would otherwise entail detailed calculation (see Sec. 7.2.10).

It is necessary to understand these ideas and principles before considering the methods of dimensional analysis described in Secs 8.2 and 8.3.

8.2 RAYLEIGH'S METHOD

Dimensional analysis is a procedure that is used to group the variables in a given situation into dimensionless parameters, these being less numerous than the original variables. The terms 'dimensionless group', 'non-dimensional parameter' and 'numeric' are all used in different texts; the former will be employed in this book. Assume there is a relationship between several variables (P, Q, R, S), of the form

$$P = f(Q, R, S) \tag{8.6}$$

Each term in the equation has the same dimensions and can therefore be written in a form in which each term is dimensionless. This is usually represented as

$$\pi_1 = f(\pi_2, \pi_3, \ldots, \pi_n) \tag{8.7}$$

where the symbol π represents a dimensionless group containing *some* of the variables $P, Q, R,$ and S.

Rayleigh's method (named after J. W. Strutt, the third Baron Rayleigh) uses dimensional

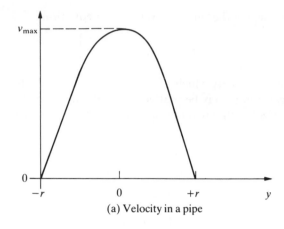

(a) Velocity in a pipe

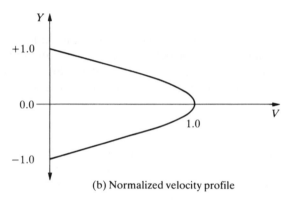

(b) Normalized velocity profile

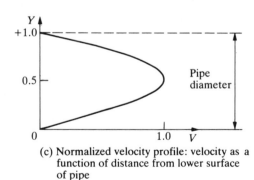

(c) Normalized velocity profile: velocity as a
function of distance from lower surface
of pipe

Figure 8.2 Velocity profiles for fluid flow in a pipe (low flowrates).

analysis to obtain an equation of the form of Eq. (8.7) (or Eq. (8.6)). Assume that Eq. (8.6) can be
written

$$P = A_1 Q^{a1} R^{b1} S^{c1} + A_2 Q^{a2} R^{b2} S^{c2} + \cdots \tag{8.8}$$

$$P = \sum A Q^a R^b S^c \tag{8.9}$$

where the coefficients $A_1, A_2, \ldots$, etc., are true constants (pure numbers), and the values of the
coefficients and the exponents (a, b, c) are unknown.

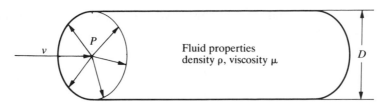

Figure 8.3 Variables used to characterize fluid flow in a pipe.

The range of possible values of a, b and c is greatly reduced if Eq. (8.8) is universally valid. This means that the dimensions of each term are the same, and the same as those of the dependent variable P. This method is illustrated by considering a particular example.

Example 8.1 Determine the dimensionless groups formed from the variables that describe the flow of fluid through a pipe, as shown in Fig. 8.3. The pressure force at any cross-section of the pipe is a function of the fluid properties and the pipe dimensions. The variables listed in Table 8.1 are thought to be significant.

Note: The units of pressure and viscosity are N/m^2 and $N\,s/m^2$ respectively, but for the purposes of dimensional analysis the dimensions in terms of mass, length and time are used here—see Chapter 1.

Select the pressure force as the dependent variable (i.e. exclude it from those variables whose effect one desires to isolate), and assume that the relationship between P and the other variables is to consist of a series of terms such as

$$P = A_1 v^{a1} \mu^{b1} \rho^{c1} D^{d1} + A_2 v^{a2} \mu^{b2} \rho^{c2} D^{d2} + \cdots$$

$$= \sum A v^a \mu^b \rho^c D^d \tag{8.10}$$

where the coefficient $A_1, A_2, \ldots$, etc., are true constants (pure numbers). Therefore

$$[P] = [v^a \mu^b \rho^c D^d]$$

$$ML^{-1}t^{-2} = L^a t^{-a} \times M^b L^{-b} t^{-b} \times M^c L^{-3c} \times L^d$$

$$= M^{b+c} L^{a-b-3c+d} t^{-a-b}$$

The powers of M, L and t on both sides of the equation must be equal, therefore:

$$M: 1 = b + c \tag{8.11}$$

$$L: -1 = a - b - 3c + d \tag{8.12}$$

$$t: -2 = -a - b \tag{8.13}$$

Table 8.1 Variables for use in Example 8.1

Variable	Symbol	Dimensions
Pressure force	P	$ML^{-1}t^{-1}$
Velocity	v	Lt^{-1}
Viscosity	μ	$ML^{-1}t^{-1}$
Density	ρ	ML^{-3}
Diameter	D	L

With three equations and four unknowns, we can only solve for three of the unknowns in terms of the fourth. Therefore:

$$c = 1 - b$$
$$a = 2 - b$$
$$d = -b$$

Equation (8.10) is of the form:

$$P = \sum A v^{2-b} \mu^b \rho^{1-b} D^{-b}$$
$$= \sum A \rho v^2 \left(\frac{\mu}{v \rho D} \right)^b$$

where the exponent b can have any value.

The form of the equation is:

$$P = \rho v^2 \left[A_0 + A_1 \left(\frac{\mu}{v \rho D} \right) + A_2 \left(\frac{\mu}{v \rho D} \right)^2 + \cdots \right]$$

where b has the value 0, 1, 2, etc., in successive terms. Alternatively:

$$\frac{P}{\rho v^2} = f \left(\frac{\mu}{v \rho D} \right)$$

The actual relationship represented by the function must be determined by experiment. However, it is now apparent that it is the relationship between the dimensionless groups $(P/\rho v^2)$ and $(\mu/v \rho D)$ that needs to be investigated, rather than experiments involving all five variables independently.

Both the dimensionless groups in this analysis should be familiar to engineering students who have studied fluid mechanics. They are given special names and have a particular physical interpretation as follows:

$$\text{Reynolds number, } Re = \frac{v D \rho}{\mu} = \frac{\text{inertia force}}{\text{viscous force}}$$

$$\text{Euler number, } Eu = \frac{P}{\rho v^2} = \frac{\text{pressure force}}{\text{inertia force}}$$

Note: It was the inverse of the Reynolds number that was obtained in the above example.

The following dimensionless group is also commonly encountered:

$$\text{Froude number, } Fr = \frac{v^2}{g D} = \frac{\text{inertia force}}{\text{gravity force}}$$

A table of conventional dimensionless groups often used and encountered in engineering is given in Table 8.2.

The use of dimensional analysis in this problem produced a solution in terms of well-known dimensionless groups. However, other solutions can be obtained if Eqs (8.11) to (8.13) are solved in terms of a different unknown, as shown:

$$b = 2 - a$$
$$c = 1 - b = -1 + a$$

Table 8.2 Conventional dimensionless groups

(Refer to BS 5775[3]. It is recommended that two letters, e.g. *Re*, are used as the abbreviation for a dimensionless parameter. This standard replaces BS 1991)

Name and symbol	Formula*	Comments
1. *Momentum transfer*		
Euler number, *Eu*	$Eu = \dfrac{\Delta P}{\rho v^2}$	
Froude number, *Fr*	$Fr = \dfrac{v}{\sqrt{(lg)}}$	
Grashof number, *Gr*	$Gr = \dfrac{l^3 \rho^2 g \gamma \, \Delta T}{\mu^2}$	
Knudson number, *Kn*	$Kn = \lambda/l$	
Mach number, *Ma*	$Ma = v/c$	
Reynolds number, *Re*	$Re = vd\rho/\mu$	
Strouhal number, *Sr*	$Sr = lf/v$	
Weber number, *We*	$We = l\rho v^2/\sigma$	
2. *Heat transfer*		
Biot number, *Bi*	$Bi = \dfrac{hl}{k}$	$\dfrac{\text{Solid thermal resistance}}{\text{Fluid thermal resistance}}$
Fourier number, *Fo*	$Fo = \dfrac{\alpha t}{l^2}$	$\dfrac{\text{Conduction of heat}}{\text{Storage of heat}}$ $(\alpha = k/\rho C_{\mathrm{p}})$
Graetz number, *Gz*	$Gz = \dfrac{\rho v d^2 C_p}{kl}$	$Gz = Re \, Pr(d/l)$ Used in forced-convection situations
Grashof number, *Gr*	$Gr = \dfrac{l^3 \rho^2 g \beta \, \Delta T}{\mu^2}$	$\dfrac{\text{Buoyancy forces}}{\text{Viscous forces}}$ Used in natural-convection situations
Nusselt number, *Nu*	$Nu = \dfrac{hl}{k}$	Basic dimensionless convective heat-transfer coefficient
Peclet number, *Pe*	$Pe = \dfrac{\rho v l C_p}{k}$	$Pe = Re \, Pr$ $\dfrac{\text{Convection heat transfer}}{\text{Conduction heat transfer}}$ Used in forced-convection situations
Rayleigh number, *Ra*	$Ra = \dfrac{l^3 \rho^2 C_{\mathrm{p}} g \gamma \, \Delta T}{\mu k}$ $= \dfrac{l^3 \rho g \gamma \, \Delta T}{\mu \alpha}$	$Ra = Gr \, Pr$ Used in free-convection problems
Stanton number, *St*	$St = \dfrac{h}{\rho v C_{\mathrm{p}}}$	$St = Nu/(Re \, Pr)$ $= Nu/Pe$ $\dfrac{\text{Heat transfer at surface}}{\text{Heat transfer by fluid}}$ Sometimes called Margoulis number, *Ms*

(*continued*)

Table 8.2 *continued*

Name and symbol	Formula*	Comments

3. *Mass transfer*

(These groups have already been defined in Sections 1 or 2 for momentum or heat-transfer situations, but they will be defined here for problems involving the transport of matter and distinguished by a dagger †)

Fourier number	$Fo^\dagger = \dfrac{Dt}{l^2}$	$Fo^\dagger = Fo/Le$
Peclet number	$Pe^\dagger = \dfrac{vl}{D}$	$Pe^\dagger = Re\ Sc = Pe\ Le$
Grashof number	$Gr^\dagger = \dfrac{l^3 \rho^2 g \beta\ \Delta T}{\mu^2}$	$\dfrac{\Delta\rho}{\rho} = \gamma\ \Delta T + \beta\ \Delta x$
Sherwood, number, Sh	$Sh = \dfrac{K_G l}{\rho D}$	Nusselt number for mass transfer $\dfrac{\text{Mass diffusivity}}{\text{Molecular diffusivity}}$
Stanton number	$St^\dagger = \dfrac{K_G}{\rho v}$	$St^\dagger = Sh/Pe^\dagger$

4. *Dimensionless groups dependent upon physical properties*

Prandtl number, Pr	$Pr = \dfrac{\mu C_p}{k} = \dfrac{\mu}{\rho\alpha}$	$\dfrac{\text{Momentum}}{\text{Thermal diffusion}}$
Schmidt number, Sc	$Sc = \dfrac{\mu}{\rho D}$	$\dfrac{\text{Momentum}}{\text{Mass diffusion}}$
Lewis number, Le	$Le = \dfrac{k}{\rho C_p D} = \dfrac{\alpha}{D}$	$Le = Sc/Pr$ $\dfrac{\text{Thermal diffusivity}}{\text{Molecular diffusivity}}$

5. *Dimensionless factors*

Friction factor, f	$f = \tau/(\tfrac{1}{2}\rho v^2)$
Heat-transfer factor, j_H	$j_H = St\ Pr^{2/3}$
Mass-transfer factor, j_M	$j_M = St^\dagger\ Sc^{2/3}$

* Definition of symbols:

c	velocity of sound
C_p	specific heat capacity at constant pressure
d	diameter
D	mass diffusivity (or diffusion coefficient)
f	frequency
g	acceleration due to gravity
h	convective heat-transfer coefficient
k	thermal conductivity (of gas or solid, as appropriate)
K_G	convective mass-transfer coefficient
l	characteristic length (e.g. diameter of pipe, length or width of flat plate, etc.)
t	time
v	velocity
α	thermal diffusivity
β	coefficient of thermal expansion $\left(= -\dfrac{1}{\rho}\left(\dfrac{\partial\rho}{\partial x}\right)_{T,P}\right)$

γ	cubic expansion coefficient $\left(= -\dfrac{1}{\rho}\left(\dfrac{\partial\rho}{\partial T}\right)_P\right)$
	Note: $-\dfrac{\Delta\rho}{\rho} = \gamma\ \Delta T + \beta\ \Delta x$
ΔP	pressure difference
ΔT	temperature difference
Δx	difference in mole fraction
μ	dynamic or absolute viscosity
ρ	density
σ	surface tension
λ	mean free path
τ	shear stress or pressure

Therefore, $e = -2 + a$. Hence, Eq. (8.10) becomes

$$P = \sum Av^a \mu^{2-a} \rho^{-1+a} D^{-2+a}$$

$$P = \sum A \left(\frac{\mu^2}{\rho D^2}\right)\left(\frac{vD\rho}{\mu}\right)^a$$

where the exponent a can have any value. Therefore

$$\frac{P\rho D}{\mu^2} = \mathrm{f}\left(\frac{vD\rho}{\mu}\right)$$

Although the solution is still dimensionally valid, and it contains the Reynolds number, the dimensionless group $(P\rho D^2/\mu^2)$ is not normally used in engineering. Results could be presented in this form but comparison with the findings of other experimenters would be more difficult.

8.3 BUCKINGHAM'S METHOD

Rayleigh's method of dimensional analysis was illustrated for a relatively simple problem. However, difficulties are encountered when more variables need to be considered. Sometimes the dimensionless groups obtained by applying the method can be predicted at the outset from a knowledge of the subject. A development of Rayleigh's method was proposed by Edgar Buckingham (see *Phys. Rev.*, vol. 2, 345, 1914) and is widely used. It is now known as *Buckingham's method*.

The number of dimensionless groups into which the variables can be combined is determined from *Buckingham's pi* (π) *theorem*, which states:

The number of dimensionless groups (i) in the final relationship is equal to, or greater than, the difference between the number of variables (V), and the number of fundamental dimensions (d) used to describe the variables.

The symbol π is used to represent a dimensionless group, so that the desired relationship between the variables can be written as

$$\pi = \mathrm{f}(\pi_2, \pi_3, \ldots, \pi_i)$$

where $i \geqslant V - d$ as defined by Buckingham's π theorem.

If the equations obtained by equating the powers of the individual dimensions, i.e. Eqs (8.11) to (8.13), are not all independent, then more dimensionless groups will be obtained. Hence, the inclusion of the 'greater than' statement. Sometimes two or more of these equations (often called *indicial equations*) may be identical. This occurs if several variables with the same dimensions are included in the analysis, e.g. pipe diameter (d), bubble diameter (d_b) and pipe length (L) can give rise to the following dimensionless groups:

$$d/d_b; \; d/L; \; L/d_b$$

In the example used to illustrate Rayleigh's method, there are five variables ($V = 5$) that were defined using three primary dimensions ($d = 3$). From Buckingham's π theorem we would expect to get a relationship between *not less than* two dimensionless groups ($i > V - d$), as obtained. The theorem does not predict the number of dimensionless groups explicitly, however, there are several ways in which this can be achieved as described below.

(a) Write down the number (n) of independent indicial equations (i.e. $n \leqslant d$), then $i = V - n$.

(b) Write down a dimensional matrix of primary dimensions (rows) and variables (columns) describing the problem, as shown below for Example 8.1 in Sec. 8.2.

	P	v	μ	ρ	D
M	1	0	1	1	0
L	-1	1	-1	-3	1
t	-2	-1	-1	0	0

The numbers in this table represent the exponent of the primary dimensions for each variable. From this array of numbers it is necessary to determine the rank of the matrix (r). This is the order (number of rows or columns) of the $largest$ non-zero determinant that can be formed from the matrix. In this example, the rank is 3 and the number of dimensionless groups to be formed is found from the expression

$$i = V - r$$

Therefore:

$$i = 5 - 3 = 2 \text{ (as obtained)}$$

Note: In order to use this method the reader must be familiar with the mathematics of determinants.

(c) The number of independent indicial equations (n) is equal to the greatest number of variables included in the analysis that *cannot* be combined into a dimensionless group. This statement is known as *Van Driest's rule*. It may be a difficult concept to understand; the other methods are probably easier to apply when initially studying dimensional analysis.

When the number of dimensionless groups has been determined, these groups can often be written down without difficulty based upon experience, intuition and reference to a table of conventional groups (such as Table 8.2). There are two rules that should be observed:

1. Each of the variables must be used at least once.
2. The dimensionless groups formed must be independent of each other, i.e. it must not be possible to form any of them by a combination of the others.

Another example of dimensional analysis that illustrates the Buckingham method will now be considered.

Example 8.2 Determine the dimensionless groups that can be used to describe the situation where heat is transferred from a pipe wall into a fluid flowing inside the pipe. This is an

Table 8.3 Variables for use in Example 8.2

Variable	Symbol	Dimensions
Heat-transfer coefficient	h	$Q/t\, L^2\, T$
Velocity (average)	v	L/t
Fluid density	ρ	M/L^3
Fluid viscosity	μ	$M/L\, t$
Fluid heat capacity	C_p	$Q/M\, T$
Fluid thermal conductivity	k	$Q/t\, L\, T$
Pipe diameter	D	L

example of forced-convection heat transfer, and more detailed discussion of this situation can be found in engineering heat-transfer textbooks.

The heat-transfer coefficient (h) is the quantity of primary interest; the dimensional analysis must also include terms that are descriptive of the thermal and flow properties of the fluid, and the system geometry. The dimensions M, L, t, T (temperature) and Q (heat) are used to represent the variables involved. Although Q can be expressed in terms of M, L and t, the analysis will be easier if Q is considered as a fundamental dimension in this situation. The important variables, their symbols and dimensional representations are listed in Table 8.3.

It will be left to the reader to verify that the rank of the dimensional matrix (or the number of independent indicial equations) is equal to 4. There are 7 variables and therefore 3 dimensionless groups are required. Equation (8.7) defines the relationship between the three π groups as

$$\pi_1 = f(\pi_2, \pi_3)$$

The four variables v, μ, k, D, are (arbitrarily) chosen to represent the core. The π groups are represented as

$$\pi_1 = v^a \mu^b k^c D^d h$$

$$\pi_2 = v^e \mu^f k^g D^i \rho$$

$$\pi_3 = v^j \mu^l k^m D^n C_p$$

Writing π_1 in dimensional form:

$$1 = \left(\frac{L}{t}\right)^a \left(\frac{M}{Lt}\right)^b \left(\frac{Q}{tLT}\right)^c (L)^d \left(\frac{Q}{tL^2T}\right)$$

Equate the exponents of the fundamental dimensions on both sides of this equation:

$$M: 0 = b$$

$$L: 0 = a - b - c + d - 2$$

$$t: 0 = -a - b - c - 1$$

$$T: 0 = -c - 1$$

$$Q: 0 = c + 1$$

Solving these equations for the four unknowns:

$$a = 0 \qquad b = 0$$

$$c = -1 \quad d = 1$$

Therefore, π_1 becomes:

$$\pi_1 = \frac{hD}{k} = \text{Nusselt number, } Nu$$

Similarly:

$$\pi_2 = \frac{vD\rho}{\mu} = \text{Reynolds number, } Re$$

and

$$\pi_3 = \frac{\mu C_p}{k} = \text{Prandtl number, } Pr$$

Therefore, a possible relation correlating the important variables is of the form:

$$Nu = f(Re, Pr)$$

The Dittus–Boelter equation is an empirical relationship of this form for forced convection-heat transfer in a circular conduit for a turbulent flowing fluid. It is written:

$$Nu = 0.023 Re^{0.8} Pr^n$$

The following restrictions apply to the use of this equation:

$Re > 10^4$;

$0.7 < Pr < 100$;

$n = 0.4$ for a heated fluid, 0.3 for a cooling fluid;

length:diameter ratio for the pipe > 60;

all fluid properties are evaluated at the arithmetic mean bulk temperature.

Note: If the core variables ρ, μ, C_p and v had been chosen, the dimensional analysis would have produced the Reynolds and Prandtl numbers as two dimensionless groups; the third group would be:

$$\text{Stanton number, } St = \frac{h}{\rho v C_p}$$

An alternative correlation is

$$St = f(Re, Pr)$$

The Stanton number can also be formed by taking the ratio $[Nu/(RePr)]$; check this is correct!

Example 8.3 The velocity of sound in a gas (v; Lt^{-1}) is thought to be dependent upon the secondary quantities listed in Table 8.4.

There are four variables and three primary quantities, therefore we would expect one dimensionless group to be formed. Therefore:

$$f(\pi) = 0$$

or

$$\pi = \text{constant}$$

Write the dimensionless group (π) as

$$\pi = v^a \rho^b P^c \mu^d$$

Table 8.4 Variables for use in Example 8.3

Variable	Symbol	Dimensions*
Density	ρ	$F t^2 L^{-4}$
Pressure	P	$F L^{-2}$
Viscosity	μ	$F t L^{-2}$

* In this example, force (F), length (L) and time (t) are chosen as the primary quantities.

Writing in dimensional form:

$$\pi = \left(\frac{L}{t}\right)^a \left(\frac{Ft}{L^4}\right)^b \left(\frac{F}{L^2}\right)^c \left(\frac{Ft}{L^2}\right)^d$$

Since there are four unknowns and three equations (obtained by equating the exponents of the fundamental dimensions), it is appropriate to select the value $a = 1$ (arbitrarily). The solution is then obtained in terms of the velocity of sound, as required.

It is left as an exercise to the reader to show that:

$$b = \tfrac{1}{2}; \quad c = -\tfrac{1}{2}; \quad d = 0$$

Therefore:

$$\pi = v\rho^{1/2} P^{-1/2} = \text{constant}$$

or

$$v = \text{constant}\sqrt{(P/\rho)}$$

(What solution is obtained if the value of a is chosen as equal to 3?)

The constant is evaluated using experimental results and the equation can then be applied to other gases, or the same gas under various conditions of pressure and density.

In the derivation, the exponent d is equal to zero and the viscosity of the gas is not included in the final equation. This could probably have been predicted since the propagation of sound involves compression waves and depends upon the mass (i.e. density) and elastic properties of the gas. The gas viscosity is only important if shearing occurs.

8.4 PROBLEMS

Problem 1 A non-linear inductance has an iron core for which the saturation curve can be expressed approximately by the ninth-power equation:

$$i = K\psi^9$$

where K depends upon the number of turns and the size of the core.

Show that the relationship between the transient current (i) resulting when a resistance (R) is connected in series, and a d.c. voltage (E) is impressed at time $t = 0$, is given by

$$i\frac{R}{E} = f\{(KRE^8)^{1/9}\}t$$

The variables can be expressed in terms of the dimensions of voltage, current and time. From the equation

$$e = \frac{d\psi}{dt}$$

the dimensions of K are

$$[K] = \text{current(voltage . time)}^{-9}$$

Problem 2 The lift and drag on the wing of an airplane depend (mainly) upon:

(a) the shape of the wing;
(b) the size of the wing;
(c) relative velocity with respect to the air;
(d) angle of attack, i.e. angle of inclination of the wing with respect to the relative wind direction;
(e) density of air;
(f) viscosity of air.

Assume that a particular shape of wing is to be investigated and this variable can therefore be omitted from the analysis. The angle of attack is also fixed (i.e. specified) and the relevant variables are given in Table 8.5.

Determine the number of independent dimensionless groups and show that the solution is of the form:

$$\frac{F}{(lv)^2\rho} = f\left(\frac{lv\rho}{\mu}\right)$$

Note: Dimensional analysis does not define the form of the functional relationship between the groups. However, in this problem the relationship depends only on the shape of the wing and the angle of attack, and it could be determined by performing experiments with suitable models.

In such tests, values of one or more parameters are varied and appropriate measurements are taken. The values of the dimensionless groups (π_1 and π_2) are calculated and plotted against each other. The shape of the curve defines the functional relationship between the groups.

If the angle of attack is changed, the function and its curve also change. A family of curves can be plotted; each curve defines the function for a particular angle of attack. In many practical situations only one or two operating conditions are applicable, and it is not necessary to determine the entire functional relationship.

Problem 3 Use dimensional analysis to find the equation for the period of a pendulum when oscillating through a small angle. Assume that the period (T) depends upon the moment of inertia about the axis of suspension (I), the weight (W) and the distance between the centre of gravity and the point of suspension (L).

Problem 4 Find an expression for the fundamental frequency of free vibration of a stretched string, if it is assumed that the frequency depends on the length, the tension and the mass per unit length.

Problem 5 Use dimensional analysis to find the equation describing the velocity of sound along a solid rod.

Table 8.5 Variables for use in Problem 2

Variable	Symbol	Dimensions
Force on the wing (either lift or drag)	F	F
'Characteristic' dimension, e.g. chord of the wing	l	L
Relative velocity	v	$L t^{-1}$
Density of air	ρ	$F t^2 L^{-4}$
Viscosity of air	μ	$F t L^{-2}$

EXERCISES

The reader should apply the methods of dimensional analysis described in this chapter to the analysis of particular experimental results.

8.1 Define the terms:
   (a) dimensionally homogeneous;
   (b) dimensionally inhomogeneous;
   (c) true constant;
   (d) suppressed constant.
8.2 Explain the principle of dimensional homogeneity of equations.
8.3 Consider examples where the normalization of equations assists the interpretation of the particular situation.
8.4 Describe and apply both the Rayleigh and Buckingham methods of dimensional analysis.

REFERENCES

1. Kreith, F. and W. Z. Black, *Basic Heat Transfer*, Harper and Row, New York, 1980, pages 255–261.
2. Dean, J. A. (ed.), *Lange's Handbook of Chemistry*, 13th edn, McGraw-Hill, New York, 1985.
3. *BS 5775: Specification for quantities, units and symbols (Parts 0 to 13). Part 12: Dimensionless parameters* (1982). This standard replaces BS 1991.

BIBLIOGRAPHY

Becker, H. A., *Dimensionless Parameters: Theory and Methodology*, Elsevier Applied Science, Oxford, 1976.
Bridgman, P. W., *Dimensional Analysis*, A.M.S. Press, New York, 1931 (reprinted).
Langhaar, H. L., *Dimensional Analysis and Theory of Models*, Robert E. Krieger Publishing Co., Melbourne, Fla, 1980.
Pankhurst, R. C., *Dimensional Analysis and Scale Factors*, Chapman and Hall, London, 1964.
Staicu, C. I., *Restricted and General Dimensional Analysis*, Heyden, Philadelphia, 1982.

FOUR

PRESENTATION OF INFORMATION

SCOPE

Part Four includes the visual presentation of data (Chapter 9) and the use of oral and written communications (Chapter 10). The reader is encouraged to consider the alternative forms of visual presentation, when each type should be used, and how the information should be presented.

Oral reports need special preparation and they must offer advantages over an equivalent written report. Different types of written reports are considered, e.g. formal reports, short-form reports and memoranda; the requirements of each type and guidelines for their presentation are given.

Examples of different types of reports are not included; it is considered that 'doing' is more useful than 'reading'. Readers should prepare and evaluate their own written (and oral) reports—and learn from their mistakes.

VISUAL PRESENTATION

CHAPTER OBJECTIVES

1. To encourage the reader to consider alternative visual ways of presenting information.
2. To provide examples of, and guidelines for, the presentation of:
   Pictorial representations
   Graphs
   Tables
   Flowcharts
   Engineering diagrams
   Engineering drawings

QUESTIONS

- How is information (especially numerical data from experiments) presented, i.e. the visual forms of presentation rather than the actual report presentation?
- In which situations are the different forms of presentation more appropriate?

COMMENTS

This chapter discusses the visual presentation of information. Guidelines for oral and written communications are given in Chapter 10. Whatever the nature of the work that has been undertaken, ultimately the results of this work must be presented to an audience. This is true for student projects, laboratory investigations, the critical analysis of a published paper, tutorial problems and examination questions. For the *effective* presentation of information, two factors need to be considered. These are:

(a) the audience receiving the information;
(b) selection of an appropriate form of presentation.

If both of these factors are correctly identified, then the outcome of the presentation is successful communication.

The way in which any information is presented should depend primarily upon the audience for which it is intended. Examples of 'audience identification' can be seen in a comparison of several daily newspapers or current affairs programmes on television. In each case the newspaper or the programme is aimed at a particular audience and is presented accordingly. In the engineering profession, the recipient of the information is usually easily identified at the outset of the work. In teaching institutions, the lecturer presents information to students who are less well qualified and knowledgeable in that subject.

The lecturer's performance in terms of the effectiveness of the communication is rarely assessed. However, a student's performance in a project is determined not only by what has been achieved but *also* by the standard of presentation of the information.

In industry, your associates will form opinions of you, your ability and your character from the way in which you present your communications—however unfair this may appear to be. You will be required to communicate with your superiors, your peers, people who work for you, people from other companies and sometimes the 'general' public. It is essential, wherever possible, to identify your audience and prepare your presentation accordingly.

It is necessary to establish the level of technical expertise of the audience. Information presented either too simply or with too much background description will be seen as boring or inappropriate. Information that is too technical may alienate the audience and lose their attention and interest, and once lost it may be difficult to recapture. Engineers and scientists who are well qualified or have wide experience may have particular insights into the fundamental principles, despite a lack of expertise in the particular area of the presentation. Such people may be impressed by the technological achievement but critical of the theoretical basis of a project; in this case particular attention should be paid to a justification of this aspect in a presentation.

Technicians may not be highly qualified but their level of practical experience often surpasses that of the engineers and scientists who direct their work. Communications with craftsmen need to be directed towards a thorough explanation of the technical aspects of a problem; it is important not to assume that they possess an understanding of the theory or basic mathematical principles.

In certain situations, an audience may include persons having a wide variety of experience and expertise. For example, a meeting to explain a new technical development may include technicians, scientists, engineers, marketing and sales personnel, etc. In this situation it may be necessary to present the information in several different forms, taking into account the background and experience of different people. If this approach is too costly or time-consuming, it may be possible to make one presentation and draw attention to appropriate features for particular persons.

If there is any doubt about the experience of your audience, it is probably safer to assume they possess less rather than more knowledge of the subject. Someone who becomes bored may wait for the presentation to 'catch up' or progress quickly to more in-depth material, whereas the person who becomes lost will probably give up and be lost forever.

This chapter is mainly concerned with the second factor mentioned earlier, namely selection of an appropriate form of presentation. Aspects of oral and written presentations are dealt with in Chapter 10; this chapter is concerned specifically with describing particular types of visual

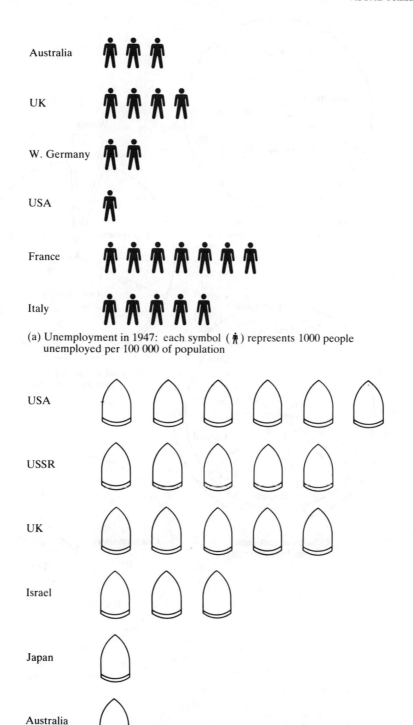

(a) Unemployment in 1947: each symbol ( ) represents 1000 people
unemployed per 100 000 of population

(b) Defence spending in 1980: each symbol (◯) represents 10^8 US

Figure 9.1 Examples of pictographs

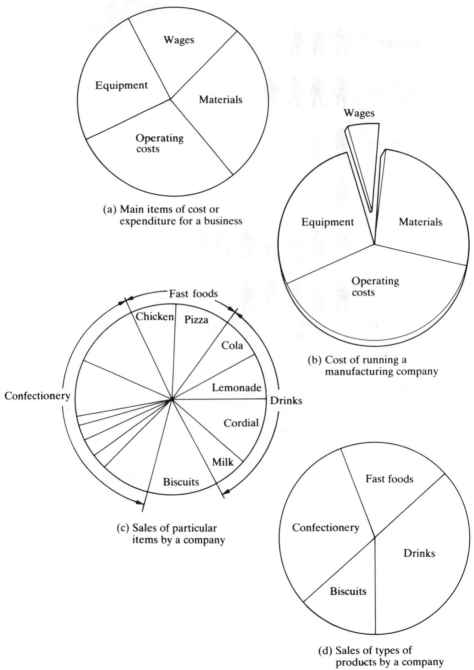

(a) Main items of cost or
expenditure for a business

(b) Cost of running a
manufacturing company

(c) Sales of particular
items by a company

(d) Sales of types of
products by a company

Figure 9.2 Examples of pie charts.

presentations. These means of presenting information include pictorial representations, graphs, tables and flowcharts, and the use of standard symbols and accepted practices when preparing engineering diagrams and engineering drawings. Each of these types of visual presentation will be discussed.

9.1 PICTORIAL REPRESENTATIONS

Pictorial representations provide a convenient way of communicating information quickly. They are particularly useful when an audience has non-technical skills or a wide variety of expertise. The salient feature(s) of the information can be grasped immediately and the presentation should convey a particular message. Examples of pictorial representations are the pictograph (Fig. 9.1), the pie chart (Fig. 9.2), the bar chart (Fig. 9.3) and the block graph (Fig. 9.4). These types of presentation often do not provide exact data; their aim is usually to create an overall impression, to make a particular point or to make comparisons. For example, after World War II unemployment was lowest in the USA and highest in France (Fig. 9.1a); the majority (70 per cent) of the working population is qualified to at least school-leaving certificate level and of these 40 per cent are qualified technicians (Fig. 9.4).

Pictorial representations are not particularly useful where differences between the variables are small. For example, Fig. 9.2(a) shows that there is little difference between the four main areas of cost/expenditure for this business; a pie chart is probably not really necessary to illustrate this

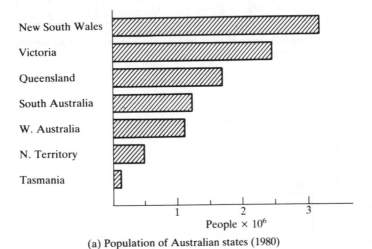

(a) Population of Australian states (1980)

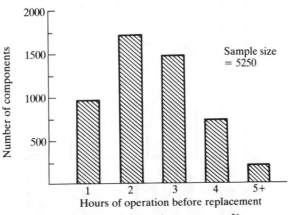

(b) Reliability/operating data for component X

Figure 9.3 Examples of bar charts.

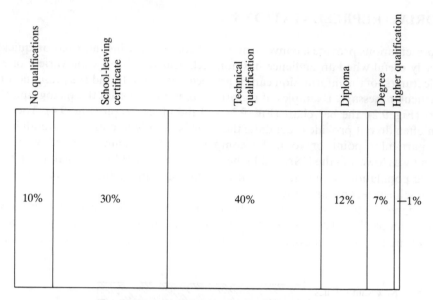

Figure 9.4 Example of a horizontal block graph (qualitative presentation): percentage of the working population having formal qualifications (West Germany, 1985).

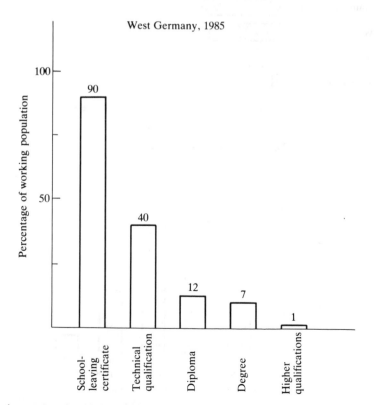

Figure 9.5 Alternative presentation of data (from Fig. 9.4).

point unless it is unexpected and requires emphasis and reinforcement. Conversely, Fig. 9.2(b) shows that for this company wages constitute only a small percentage of total costs, and operating costs are the main expenditure.

Pictographs are usually the least precise means of transmitting information. Exact data can be provided by all these types of presentations, however to do so requires the inclusion of more information which may detract from the immediate impression (and this is usually the reason for choosing this particular form of presentation). If the reader is likely to want (or need) further clarification of the figures presented, it may be more appropriate to collect all the data used in tabular form and include a reference to this table (often in an Appendix to a report). The use of tables for presenting information is discussed in Sec. 9.3.

These types of presentations are inappropriate where a large number of small values occur (see Fig. 9.2c), especially if these are relevant to the discussion. In this case, graphs (see Sec. 9.2) or tables (see Sec. 9.3) may be a more useful form of presentation. Alternatively, the minor items may sometimes be grouped together under an appropriate heading (see Figs 9.2c and 9.2d).

Pie charts are frequently used by governments to explain the effects of a budget, or how taxes are allocated. Hence the common phrases 'dividing up the cake' and 'taking a smaller slice of the pie'.

As with all presentations it is important that the information is presented clearly and in an unambiguous manner. The data in Fig. 9.4 could be misinterpreted as showing that only 30 per cent of the working population attained a school-leaving certificate, whereas the actual figure is 90 per cent (all higher qualified persons must also have obtained this basic qualification). The reader who does not read explanatory notes or headings carefully could obtain a wrong impression. The information presented in Fig. 9.4 could be presented in an alternative form as shown in Fig. 9.5, in an attempt to avoid this possible misinterpretation.

9.2 GRAPHS

Most engineering data are stored in tables (see Sec. 9.3) and displayed in line graphs. Graphs provide a means of representing quantitative information and they indicate the precision and/or reproducibility of the measurements. Sometimes a possible relation between the variables can be obtained from a graph. The data are shown as points on the graph; the distances of the points relative to two coordinate axes represent their values. The axes usually intersect at right angles, and the vertical and horizontal reference lines are referred to as the ordinate and the abscissa respectively. The independent variable is normally plotted along the abscissa and the dependent variable along the ordinate.

Before plotting data on a graph, the data should be arranged in tabular form and the types of scales that emphasize the functional relationship between the variables should be determined. The scale for each axis can be estimated by dividing the range in the variable by the scale length available. The actual scale to be used should be suitable for the graph paper available. Many different types of rectangular graph paper are available on which grids are printed for easy plotting. This paper is available with 5 or 10 lines per cm (or between 4 and 16 lines per inch) in different grid sizes, e.g. 18 × 25 cm or 7 × 10 in.

A scale should be chosen that will not require awkward fractions in the smallest calibration on the paper. The scale and the distances between grid lines should be consistent with the precision of the data. The coordinate scales and their value ranges should be chosen so that the data points cover as large an area of the graph paper as possible. If the precision of the data is approximately the same in both directions (ordinate and abscissa), or if it is unknown, the units and the range of the scales should be chosen to ensure that the data points cover a nearly square area. Alternatively, if the precisions differ in each direction, the ratio of height-to-width for the area covered by the data points should correspond (if practicable) to the ratio of the precisions in each direction. Very large or very small numbers may be written in standard index form, e.g. 2.5×10^6 or 3.9×10^{-3}.

The beginning of the ordinate and abscissa quantities is normally zero, unless this results in compression of the curve. The origin is usually located in the lower left corner of the graph except where negative values are to be plotted, in which case it is then located so that all values can be shown. The main grid lines must be marked, and the values assigned are written along the coordinate axes.

The ordinate and abscissa variables *must* be labelled, and their respective units of measurement included. A title *must* be included; the ordinate values are usually referred to first and the abscissa values second. The title should provide as full a description as possible of the information presented and how it was obtained. Numbers labelling the divisions along the axes should have a small number of digits. An axis should not be labelled thus:

0.001	0.002	0.003	Time (seconds)

but written as:

1	2	3	Time $\times 10^3$ (s)

or

1	2	3	Time (s) $\times 10^3$

The notation: 'Time, seconds $\times 10^3$' can be ambiguous and should be avoided.

The data points are plotted as fine, tiny dots and once located they are identified by a small circle, triangle, square, etc. Different identification symbols can be used to distinguish the results of different experiments or conditions. After the data points are located, either a smooth curve or a straight line (as implied by the data) is drawn to indicate an average of the plotted points. The method by which a best straight-line fit is obtained for a set of data is described in Sec. 7.5. Techniques that can be used to obtain a linear relationship for a set of data points are described in Sec. 7.7. The dispersion of the data points about an average value occurs because of the many variable factors that influence experimentally determined measurements.

When the final straight line or curve is drawn, the line should stop at the perimeter of the symbol used to identify the point rather than passing through the symbol. Graphs are also drawn for theoretical relationships and empirical relationships. These relationships are usually represented as smooth curves or straight lines depending upon the form of the mathematical expression; point designations are not normally included. Distinctive lines can be used to distinguish between different sets of experimental data, theoretical relationships and empirical relationships, e.g. solid line, broken line, dash–dot–dash line, etc.

Graphs are drawn to provide both a visual appreciation of the results and also to function as a data source. Although actual measurements are usually stored in tabular form, the convenience and compact nature of graphs means that they are often used as a source of data. Graphs are also more convenient than tables when interpolation or extrapolation of data is required (see Sec. 7.6). The degree of precision of the results requires some consideration. If significant effort and cost have been expended to obtain high precision, then results should be plotted such that the precision is maintained for any data subsequently obtained from the graph. The number of

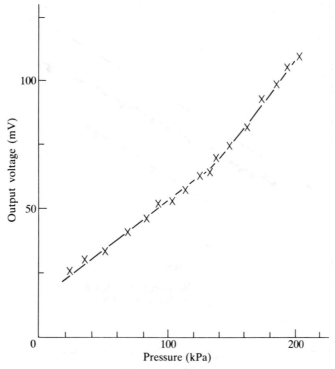

Figure 9.6 Graphical presentation of data (relationship between two variables): calibration curve for a pressure transducer.

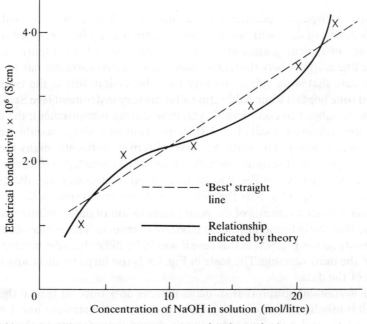

Figure 9.7 Data subject to variation during measurement (too few data points).

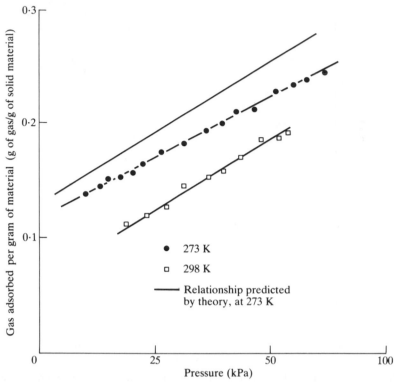

Figure 9.8 Graphical presentation of data showing the relationship between three variables: adsorption isotherms, quantity of gas (air) adsorbed by granular carbon.

variables that can be displayed effectively on a single graph is often restricted to three, and sometimes only two. Sets of data with several variables are usually best reported in tabular form.

Some examples of typical graphs are given in Figs 9.6 to 9.10. Figures 9.6 and 9.7 are examples of single line graphs where the relationship between two variables has been investigated. Figure 9.6 shows data that were subject to very little dispersion; this is the type of graph that would be obtained (one hopes!) when calibrating a laboratory instrument (see Sec. 2.7). Figure 9.7 shows data that were subject to considerable variation during measurement; the curves that are drawn are only approximately located and their position and shape would be suggested by reference to appropriate theory. To locate the curve with more certainty, many more data points would be required, especially in regions where the slope of the curve 'appears' to be changing. The effect of three variables is shown in Fig. 9.8; the theoretical relationship is indicated by the solid line. The use of a key or legend avoids the need to write on the graph itself.

Figures 9.9 and 9.10 are examples of the poor presentation of data; unfortunately these types of presentation are frequently found in student laboratory reports. The choice of three variables is too many for presentation in Fig. 9.9; in this case it would be difficult to draw reliable conclusions about the effect of the third variable. The scale in Fig. 9.9 is too large to allow any meaningful use or interpretation of the data.

Figure 9.10 is a 'classical' graph that students present only once, at least if they are sensible! This type of graph is usually presented with either the curve or the straight line; I usually draw in the other one in red ink and put a line through any discussion referring to this figure. The only valid point that can be made from Fig. 9.10 is that too few data points are available to consider

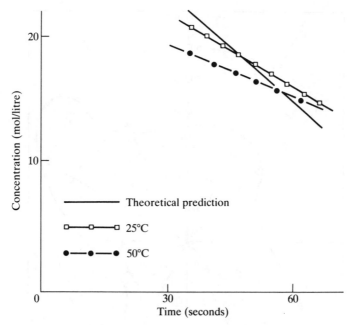

Figure 9.9 Graphical presentation of data (poor choice of scale, unsuitable for three variables).

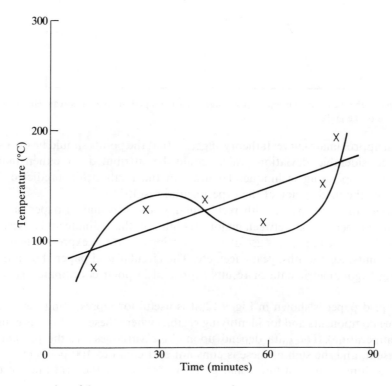

Figure 9.10 Poor presentation of data.

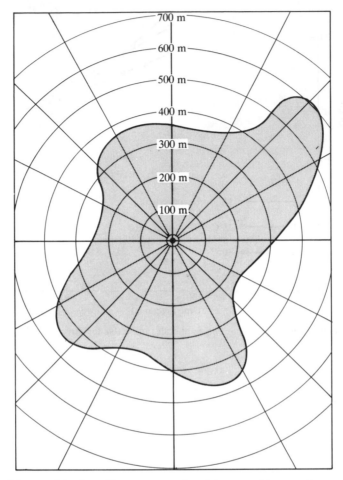

Figure 9.11 Example of the use of circular grid paper: range of particles present in a gas stream emitted from a chimney (located at the centre of the grid).

drawing even an approximate curve. If theory suggests that the points should lie on a straight line, then there is considerable deviation which could be attributed to either malfunctioning equipment or poor operating technique. To draw in the relationship predicted from theory merely emphasizes the inadequacy of the experimental results!

A wide variety of graph paper with rectangular grids is available; paper having different distances and/or number of lines between major grid lines on the ordinate and abscissa directions is also produced. This type of paper is useful when plotting quantities expressed in units that have non-decimal subunits, e.g. months, years, feet, etc. The circular grid paper shown in Fig. 9.11 is used for plotting trigonometric data or results expressed in polar coordinates, i.e. distance and angle.

Triangular grid paper is shown in Fig. 9.12; it is useful for representing the composition of mixtures of three components and for identifying regions where these mixtures are immiscible (at a particular temperature). The data depend upon three variables, i.e. the percentage of each component present, and the sum of these is constant and equal to 100 per cent.

The distinguishing feature of functional graph paper is that the grid lines in one or both directions are spaced over distances proportional to the values of a given function, i.e. $y = f(x)$,

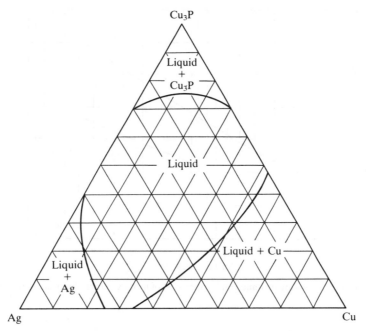

Figure 9.12 Example of the use of triangular graph paper: isothermal section at 800°C for the Ag–Cu–Cu$_3$P system.

for a conventionally chosen series of values of x. The most common types of functional graph paper are semi-logarithmic and log–log plots. The semi-logarithmic plot is used where an exponential relationship is expected between the variables, e.g. $y = a \exp(bx)$ where a and b are constants. Then, $\ln y = bx + \ln a$, or $\log y = 0.4343bx + \log a$. Semi-logarithmic graph paper has the distances from the origin to the intersections of the grid lines proportional to the logarithm of the numbers marked along one axis. The other axis has equidistant, arithmetic divisions. An example of a semi-logarithmic plot is given in Fig. 9.13; graph papers with one to seven logarithmic cycles are available.

Power law relations of the form $y = ax^b$ can also be written as: $\log y = b \log x + \log a$. Data related to this type of equation can be plotted on log–log graph paper (Fig. 9.14) where both axes are marked with divisions in logarithmic proportions. Graph paper with grid patterns from 1×1 to 3×5 logarithmic cycles is available. Other special graph paper includes relative frequency and cumulative frequency papers (used to test whether a series of results from repeated experiments follow the normal distribution) and probability or normal curve paper (having the abscissa presented as a cumulative frequency from 0.01 to 99.99 per cent). Special graph paper is not available for every type of functional relationship, however correct choice of appropriate functions of the prime variables allows a relationship to be tested, as described in Sec. 7.7.

9.2.1 Nomographs

A *nomograph* (sometimes called a nomogram) is a pictorial representation consisting of several scales graduated so that distances are proportional to the variables involved. A simple example is shown in Fig. 9.15; this consists of a single line having graduations corresponding to centimetres and inches on opposite sides. Nomographs are a useful method of solving problems involving equations of various types. Solutions requiring repeated readings of data may be obtained from nomographs. The construction of nomographs is beyond the scope of this chapter, but since they

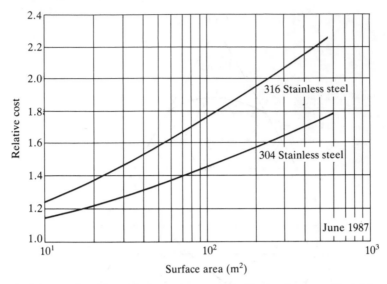

Figure 9.13 Example of the use of semi-logarithmic graph paper: cost of heat exchangers with stainless steel tubes relative to carbon steel construction.

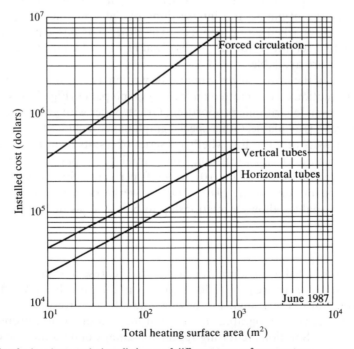

Figure 9.14 Example of a log–log graph: installed cost of different types of evaporators.

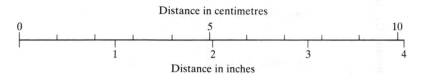

Distance in centimetres

Distance in inches

Note: 1 inch = 2·54 cm

Figure 9.15 Example of a simple nomograph: a functional chart.

are frequently used by engineers and scientists, a brief description of the common types of charts and their uses is included.

A *functional chart* is illustrated in Fig. 9.15; this type of chart is used when two variables are related by a constant coefficient. An *alignment chart* in its simplest form is shown in Fig. 9.16, consisting of three graduated parallel lines. If points are fixed on two of the graduated lines, then a straight line passing through these points will intersect the third graduated line at a point that satisfies the relations between the variables. The use of the chart is illustrated by the dashed line in Fig. 9.16. A *Z-chart* is an alternative form of the alignment chart. This type of chart has the centre graduated line running diagonally. It is used for the solution of equations of the form $f(z) = f(x)/f(y)$.

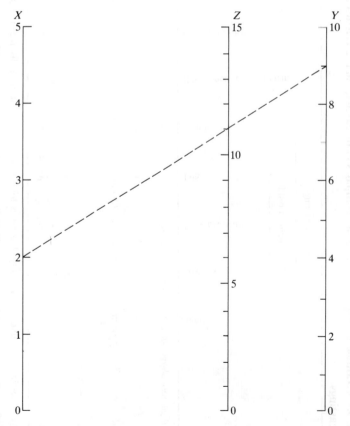

Figure 9.16 Example of an alignment chart. This nomograph can be used to satisfy the relationship $X + Y = Z$. An example is shown by the broken line.

Table 9.1 Example of a qualitative table: comparison of casting characteristics and other properties of aluminium and aluminium alloy ingots and castings

| BS designation | Form of casting | | | | Fluidity | Resistance to hot tearing | Pressure tightness | Machinability | Resistance to corrosion | Strength at room temperature | Strength at elevated temperature | Shock resistance | Electrical conductivity | Decorative anodizing |
| | Permanent mould | | | Die | | | | | | | | | | |
	Sand	Gravity	Low pressure											
General purpose alloys: die casting (pressure) alloys														
LM2	—	—	—	4	3	4	4	3	3	4	3†	2	2	1
LM20	—	—	—	4	4	4	4	3	3	4	1†	2	3	1
General purpose alloys: permanent mould and sand casting alloys														
LM4	4	4	3	—	3	3	4	3	3	2	3	2	2	1
Special purpose alloys														
LM5	2	2	2	n	2	2	1	3	4	2	2	2	2	4
LM9	3	3	4	n	3	4	3	2	3	3	3	2	2	1
LM10	2	2	n	n	2	3	1	3	4	4	n	4	1	2

* 4 denotes highest value or suitability; 1 denotes lowest; n indicates not normally recommended in this form or condition.

† The use of die castings is usually restricted to only moderately elevated temperatures.

9.3 TABLES

A table is an orderly arrangement of corresponding descriptions of dependent and independent variables. Tables may contain qualitative or quantitative descriptions. Table 9.1 is an example of a qualitative table. The use of a qualitative score or grading between 1 and 4 is subjective and is intended to assist the selection of a process based upon this qualitative comparison. Simple comparative terms such as high-low, good/bad, simple/complex, etc., can also be used. The advantage of using a number scale, even though it it not based upon well-defined rules, is that totals can be obtained for all the processes, making comparisons somewhat easier. However, choices are often made based upon particular requirements rather than consideration of a range of advantages. For example, a pump required to handle oxygen gas *must* be oil-free for all its working parts in contact with the gas, therefore a pump that was rated most suitable on an overall qualitative points basis would be completely unsuitable if it scored zero on the 'freedom from oil' section.

Tables 9.2 and 9.3 are sometimes referred to as semi-quantitative tables because the body of the table contains some qualitative descriptions. The independent variable is normally listed in

Table 9.2 Examples of semi-qualitative tables: conversion factors

(a) *Conversion of work, energy and heat*

	J	kWh	kcal	Btu	ft lbf
J	1	2.778×10^{-7}	2.39×10^{-4}	9.48×10^{-4}	0.7376
kW h	3.6×10^6	1	860	3413	2.655×10^6
kcal	4187	1.163×10^3	1	3.9685	3087.4
Btu	1.055	2.93×10^4	0.251 98	1	777.97
ft lbf	1.3558	3.766×10^{-7}	3.239×10^{-4}	1.285×10^{-3}	1

(b) *Metric conversion factors*

Property	Exact conversion		
	From	To	Multiply by
Length	inch	mm	2.540×10
	foot	mm	3.048×10^{-4}
Mass	ounce	g	2.835×10
	pound	kg	4.536×10^{-1}
	ton (2000 lb)	kg	9.072×10^2
Density	lb/ft^3	kg/m^3	1.602×10
Temperature	°F	°C	$(°F - 32) \times (5/9)$
Area	inch^2	mm^2	6.452×10^2
	ft^2	m^2	9.290×10^{-2}
Volume	inch^3	mm^3	1.639×10^4
	ft^3	m^3	2.832×10^{-2}
Force	ounce (force)	N	2.780×10^{-1}
	pound (force)	kN	4.448×10^{-3}
Stress	lbf/in^2 (p.s.i.)	MPa	6.895×10^{-3}
Torque	lbf in.	N m	1.130×10^{-1}
	lbf ft	N m	1.356

Table 9.3 Example of a semi-qualitative table: properties of some pure metals

Element and symbol	Atomic number	Atomic mass	Melting point (°C)	Density × 10^{-3} (kg m^{-3})
Aluminium (Al)	13	26.98	660	2.699
Copper (Cu)	29	63.54	1083	8.96
Iron (Fe)	26	55.85	1535	7.87
Lead (Pb)	82	207.21	327	11.36
Magnesium (Mg)	12	24.32	651	1.74
Nickel (Ni)	28	58.71	1458	8.90
Tin (Sn)	50	118.70	232	7.30
Titanium (Ti)	22	47.90	1660	4.51
Tungsten (W)	74	183.86	3410	19.3
Zinc (Zn)	30	65.38	420	7.13

the far left column of the table (known as the 'stub'). This type of table is particularly useful for presenting conversion factors (e.g. Table 9.2), atomic masses of the elements and compounds, and the physical constants of various types of materials (compare Table 9.3). When a large number of different types of measurements have been carried out on a wide range of materials, there is usually no alternative but to list all the results. Handbooks contain many tables of this type. However, if a measurement has been repeated many times, it is not usually necessary to record all the results obtained (or even possible in a reasonably compact form). In this case it is usually sufficient to obtain and record an appropriate average value and some indication of its precision, e.g. the standard deviation (see Sec. 7.2 concerned with descriptive statistics).

Tables 9.4 and 9.5 are examples of quantitative tables; data for all the variables are listed in numerical form. Tables are often used for recording information rather than direct presentation and comparison of data; Table 9.4 contains data for the relationship between two variables whereas Table 9.5 contains data for several variables. Functional tables are those that tabulate

Table 9.4 Example of a simple quantitative table (relationship between two variables): conversion of temperatures

	(a)		(b)
°C	°F	°F	°C
−40	−40	−40	−40
−20	−4	−20	−28.9
−10	14	−10	−23.33
0	32	0	−17.78
10	50.0	10	−12.22
20	68.0	20	−6.67
50	122.0	50	10.00
100	212.0	100	37.8
250	482	250	121.1
500	932	500	260.0
1000	1832	1000	538
2000	3632	2000	1093
3000	5432	3000	1649

Table 9.5 Example of a quantitative table (relationship between three variables): density (g/cm^3) of aqueous copper sulphate (CuSO$_4$)

Weight % CuSO$_4$	0°C	20°C	40°C
1	1.0104	1.0086	1.0024
4	1.0429	1.0401	1.0332
8	1.0887	1.0840	1.0764
12	1.1379	1.1308	1.1222
16	—	1.1800	—
18	—	1.2060	—

relations of the type $y = f(x)$. Tables of logarithms, trigonometric functions, etc., are examples of this type of table.

The number of digits recorded for each measurement should reflect the precision inherent in the measurement. However the choice is usually based upon aesthetics as the confidence interval reflects the reliability, not the number of digits included. When tables are prepared specifically for presentation with the intention of conveying a particular aspect of the work, then the data can usually be rounded off (and stated as such) to a number of figures appropriate for easy assimilation of the meaning (rather than presenting exact numerical values). Furthermore, tables for presentation and comparison should contain only the information that is required; any additional information will tend to distract or confuse the reader. Table 9.6 is an example of a table prepared merely for comparison and presentation; contrast this table with Table 9.5 which is intended for the storage of information.

Very large or very small measurements (i.e. numbers) should not be written in the table, instead the data should be written in standard index form, e.g. 2.05×10^3 or 7.6×10^{-2}. The exponent should be included in the column heading. However, the meaning is often misinterpreted; for example, a column heading of '*Flowrate* $\times$ 10^3 (m^3/s)' means that a value of 6.5 taken from the table is equivalent to a flowrate of 0.0065 m^3/s *not* 6500 m^3/s. The interpretation is that the flowrate (in the table) *has been* multiplied by 10^3. This is often a source of confusion and in some books it is actually expressed incorrectly. The problem can easily be overcome by interpreting the data based upon an understanding of the system, i.e. asking whether a flowrate of 0.0065 or 6500 m^3/s is appropriate or possible for that particular reading. This requires under-

Table 9.6 Example of a quantitative table for comparison of data: densities (g/cm^3) of aqueous inorganic solutions

Weight %	NH$_4$Cl	NH$_4$NO$_3$	(NH$_4$)$_2$SO$_4$
(a) *Temperature = 10°C*			
4	1.0126	1.0168	1.0234
20	1.0596	1.0870	1.1184
(b) *Temperature = 20°C*			
4	1.0168	1.0144	1.0220
20	1.0870	1.0827	1.1154

standing rather than mere rote learning! (As already noted in Sec. 9.2 regarding the labelling of axes of graphs, the notation 'Flowrate $(m^3/s \times 10^3)$' should be avoided.)

Any type of presentation, whether a table, graph, flowchart, etc., should be clear, accurate and unambiguous. Mistakes often occur during typing of the hand-written copy of a table; however, ambiguities can also be introduced at the typing stage especially if the left-hand column contains descriptive information which requires more than one line. The typist rarely possesses the technical knowledge to appreciate where spaces should be left in a table. Careful proof reading is essential if accuracy is to be maintained.

9.4 FLOWCHARTS

Flowcharts are visual aids used to explain the nature of a process that consists of a set of logical and sequential stages. They can provide the broadest overview of a process or an explanation of the smallest detail of calculation. The choice depends upon the needs of the reader. Flowcharts are a particularly useful tool when developing computer programs. They show the nature of the computational process; their understanding should not depend upon a knowledge of the particular computer language that is used. Students are often reluctant to spend time preparing a flowchart before developing a computer program; however, it is usually difficult to determine programming errors based on logic without a flowchart. This is especially true for the teacher who has not written a comparable program, and who may not be an expert computer programmer.

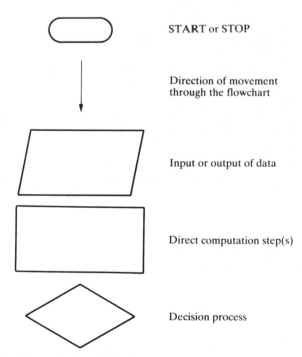

START or STOP

Direction of movement through the flowchart

Input or output of data

Direct computation step(s)

Decision process

Figure 9.17 Symbols used in the construction of flowcharts.

Flowcharts and computer programs are particularly useful for solving *algorithms*; this is a general term that is used in various ways when referring to computational processes. A broad definition of an algorithm is a step-by-step procedure leading to a solution of a computational problem in a finite number of steps. A narrower definition adds the requirement that there must be some sort of loop in the procedure. It must also have one or more rules for stopping the calculation when an acceptable result is obtained, or when a solution cannot be obtained.

The basic symbols or blocks used in the construction of flowcharts are shown in Fig. 9.17. Steps in an algorithm are shown using lines to connect the blocks. The direction of flow through a flowchart may be indicated by arrows on the lines; however, arrows may be omitted if the convention is adopted that flows are downward or to the right unless otherwise indicated.

The three main types of flowchart that will be discussed are simple, branched and looped flowcharts. A *simple flowchart* (without branching or looping) does not have decision points and is used to describe a straight-through computation. The flowchart contains complete details of the process but, because of the absence of branching or looping, its only advantage over a numerical list is the visual presentation. However, flowcharts have definite advantages when decisions have to be made.

A *branching flowchart* has several branches containing different steps; the branches originate at a decision point and the branch to be followed depends upon the value of the decision variable. The use of a branching flowchart to obtain the solution of a quadratic equation is shown in Fig. 9.18.

A *looped flowchart* contains branches that cause a return to an earlier part of the process; these branches are called loops. The use of loops provides a significant increase in the ability of the algorithm to solve problems. Loops also enable flowcharts and computer programs to become more compact. An example of the use of a looping flowchart is shown in Fig. 9.19; this is used to test whether a triangle is right angled. The use of loops can be disadvantageous if they are not used correctly (and carefully). It is possible to create an infinite loop such that the process has no way of terminating. Inequalities are often used to prevent infinite looping, provided that an appropriate difference is used, e.g. $(a - b) > 0.01$ or $(a - b) > 0.0001$. If the difference is too small, a solution may never be obtained; alternatively, the specification of too large a difference may mean that an inaccurate solution is obtained prematurely.

An additional decision variable and branch are often included to ensure that the looping will terminate. This may include a statement such as 'when $n = 100$, print OUT OF DATA, then STOP', where n is a counter, set initially to zero and increased by 1 each time the loop is traversed. A procedure may be repeated until all data have been used (assuming these data are not self-generating) and an appropriate message is printed if no solution is obtained. With computer programs it is essential to incorporate appropriate print statements that indicate what the computer has done when it finally stops running the program. However, too much print-out can also be a disadvantage. The advantage of flowcharts with branching and/or looping is that they are more easily visualized than a listing or numbered instructions or statements.

Finally, a few general observations regarding the construction and interpretation of flowcharts:

(a) The flowchart must not be ambiguous; the path to be taken must be clear.
(b) Only one starting point is possible, usually at the top of the flowchart.
(c) The only block that may have more than one exit line is a decision block.
(d) The stop block cannot have an exit line; more than one stop block is permissible.

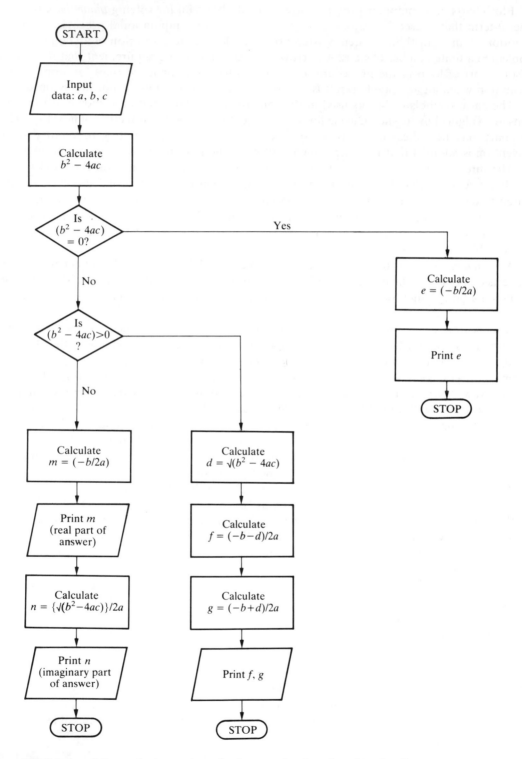

Figure 9.18 Solution of the quadratic equation $ax^2 + bx + c = 0$, using a branching flowchart.

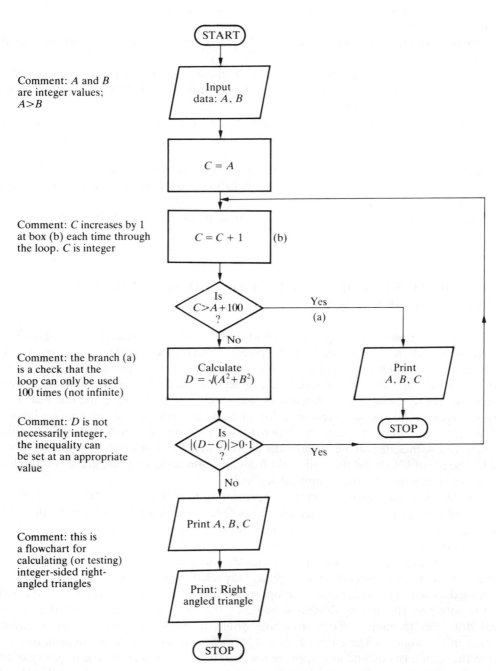

Comment: *A* and *B* are integer values; *A>B*

Comment: *C* increases by 1 at box (b) each time through the loop. *C* is integer

Comment: the branch (a) is a check that the loop can only be used 100 times (not infinite)

Comment: *D* is not necessarily integer, the inequality can be set at an appropriate value

Comment: this is a flowchart for calculating (or testing) integer-sided right-angled triangles

Figure 9.19 Use of a looping flowchart to determine whether a triangle is right angled. *Note:* Real values of *C* can be considered by making $C = C + 0.1$ (for example) at point (b). The following step in the flowchart then becomes 'Is $C > A + 10$?' for a restriction of 100 on the number of times through the loop.

(e) Flow path lines can join, e.g. looping.

(f) Decision blocks must specify the conditions under which *each* exit branch will be followed.

(g) Draw an arrow at the end of every line (optional) or where two lines join (highly recommended).

(h) STOP statements should be situated where they logically occur, not necessarily at the bottom of the flowchart.

(i) Information inside the blocks should be brief but descriptive, using combinations of English and algebra.

(j) Lines leading from decision blocks must be definitive in terms of the decision statements within the block. Words such as TRUE/FALSE or YES/NO may be used (appropriately) to designate the departing lines.

(k) A single line should enter each block (except the START block) from the top. If more than one line enters a block, their paths should be joined outside the block, as shown in Fig. 9.20(b).

(l) Flowcharts should be prepared to minimize the crossing of lines. If crossovers are unavoidable, the symbol shown in Fig. 9.20(c) should be used to show that lines are not connected.

(m) It takes a lot less time to redraw a flowchart than to 'de-bug' an infinite loop!

9.5 GRAPHICAL SYMBOLS, ENGINEERING DIAGRAMS AND ENGINEERING DRAWINGS

Engineers of all disciplines are required to build things; they apply scientific principles for the benefit (it is hoped) of the human race. To perform these tasks effectively it is necessary to communicate results, ideas, information, etc., between different groups of people in an unambiguous manner. The use of standard engineering diagrams and engineering drawings enables engineers to communicate information accurately and in a simple and compact form.

An *engineering diagram* is a representation of an engineering system, equipment layout or circuit arrangement. It is not usually drawn to scale or dimensioned (although significant dimensions can be included); it merely provides a record of the way in which a system functions. It would not be possible to build the equipment from an engineering diagram alone, although that diagram would explain how the equipment works.

An *engineering drawing* is a pictorial representation of an object, drawn to scale and containing all relevant dimensions. It should be possible to construct an object from an accurate engineering drawing, or in some cases several drawings showing the object as viewed from different positions.

Graduate engineers have usually undertaken some formal instruction in engineering drawing. Graduate mechanical engineers are also expected to be capable of producing accurate engineering drawings for a wide range of equipment. Other engineering graduates are expected to be able to interpret engineering drawings; however, they are not usually expected to prepare detailed drawings themselves. Such drawings would normally be prepared by a qualified draughtsman (or woman). The engineer should be able to communicate the information to be drawn to the draughtsman, and also approve the final drawing as correct. The majority of engineers are therefore more concerned with producing engineering diagrams and suitable sketches than with preparing accurate engineering drawings. The ability to produce a sketch for an item to

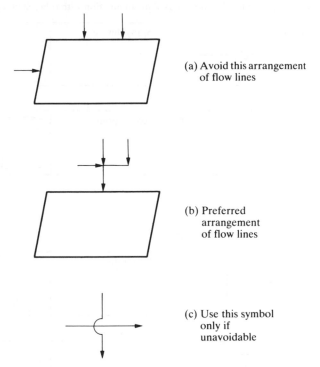

(a) Avoid this arrangement
of flow lines

(b) Preferred
arrangement
of flow lines

(c) Use this symbol
only if
unavoidable

Figure 9.20 Directions indicated on a flowchart.

be made in the workshop, or to be redrawn by a draughtsman, is not a skill that is usually taught. It is developed and improved by practice, by discussion of the merits and failings of prepared sketches, and by feedback from those people who must work from these 'rough' ideas. In a degree course, the student should develop this skill and receive appropriate feedback regarding the nature of the sketches produced. Hopefully, at the end of a course the graduate will be proficient in this essential aspect of communication.

9.5.1 Engineering diagrams

The different engineering specializations have established their own terminology to specify different types of diagrams. Many terms have a different interpretation depending upon the particular field of engineering. Certain standards are available, e.g. BS 5070, that attempt to remove ambiguities and provide some uniformity between different specializations. A list of relevant British Standards is included in Sec. 9.5.3. Standards are also available that specify the graphical symbols for use in the creation of engineering diagrams, e.g. BS 1553. Reference should be made to these standards and the additional references given at the end of this section. The graphic symbols used in engineering diagrams need to be relatively simple and versatile, so that they can be easily modified to suit a particular design requirement; also their form should be representative of the equipment that they describe. Some examples of the more common symbols are given in Fig. 9.21.

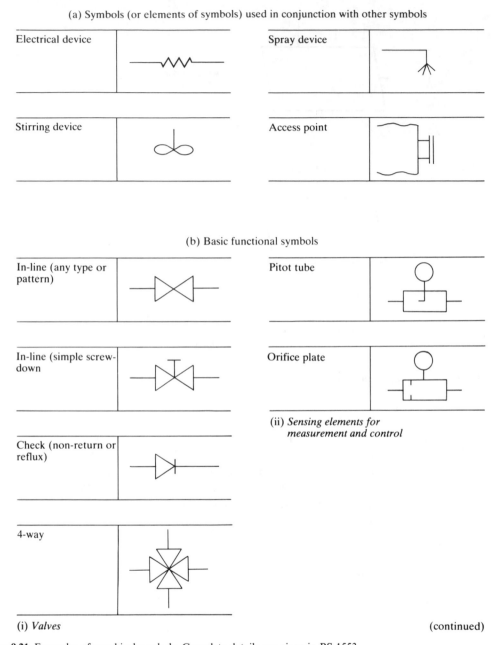

(a) Symbols (or elements of symbols) used in conjunction with other symbols

Electrical device		Spray device	

Stirring device		Access point	

(b) Basic functional symbols

In-line (any type or pattern)		Pitot tube	

In-line (simple screw-down		Orifice plate	

(ii) *Sensing elements for measurement and control*

Check (non-return or reflux)	

4-way	

(i) *Valves*   (continued)

Figure 9.21 Examples of graphical symbols. Complete details are given in BS 1553.

Diagrams that illustrate a *function* include block, system and circuit diagrams, and those illustrating *assembly* include wiring, piping and installation diagrams. Descriptions of each general type of diagram and examples of each are given in Table 9.7. These diagrams are normally

(c) Basic and developed symbols for plant and equipment

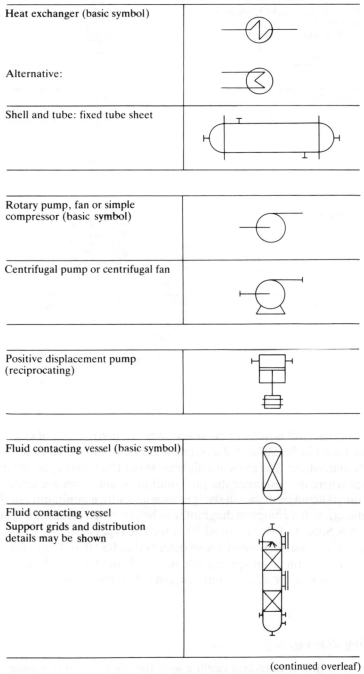

Heat exchanger (basic symbol)	
Alternative:	
Shell and tube: fixed tube sheet	
Rotary pump, fan or simple compressor (basic symbol)	
Centrifugal pump or centrifugal fan	
Positive displacement pump (reciprocating)	
Fluid contacting vessel (basic symbol)	
Fluid contacting vessel Support grids and distribution details may be shown	

(continued overleaf)

Figure 9.21 *continued*

used in successive stages of a project. The definition, principles and presentation of each type of diagram are given in BS 5070; this standard also includes 46 diagrams as examples of the different types. Three examples of engineering diagrams are included here as Figs 9.22 to 9.24.

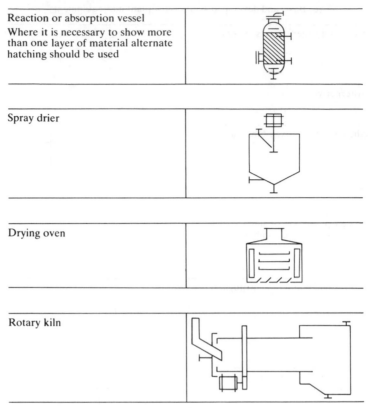

Reaction or absorption vessel Where it is necessary to show more than one layer of material alternate hatching should be used	
Spray drier	
Drying oven	
Rotary kiln	

Figure 9.21(c) *continued*

Some relevant points will now be made concerning the preparation of engineering diagrams; full details can be found in BS 5070. A diagram should be presented so that its meaning can be quickly and easily understood. The parts of a diagram should be evenly spaced; large spaces should be avoided except where they enhance the presentation or additions are anticipated. Wherever possible, connections should be drawn in straight lines and with a minimum number of crossovers and changes of direction. If a complete diagram is to be presented on several sheets, then a useful numbering system is Sheet 1 of 10, ..., Sheet 10 of 10. The types of lines, e.g. continuous, broken, etc., to be used in engineering diagrams are also described in BS 5070. The graphical symbols used on a diagram should conform to an appropriate standard, and this standard should be stated on the diagram. Special symbols or adaptations of standard symbols should be supported by a note or legend.

9.5.2 Engineering drawings

To reiterate, the main British Standard dealing with the preparation of engineering diagrams is BS 5070; other standards are available which describe the graphic symbols to be used, e.g. BS 1553. As already discussed in Sec. 9.5.1, engineering diagrams are normally associated with a type of flow and relate components (usually indicated by symbols) functionally to each other by the use of lines. Such diagrams do not depict the shape, size or form of the components and, in general, actual connections or locations are not indicated. BS 308 covers what are commonly

Table 9.7 Types of engineering diagrams, descriptions and examples

Type	Description	Examples of titles
Block and system diagrams	Simplified illustration of the main interrelationships of elements in a system, and how the system works or may be operated	Chemical process block diagram Logic diagram Control system diagram Power system diagram
Circuit and flow diagrams	Describe the full functioning of a circuit, process or installation, showing all essential parts and connections (using symbols), without regard to the physical layout	Process flow diagram (flowsheet) Control circuit diagram
Wiring and piping diagrams	Showing the detailed connections between components or items of equipment	Wiring layout Wiring schedule Piping layout Piping schedule
Installation diagram	Describing connections and installation information related to a particular site or structure	Pipework installation diagram Cabling installation diagram Installation layout
Supplementary diagram	Containing information for which none of the other types is suitable	Timing diagram Lubrication diagram Topographic diagram Sequence chart

accepted to be drawings that define shape, size and form. It contains recommendations for drawing layout, scales, types of lines, lettering, methods of orthographic projection, sections and conventional representations.

A drawing may consist of more than one sheet and is, in general, either any form of graphical representation or more specifically a graphical representation of an object that depicts its form and/or position. There are several types of drawings; these are described as follows:

(a) A *design layout drawing* represents in broad principles the feasible solutions meeting the design requirements, e.g. a design scheme.

(b) A *detail drawing* depicts a single object and includes all necessary information to define the object completely.

(c) A *tabular drawing* depicts an object typical of a series of similar such objects having their variable characteristics presented in tabular form.

(d) An *assembly drawing* shows two or more parts, or sub-assemblies, in their assembled form, and includes any dimensions and instructions necessary to effect assembly.

(e) A *combined drawing* shows an assembly, item list and constituent details drawn separately but all on the same drawing.

(f) An *arrangement drawing* of the complete finished product or equipment shows the arrangement of assemblies, including important functional and/or performance features.

(g) A *diagram* is a drawing depicting the function of a system or the relationship between component parts, using a simplified representation.

(h) An *item list* is a list of component parts required for an assembly, shown either on the drawing or on a separate list.

(i) A *drawing list* is a list of drawings associated with an assembly.

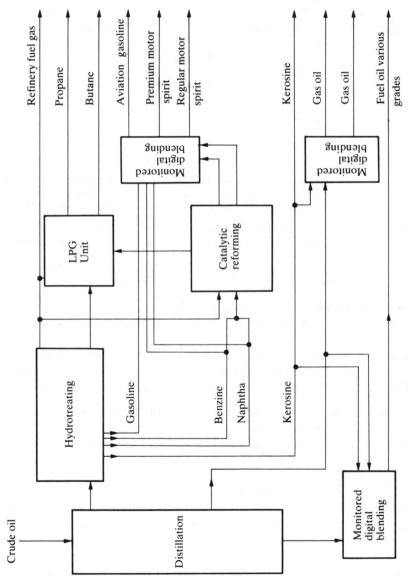

Figure 9.22 Example of a block diagram. (Reproduced from BS 5070 (1974), page 19, with permission.)

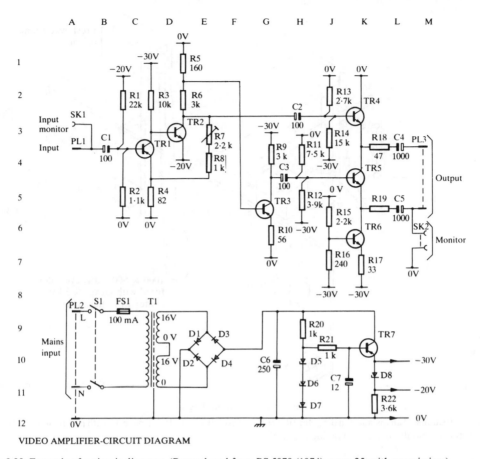

VIDEO AMPLIFIER-CIRCUIT DIAGRAM

Figure 9.23 Example of a circuit diagram. (Reproduced from BS 5070 (1974), page 25, with permission.)

Drawing sheets should conform to the International Organization for Standardization (ISO) 'A' series sizes, ranging from 841 × 1189 mm (A0) to 210 × 297 mm (A4). Drawing sheets can have two formats. Landscape format is intended to be viewed with the longest side of the drawing sheet horizontal. Portrait format is intended to be viewed with the longest side of the drawing sheet vertical.

The title block is that area of the drawing sheet containing the information required for the identification, administration and interpretation of the drawing. It is usually situated in the lower right-hand corner of the drawing frame. The following basic information should be included:

(a) name of company or organization;
(b) drawing number—this may also appear elsewhere on the drawing sheet for convenience;
(c) descriptive title;
(d) date of drawing;
(e) signature(s);
(f) original scale;
(g) copyright clause;
(h) projection symbol;
(i) unit of measurement;

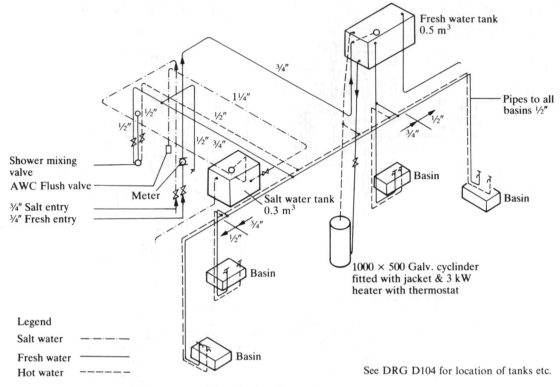

Plumbing installation diagram

Figure 9.24 Example of an installation diagram. (Reproduced from BS 5070 (1974), page 37, with permission.) Note that inch sizes are still commonly used for plumbing pipework.

(j) reference to standards;
(k) sheet number;
(l) number of sheets;
(m) issue information.

The following supplementary information may also be considered for inclusion:

(n) material and specification;
(o) treatment/hardness;
(p) finish and surface texture;
(q) general tolerances;
(r) key to geometrical tolerances;
(s) sheet size;
(t) first-used date;
(u) equivalent part;
(v) supersedes or superseded by;
(w) warning notes, e.g. 'do not scale';
(x) other appropriate information.

The scale is the ratio of the linear dimension of an element of an object, as represented in the original drawing, to the real linear dimension of the same element of the object itself. Full size is indicated by '1:1', an enlargement scale by 'X:1' and a reduction scale by '1:X'.

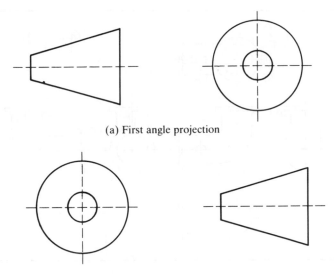

(a) First angle projection

(b) Third angle projection

Figure 9.25 Symbols used to indicate the system of projection.

Multi-view orthographic projection is predominently used in engineering drawings. There are two systems, known as first angle and third angle; both are based on a framework of planes at right angles. Both systems are approved internationally and have equal status. The first angle projection method is normally used for British Standards, e.g. BS 308, although the third angle method could be used without prejudice to the principles established. The system of projection used on a drawing should be indicated by the appropriate symbol as shown in Fig. 9.25. An example of first angle projection is shown in Fig. 9.26; each view shows what would be seen by looking on the far side of an adjacent view. In third angle projection, each view shows what would

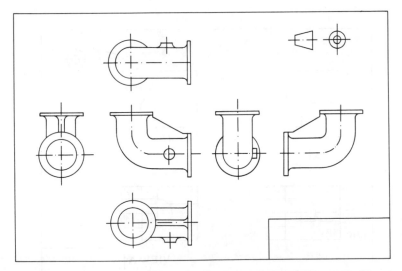

Figure 9.26 Example of first angle projection. Each view shows what would be seen by looking on the far side of an adjacent view. (Reproduced from **BS 308**: Part 1 (1984), Fig. 7, with permission.)

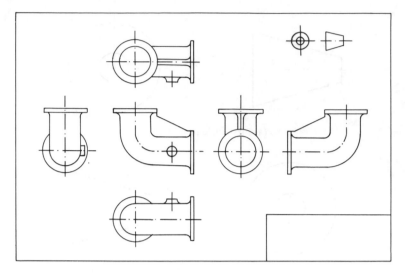

Figure 9.27 Example of third angle projection. Each view shows what would be seen by looking on the near side of an adjacent view. (Reproduced from BS 308: Part 1 (1984), Fig. 8, with permission.)

be seen by looking on the near side of an adjacent view. This method is illustrated in Fig. 9.27. A special arrangement of views is shown in Fig. 9.28 and is used where the position of a view cannot conveniently conform to the method indicated by the symbol. In this case the direction of viewing should be clearly shown. The number and choice of views should be selected so that the maximum amount of information is presented clearly, and the number of views is the minimum necessary to ensure that the drawing cannot be misunderstood.

The information contained in this section is intended to provide an introduction to engineering drawing practice. Further details can be found in appropriate standards, e.g. BS 308, or in the many texts describing this subject (see Sec. 9.5.4).

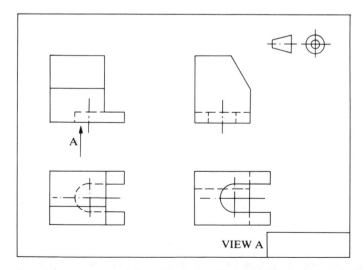

Figure 9.28 Example of special arrangement of views. (Reproduced from BS 308: Part 1 (1984), Fig. 9, with permission.)

9.5.3 Publications of the British Standards Institution

All standards are subject to periodic independent revision; it is important to ascertain that the latest edition or amendment of a particular standard is being consulted. The British Standards Institution issues a new catalogue of its publications each year. The publications listed in this section are related to the preparation of engineering diagrams and engineering drawings.

BS 308: *Engineering drawing practice*
  Part 1: Recommendations for general principles
  Part 2: Dimensioning and tolerancing of size
  Part 3: Geometrical tolerancing

BS 1192: *Construction drawing practice*
  Part 1: Recommendations for general principles
  Part 4: Recommendations for landscape drawings

BS 1553: *Graphical symbols for general engineering*
  Part 1: Piping systems and plant
  Part 2: Graphical symbols for power generating plant
  Part 3: Graphical symbols for compressing plant

BS 1646: *Symbolic representation for process measurement control functions and instrumentation*
  Part 1: Basic requirements
  Part 2: Additional basic requirements
  Part 3: Detailed symbols for instrument interconnection diagrams
  Part 4: Basic symbols for process computer, interface and shared display/control functions

BS 3363: *Letter symbols for semiconductor devices and integrated microcircuits (including two Supplements)*

BS 3641: *Symbols for machine tools (see also PD 6490)*
  Part 1: General symbols
  Part 2: Numerical control symbols
  Part 3: Additional general symbols

BS 3939: *Graphical symbols for electrical power, telecommunications and electronics diagrams*
  Part 2: Symbol elements, qualifying symbols and other symbols having general application
  Part 3: Conductors and connecting devices
  Part 4: Passive components
  Part 5: Semiconductors and electron tubes
  Part 6: Production and conversion of electrical energy
  Part 7: Switchgear, controlgear and protective devices
  Part 8: Measuring instruments, lamps and signalling devices
  Part 9: Telecommunications: switching and peripheral equipment
  Part 10: Telecommunications: transmission
  Part 11: Architectural and topographical installation plans and diagrams
  Part 12: Binary logic elements
  Part 13: Analogue elements

BS 4210: *35 mm microcopying of technical drawings*

BS 5070: *Drawing practice for engineering diagrams*

BS 5100: *Guide to the principles of geometrical tolerancing*

BSI educational publications (Published Documents PD)

PD 6490: Guide to application of additional symbols (and index) for machine tools [Supplement to BS 3641]

PD 7303: Electrical and electronic graphical symbols for schools and colleges

PD 7304: Introduction to geometrical tolerancing

PD 7307: Graphical symbols for use in schools and colleges

PD 7308: Engineering drawing practice for schools and colleges

PD 7309: An introduction to the tolerancing of functional length dimensions

9.5.4 Additional useful references

General literature

Austin, D. G., *Chemical Engineering Drawing Symbols*, George Godwin, London, 1979.
    (A reference book providing a comparative guide to over 1150 drawing symbols from British and US Standards. These symbols are used by draughtsmen and design engineers in the preparation of process flow diagrams and engineering line diagrams.)
Giesecke, F. E., A. Mitchell, H. C. Spencer and I. L. Hill, *Technical Drawing*. 8th edn, Macmillan, New York, 1986.
Giesecke, F. E., A. Mitchell, H. C. Spencer, I. L. Hill and R. O. Loving, *Engineering Graphics*, 3rd edn, Macmillan, New York, 1981.
Hill, R. G., 'Drawing effective flowsheet symbols', *Chemical Engineering*, 1 January, 1968, page 84.
Wells, G. L., C. J. Seagrave and R. M. C. Whiteway, *Flowsheeting for Safety*, Institution of Chemical Engineers, Rugby, 1976.

Standards

ASA Y32.11 (1961) *Graphical symbols for process flow diagrams*, ASME.
ASA Y32.2.6 (1962) *Graphical symbols for compression plant*, ASME.
ASA Z32.2.3 (1949) *Graphical symbols for pipe fittings, valves and piping*, ASME.
DIN 40716 (1961) *Graphical symbols for measuring, indicating and recording instruments*, Deutsches Institut für Normung.

EXERCISES

The information presented in this chapter is intended to provide ideas for the visual presentation of data. Be aware of the following comments and suggestions when presenting your *own* information:

Choose an appropriate form of presentation
Aim for clarity and avoid ambiguity

Consider the following types of visual presentation in relation to your own work. This will be more useful than performing textbook problems and exercises.

Pictorial representations
Graphs
Tables
Flowcharts
Engineering diagrams
Engineering drawings

Many alternative forms of these presentations can be adopted, in addition to those included in this chapter. Other types of visual presentations exist.

BIBLIOGRAPHY

Andrews, G. H. W., *Technical Drawing in Third and First Angle Projection*, Edward Arnold, London, 1973.

Breckon, C. J., *Graphic Symbolism*, McGraw-Hill, Sydney, 1975.

Dezart, L., *Drawing for Publication: A Manual for Technical Illustrators*, The Architectural Press, London, 1981.

Earle, J. H., *Descriptive Geometry*, 2nd edn, Addison-Wesley, Reading, MA, 1978.

Eide, A. R., R. D. Jenison, L. H. Mashaw, L. L. Northup and C. G. Sanders, *Engineering Graphics Fundamentals*, McGraw-Hill, New York, 1985.

Enrick, N. L., *Effective Graphic Communication*, Auerbach Publishers, New York, 1972.

Gill, R. W., *Creating Perspective*, Thames and Hudson, London, 1975.

Guidry, N. P. and K. B. Frye, *Graphic Communication in Science*, National Education Association, Washington, D.C., 1968.

Hartley, J., *Designing Instructional Text*, 2nd edn, Kogan Page, London, 1985, pages 80–103.

Jjalve, E., M. M. Andreasen and F. F. Schmidt, *Engineering Graphic Modelling*, Newnes–Butterworths, London, 1979.

Knowlton, K. W., R. A. Beauchemin and P. J. Quinn, *Technical Freehand Drawing and Sketching*, McGraw-Hill, New York, 1977.

McKenzie, J., L. R. B. Elton and R. Lewis, *Interactive Computer Graphics in Science Teaching*, Ellis Horwood, Chichester, 1978.

Morris, G. E., *Technical Illustrating*, Prentice-Hall, Englewood Cliffs, NJ, 1975.

Parker, D., *Basic Graphic Communication*, Macmillan, South Melbourne, 1981.

White, J. V., *Editing by Design: A Guide to Effective Word-and-picture Communication for Editors and Designers*, 2nd edn, Bowker., New York, 1982, pages 174–223.

ORAL AND WRITTEN COMMUNICATIONS*

CHAPTER OBJECTIVES

1. To encourage the reader to consider when an oral report is an appropriate form of presentation, and its advantages compared with a written report.
2. To encourage the reader to consider the different types of reports, e.g. formal, short-form and memoranda, that can be presented, and what information should be included in each type.

QUESTIONS 1

- In which ways can information be presented?
- What factors influence the communication of ideas?
- What methods of presentation and communication strategies improve the effective-ness of communications?
- In which situations are oral reports appropriate and effective?
- How are effective written reports presented?

EXERCISES

1. You are required to attend an employment interview. List the personal communication factors that will influence your performance and success.
2. Discuss the factors that will influence your success when using the telephone to obtain information concerning:
   (a) an air-conditioning unit;

* The material presented in this chapter was originally published in a shorter form in the book: *Elements of Engineering Design: An Integrated Approach* by M. S. Ray, published by Prentice-Hall International (UK) Ltd, 1985. It is reproduced with the permission of the publisher.

(b) travel arrangements between London and San Francisco, including a stop-over in New York.

3. You are about to launch a new type of 'do-it-yourself' double glazing unit. Consider details of the requirements for a promotional video or television film.

4. What are the important factors when advertising a car telephone in a national newspaper or business magazine?

5. You have completed the 'on-paper' design of a wheelchair. Now consider possible uses and advantages of scale models to assist an evaluation of the ideas proposed.

6. You have designed a domestic cooker. Make recommendations for building and testing a prototype unit.

7. What documents should be available when a project is completed?

8. What recommendations would you make for the presentation of reports? How would these recommendations change for different types of material to be recorded and presented? Compare your ideas with the requirements for laboratory and project reports in your course.

9. What are the differences between an abstract and a summary? Prepare an abstract. Let a colleague criticize it and then discuss it together.

10. Recommend the packaging to be used for a 20-piece construction toy, suitable for children aged 9 to 12 years.

11. You manufacture a multi-purpose aerosol lubricant. Prepare technical literature for purposes of advertising, marketing and sales.

12. Make recommendations for the contents of the maintenance and repair manual for a washing machine. Identify specific communications that relate to the servicing of the machine by an engineer.

13. Assess the effectiveness of this book, and any lectures and tutorials for an associated course concerning the communication of particular information.

QUESTIONS 2

After completing a task or assignment always ask the following questions.

- Am I required to prepare a report?
- Is it necessary to produce a report?
- Is the report intended to provide a record of the work, or communication/information, or both?
- Is the report to be presented in written form or orally, or both?
- Are standard formats to be followed?
- Is a written report to be a short-form or formal presentation?
- What are the specifications/suggestions for the presentation of the report, e.g. length, style, etc.?

10.1 GENERAL COMMENTS

Communication is nearly always symbolic; the common symbols are words (written and spoken) and pictures (graphs, diagrams, etc.). The symbols are then combined to convey information, i.e. to formulate a message. Information science is concerned with studying the message used in a communication. However, if the message is viewed in isolation without consideration of both the sender and the receiver, the result may be a good message but poor communication.

The engineer must be aware of all aspects of the communication system and must view it as an activity involving people. Although machines can record, relay and act upon messages, they must originally be programmed by humans. It is the human transfer of information and the response that are important in engineering. Surprisingly, many messages are prepared and sent without any thought at all for the recipient. The sender often prepares the information in a form that is considered to be appropriate, i.e. as the individual would prefer to receive it, and this is the reason why so many communications are at best inefficient and at worst completely useless. Even some lecturers make the mistake of ignoring the requirements and capabilities of their audience, when part of their job is supposed to be information transfer! Always prepare information bearing in mind the person who will act upon it. Both students and lecturers would benefit if some of their time were devoted to 'effective' communications! Communication is the *efficient* transfer of information. If it is not efficient, then it is not communication!

Students often ask why they have to produce a report concerning their experimental work, and also why they have to spend so much time writing reports. These questions are concerned with two separate issues: the need to write reports and the time that this requires. Reports are written for two main reasons: first, the student needs to acquire the skills necessary for effective report writing as this is the main means of communication in their future career; second, the report is a permanent record of the work that was performed and it provides a convenient and objective means of assessment and grading. The time that needs to be spent writing reports depends upon the specifications provided. Students often spend unnecessary time preparing complete formal reports when a shorter version would be sufficient (and acceptable!). Open-channels of communication between the student and the lecturer should alleviate this problem. Do not use your previous reports as a model for all situations, and do not necessarily adopt the format that others have used. Assess each situation independently and produce a report that is appropriate (and required!). Do not assume that marks awarded are proportional to the length of a report. In many cases, the longer a report becomes then the more material it contains that can be attributed to external sources rather than to the author!

There are many guidelines that could be proposed and entire books have been written describing just one aspect of communication, e.g. report writing or effective public speaking. Some suggestions will be discussed in this chapter. However, the best way to improve your communicating ability is to be widely read, generally observant and interested in a wide range of topics. In this way, you will improve your knowledge and awareness of the alternative means of presenting information, and be able to appreciate examples of good and bad communications.

10.2 ORAL COMMUNICATIONS

Purely oral communications occur if two people are working in the dark or with an intervening partition; the most common example is talking on the telephone. The efficient transfer of information then depends upon both speech and hearing, as well as the way in which the message is constructed.

Oral communication can also occur by personal contact, but it is then influenced by facial expressions, tone of voice and mannerisms. Transfer of information using spoken words can be in the form of questions and answers, or as direct statements. For example:

'*Can* I have component X?'
'Not until next week.'
'I *want* six of these.'
'They are out of stock.'

Problems can arise if the two people who are conversing have different native tongues or different accents or dialects. The manner in which the words are spoken will also affect their understanding, as will any background noise. Ultimately, if the person receiving the message does not possess satisfactory hearing or is not concentrating, then the interpretation of the message will be incorrect.

10.2.1 Oral presentations

An oral presentation can be used to inform, advise or instruct; the first of these is the most common use. Attending oral presentations requires a greater contribution of time and effort for the audience than is required when reading a written report on the same topic. If oral communication is to be effective then the speaker must also prepare his material and approach thoroughly; the time spent should be at least equal to that required for a written report. Oral presentation should not be considered a quick, easy, 'off-the-cuff' way of communicating information. The audience will expect additional benefits by attending an oral presentation, otherwise they would merely ask for a written summary.

A reader who becomes bored with a written report can put it aside, but retains the option of giving it further attention later. An audience that becomes bored with an oral report loses both the content and the opportunity for a second attempt. Audiences are generally polite but by definition they are human, and if a message is not conveyed in an appropriate and satisfactory manner, they will probably show their dissatisfaction. Ultimately they may 'vote with their feet', i.e. leave.

There are several disadvantages that may arise with oral reporting as compared with written reports. Less material can be presented, although additional printed material can be distributed; this, however, may act as a distraction for the audience. The listener cannot easily refer back to material presented earlier, and first impressions are particularly important, especially as corrections and changes reduce the effectiveness of the presentation. The ability to revise written material makes it effective in its final form; revision is not possible once an oral report has begun. If the listener loses interest, it will be difficult to recapture. If the oral report is part of a series, it may be difficult to maintain full audience attendance for the other reports.

However, there are also advantages to be gained. The main advantage is the opportunity to influence the audience by the speaker's delivery, charisma, intonations, gestures, personality, etc. If this does not work, the advantage rapidly becomes a disadvantage! A more thorough coverage of a particular topic and increased audience participation can be achieved by the use of questions and answers, tutorials, problem-solving sessions and group discussions. Oral reporting *should* encourage the speaker to concentrate only upon the essential details, and it may also be more economical than report preparation and distribution. The speaker will have the opportunity to involve the audience directly and to use its abilities and experience during the presentation and any subsequent discussion session. The opportunity also exists of improving and enhancing material and its presentation by using appropriate audio and visual aids.

One final important point: both oral and written reports should be prepared for different audiences and different situations. The oral report has the added dimensions of vocal and visual expression, but it is also normally necessary to present it as a continuous single unit. Written reports prepared for readers should *never* be read aloud to an audience; they do not produce good communication. Consider this point in relation to the lectures in your course.

10.2.2 Summary

An oral presentation must be thoroughly prepared and *not* considered a quick and easy form of communication. Care should be taken regarding the material to be included *and* its presentation. There should be advantages from the oral presentation compared with the equivalent written report.

An oral presentation should be no longer than is necessary to present the essential information. There will be no opportunity for revision of material at a later time. The presentation *must* be interesting; the speaker's delivery and appropriate visual aids can be used to achieve this objective. Direct audience participation may be possible.

Written reports prepared for readers should never be read aloud to an audience. Prepare an oral report specifically for the audience.

10.3 WRITTEN COMMUNICATIONS

An opinion that is sometimes expressed in schools, universities and industry is that engineering students need to be taught how to 'write', or in particular how to express themselves in writing. There are no reliable statistics available to confirm or refute this opinion, which anyway is a subjective view. However, it is important to realize that there are two aspects of writing: literacy and competence. Literacy involves correct spelling, grammar and a reasonable vocabulary. The associated skills are not difficult, and therefore poor performance is often severely criticized. Competence is the ability to express thoughts, ideas, logical arguments, etc., and depends upon the subject matter being presented, whereas literacy is independent of the subject. A competent writer uses language in a clear, concise and easily understandable way when communicating with others. Someone who is literate will not necessarily be competent, and vice versa, although competence requires some grammatical ability.

The skills of literacy should be acquired mainly in schools, whereas competence will be developed and refined at the college level and thereafter. Those who are critical of the written work of engineers (and students) will need to decide which aspects are unsatisfactory before they can begin to identify the causes and solutions to the problem.

Lecturers should attempt to make the work performed by the students typical of the real world. Maximum limits should be stated for written work, and any 'excess' included in appendices. Peters and Waterman[1] conducted a detailed study of US companies and found that Proctor & Gamble and United Technologies, two of the most successful companies, had introduced the 'mini-memo', requiring that all written communications be summarized in one page. The efficient communication of information was identified as a common feature of other successful US companies.

The remainder of this section will consider different types of written material which the engineer will be required to prepare or evaluate. Ideas and suggestions will be mentioned briefly as the student will only become more adept in written communications by producing individual material, and evaluating its effectiveness.

10.3.1 Project documentation

The engineer will be required to record all data and decisions associated with a project as the work proceeds. This will provide the full product specification and documentation and will consist mainly of data and records, including relevant standards, calculations, quotations, technical literature, patents, etc., but will also contain details of why and how certain decisions and choices were made. The documentation will contain full sets of detailed drawings associated with the product and its subsequent manufacture. This will include drawings of all components, sub-assemblies and the general overall layout or assembly. There will be an extensive amount of cross-referencing, and an appropriate system of identification and drawing number sequencing should be adopted as early as possible. An efficient filing system should also be implemented which includes all aspects of the work, all references including those that are not readily available as copies, suppliers' names, addresses and telephone numbers, and all relevant costs and quotation information and dates of acquisition and/or use. This should not be left until the project documentation has grown significantly, otherwise it will be an onerous, if not impossible, task. A system should be implemented that is easy to understand, use and cross-reference, so the new staff can easily appreciate its functions.

Report writing is one of the most important tasks that the engineer performs. The report may be a policy document, a technical document, results of a feasibility study, recommendations for a strategy to be adopted, or the record and conclusions of a particular stage of a project. Company reports are usually confidential and are intended to provide a permanent record of a project. Report writing is a necessary aspect of the engineer's work. It is important that a report is presented in the correct format and that the information it contains is both relevant and complete. Some general guidelines that will be useful in many situations are presented in Secs 10.3.2 and 10.3.3, and the requirements of different types of reports are considered in Secs 10.3.4 to 10.3.8.

10.3.2 General guidelines for report writing

Questions to be answered
Who are the readers, e.g. engineers, salesmen, customers?
What is their level of knowledge of the subject?
What is the purpose of the report, e.g. to provide information, to be used (a servicing manual), an instruction (to be acted upon)?
Is it confidential?
What type of material does it contain, e.g. technical or non-technical, facts or conjecture?

Features of the presentation
(a) Assign a brief but meaningful title.
(b) Include an abstract or summary, or both. (What is the difference between them?)
(c) Include only necessary background information with a complete reference list.
(d) Describe material in a depth suitable for the particular reader.
(e) Describe the background before the details.
(f) Plan the contents of the report as a logical sequence.
(g) Emphasize essential points.
(h) Distinguish clearly between facts and opinions.
(i) Use clear, concise language.
(j) Quantify statements wherever possible, e.g. 'mild steel is cheaper than aluminium' should be followed by '— and the prices are $X and $Y per kg, respectively'.
(k) Plan the report so that it is easy to read and can be read in stages, e.g. appropriate headings, graphical presentation of data. When numbering sections do not use more than two subsections, e.g. 5.3.2 *not* 5.3.2.1. Organization of material and choice of appropriate headings and sub-headings should enable unwieldy numbering to be avoided.

(1) Choose the appropriate type of report. This may be a memo, letter, informal short report, instruction, specification, formal report, proposal, article, or technical paper. What particular features or functions relate to each of these types of report? These aspects of the presentation have been expanded in more detail by Rathbone (see Rosenstein *et al.*)[2] Different types of reports are considered in Secs 10.3.4 to 10.3.8.

Preparation of the report

(a) Prepare an outline.
(b) Write a draft copy.
(c) Set the draft aside for a while.
(d) Then revise the draft by observing the questions to be answered (see above) and the required features of the presentation (also see above).
(e) Repeat these preparation stages (c) and (d) as necessary.
(f) Obtain a constructive critical review of the draft report from someone who is knowledgeable in that field and who is also able to correct the style and the grammar.

Data presentation

- Only include *necessary* data in the body of the report; include all other data in appendices or as a separate volume.
- Present the data in a form that emphasizes the points to be made in the text.
- In general, graphs, charts and diagrams are preferable, but there are situations where tables are more appropriate.
- If using tables include only relevant information, but present it in a clear, complete and unambiguous manner. Refer to Woodson[3] (pages 338–342) and Vidosic[4] (pages 145, 148–151) for examples of the alternative ways in which data can be presented.
- Remember that data can be presented in many forms, not just as graphs and tables. Alternatives are charts, block diagrams, flowcharts, schematics, nomographs, curves, sketches, drawings, maps and photographs. These types of visual presentation are considered in Chapter 9.
- Only include visual information that is necessary and relevant.

Avoid

- Mistakes in wording and spelling.
- New, unusual or undefined symbols or abbreviations.
- Incomplete diagrams or tables.
- Incorrect references.
- Ambiguous or unfamiliar words.
- Personal or biased opinions and comments.
- Disordered or wrongly numbered pages.
- Marked or damaged paper.
- Poor reproduction.
- Typing errors and untidy corrections.
- Inadequate binding.
- Damage during postage.

10.3.3 Summary of important points

(a) When writing a report always keep the following questions in mind:

> *What is the purpose of the report?*
> *What actions should it initiate?*

(b) The reader may ask:

Why should I read this report?

The report should be written so that the answer becomes:

Because I require the information contained in the report and it is interesting, well written, well presented, and the structure of the report makes it easy to read.

(c) The body of a report should contain:

  (i) the ideas and points that the author wants to communicate;
  (ii) relevant data/information;
  (iii) results and their discussion/explanation.

These aspects of the report should be contained within a logical framework.
(d) A report should be:

  (i) factually correct (distinguish clearly between facts and opinions/conjecture);
  (ii) complete.

However, a report should be relevant, interesting and 'reader friendly', i.e. easy to read and understand.
(e) A report should be written in 'good' English. This refers to the style of writing, something that is not easily taught (and certainly outside the scope of this chapter). Writing style *should* be taught or developed in schools; the role of college and university courses should be to adapt the student's style to the requirements for written communications of the professional bodies. Several books have been written that offer advice regarding style of writing (see Bibliography), and *Roget's Thesaurus*[5] is always a useful companion when preparing a report. The style used in a technical report should be:

  (i) brief and simple;
  (ii) clear and informative;
  (iii) formal.

Although writing in the first person is not wrong, it is generally more acceptable to write impersonally, i.e. 'the authors found . . .' rather than 'we found . . .', 'it was impossible to . . .' rather than 'I could not . . .', etc. After starting to write a section of a report always check the style after a few sentences.

It is particularly important when writing technical reports to avoid using unnecessary phrases or 'flowery' language. The following phrases can be replaced by single words: 'in the fullness of time' means 'eventually' and 'at this moment in time' means 'now'. Phrases such as 'it can be shown that', 'it is my opinion', 'it is evident from the previous discussion', etc., can usually be omitted entirely from a presentation or replaced by one or two appropriate words. Technical reports should be concise and only contain necessary information. The effect of excess phrases is usually to detract from the presentation. This creates the impression that the author feels a need to 'package' the information, i.e. that the work itself is not completely acceptable. Reread written work, remove any 'obsolete' phrases so that only essential information is presented, and develop a technical-writing style.
(f) Some particular points regarding presentation:

  1. Abbreviations are useful provided the reader understands what they mean. If the abbreviation used is not a standard form, it should be explained when first used. The period or full stop may usually be omitted after an abbreviation except where a word is formed, e.g. 'in.' for 'inch'. Abbreviations for units in the text should be preceded by a

number, e.g. 9.8 m/s^2. The plural of an abbreviation is the same as the singular, e.g. 'm' for metres. Abbreviations are acceptable in tables, graphs or figures where space is limited.

2. Statements that are generally true are written in the present tense, e.g. 'acceleration due to gravity is 9.81 m/s^2', whereas the past tense is used for events in the past, e.g. 'the vessel was specially constructed'. In general, do not change the tense within a section of the report.

3. Avoid the use of informal or colloquial phrases, and contractions, e.g. don't and can't. Humorous and witty phrases are not appreciated or required; the report will be interesting to the reader because it is clear, concise, informative and well presented.

4. Hyphens are used to connect words that are compounded into adjectives, e.g. 'close-packed hexagonal', 'a high-temperature instrument', but there is no hyphen in 'a slowly moving fluid' (i.e. not following words ending in 'ly').

5. *That* is used when the clause it introduces is necessary to define the meaning of its antecedent. *Which* introduces additional or incidental information. *Which* can often be substituted for *that* even though it is not strictly gramatically correct.

6. *Data* is usually plural, i.e. 'data are', not 'data is'.

7. *Affect* is a verb meaning 'to influence; *effect* is a noun meaning 'result'.

8. Use 'because of' instead of 'due to'.

9. Use 'different from' instead of 'different than'.

10. Use 'remainder' instead of 'balance'.

11. 'Fewer' refers to numbers and 'less' refers to quantity or degree.

12. 'Farther' refers to distance and 'further' indicates 'in addition to'.

10.3.4 Types of written reports

In this section the requirements of different types of written reports are considered. The information contained in this section, and in this chapter, should be considered as a set of guidelines and examples rather than as definite rules to be followed in every situation. Most companies and organizations have their own 'in-house' requirements for the presentation of written communications. The same situation applies in educational institutions except that student projects and laboratory reports are usually better received if they satisfy both an individual lecturer's preferences (i.e. the marker) and the departmental requirements. Technical journals and book publishers also have their own individual requirements for the submission and publication of manuscripts. Despite these often conflicting views and requirements, this section attempts to present suggestions for the preparation of good technical reports; it should not be considered a 'blue-print' for all situations.

In Secs 10.3.5 to 10.3.8, the general aspects of report presentation will be considered, and the requirements of particular types of reports, e.g. the formal report, the short report and the memorandum. I was tempted to include examples of both good and bad reports; however I did not do so for two reasons. First, much space would be required to do this adequately. Second, whenever an example of a 'good' report is given to a class of students, several people adopt this as their model for future presentation, whether or not it is appropriate to the situation and the information to be presented. Some books, e.g. Bragg[6], Ulrich[7], include examples of reports, although specific examples such as projects, laboratory reports, company reports, etc., are usually available and more useful.

The best way to develop good report writing skills is to write your own reports and have them critically appraised – by either a lecturer or a peer. Learn from your mistakes. Also choose a presentation that is appropriate to the particular situation; take into account the guidelines presented in Sec. 10.3.2 and the questions posed in Sec. 10.3.3. The 'best' reports are usually written by people who are widely read and have many broad interests. Make life a learning experience; learn from everything that you do, see and read. Evaluate, criticize and use all information that you receive.

10.3.5 General aspects of report presentation

(a) Choose an appropriate descriptive *title*. This should be a single sentence, however not so long as to be unmanageable when cited. Many titles are either too long because too much information has been included, or too short (one or two words) and therefore do not provide an adequate description of the work contained in the report.

(b) The *title page* should also include:
   (i) the department of origin, and/or course name and number;
   (ii) name of author(s);
   (iii) names of persons who assisted with the work;
   (iv) name of person to receive the report;
   (v) date of submission.

(c) *Abstract*. In general, write an abstract *or* a summary but not both. The difference between an abstract and a summary is not always clear; hopefully the description here and in part (d) following will clarify any confusion.

   A copy of the abstract is often filed separately and it is therefore useful to include it on the title page, or to include certain essential details such as the title, author, etc., on the same page.

   The abstract acquaints diverse readers with the essential findings of the report, i.e. it is a concise mini-report. It requires the most careful writing of the whole report.

   The abstract tells the readers whether the report falls into their area of interest. It must be comprehensive, but not detailed. Precise quantitative expression of the main findings is required. An abstract should include:
   (i) a statement introducing the subject;
   (ii) what was done;
   (iii) major results (preferably numerical).
   Discussion of results is rarely included in the abstract. An abstract usually consists of a few well chosen sentences (in one paragraph).

(d) A *summary* conveys the entire contents of the report in brief form. It is therefore necessarily longer than an abstract. A summary usually includes:
   (i) a statement introducing the subject;
   (ii) what was done;
   (iii) how it was done;
   (iv) *all* important results.
   The summary usually comprises a few paragraphs, typically half a page and rarely more than one page in length.

(e) *Contents page*. The contents page shows the structure of the report; it may also act as a useful reference source if an index is not included. The contents list should also help the reader answer the question: 'Why should I read this report?'

   The contents page lists the divisions and subdivisions of a report in order of appearance; reference should be given to page numbers. The use of indentation in the contents may be used to denote a division, subdivision, etc., within the report. Lists of tables and/or figures are included at the end of the contents, *if* this is necessary (i.e. useful).

(f) Sections comprising the *body* of the report will be dicussed in detail in Sec. 10.3.6 describing the formal type of report.

(g) *Conclusions.* This section consists of a series of *brief*, concise sentences that answer the questions relating to the purpose of the work. Pertinent remarks regarding particular aspects may also be included. Numbering the conclusions helps to clarify the points made.

This section may also contain *recommendations* or *actions required*, or these aspects may be listed separately. Recommendations are specific suggestions for future work on any aspect of the project. Actions required identify the person(s) to be responsible and should specify the timescale. Individual numbering is useful in both these sections.

The conclusions should help the reader answer the question: '*What needs to be done now?*'

Note: The title, contents page, abstract and/or summary, and conclusions should be able to stand alone as a useful mini-report of the work. Busy executives often have only enough time to read these sections of a comprehensive report, and expect to gain an understanding of the work and its findings from this cursory appraisal. Important decisions may sometimes be made based only upon these particular sections, and they are strongly influenced by the presentation.

(h) Sections of a report that usually follow the conclusions are described in detail in Sec. 10.3.6.

(i) *General presentation*

   (i) Adequate margins for reports are 25 mm from the top and right side, 30 mm from the left side and 40 mm from the bottom.

   (ii) Paragraphs should be separated by an extra line, and the first line may be indented (5 spaces if typed).

  (iii) Use black ink as other colours do not photocopy clearly, and avoid the use of red ink (useful for later additions and comments, or marking).

  (iv) Use A4 size paper, either wide line spacing if hand-written or single (or $1\frac{1}{2}$) spacing between lines if typed (on unlined paper).

   (v) Write or type on *one* side only.

  (vi) All pages including tables, graphs, photographs, etc., are numbered, starting after either the title page or the contents page. It is preferable to begin numbering after the abstract as this may be removed for separate filing.

(j) *Letter of transmittal*

A letter of transmittal may be attached to a report and it has two functions. First, the recipient is informed that the project is completed. Second, it records the time spent on the project by the author and assistants, e.g. laboratory time, library research time, report writing, etc.

The letter is usually detached and sent to the Accounts Department for appropriate cost allocations. Therefore, the letter should make reference to the title of the project and (briefly) describe the work that was performed (as specified in the project directive).

10.3.6 The formal report

The formats presented here for different types of reports are not the only ones that could be used; many companies specify their own requirements. However, the reports described should be acceptable in situations where no format is specified. It is important to choose the type of report, e.g. short or formal, that is appropriate to the particular reader and suitable for the information to be presented. The length of any section of a report will vary widely and the author's judgement will determine the information that is necessary and should be included.

A formal report contains the full documentation for a project; it aims to inform the reader of the relevant findings and to provide a complete reference source for other workers. Shorter reports can be considered as a condensed version of the formal report, containing information appropriate to a specific audience.

A formal engineering report usually consists of the following parts:

(a) Sections providing an introduction or overview of the project, e.g. Title Page, Abstract, Contents, Letter of Transmittal, Introduction.
(b) Sections providing detailed information of the study, usually referred to as the 'body' of the report. This includes Theory, Apparatus, Experimental Procedure, Results and Discussion.
(c) Sections that complete the report and provide supportive information, e.g. Conclusions, Acknowledgements, References, Appendices.

The presentation of a formal report is discussed in detail in this section, followed by the requirements for a short report (Sec. 10.3.7) and for a memorandum (Sec. 10.3.8).

A formal report usually consists of the following sections:

Letter of Transmittal
Title Page
Abstract *or* Summary
Contents List
Introduction
Literature Review
Theory
Equipment
Experimental Procedure
Results
Discussion
Conclusions
Recommendations
Actions Required
Acknowledgements
Nomenclature
References
Appendices
   Data (e.g. physical properties)
   Sample calculations
   Computer programs
   Tables of results
   Manufacturers' literature
   Instrument specifications

Some of these sections may be combined, e.g. the Literature Review may be included in either the Introduction or the Theory sections as appropriate, the Results and Discussion sections are often combined as are Equipment and Experimental Procedure. Other headings may be used, although the ones used here are probably the most common in engineering reports. The format presented here represents one alternative; other types of presentation may be more appropriate in particular situations.

Certain sections of the formal report, e.g. Abstract, Conclusions, etc., have been discussed in Sec. 10.3.5. This section discusses the remaining parts of the report.

(A) *Introduction*

The purpose and scope of the work are presented in the Introduction. It is often similar to the Abstract although usually longer; however the Abstract may be detached. The project directive may be rewritten to serve as part of the Introduction.

The title 'Introduction' is not very informative and with a little thought a better title may be found, depending upon the subject matter. The introductory section may describe the aims of the report, the project directive, how the report is structured or the historical background to the subject under investigation. A descriptive title would be more useful. If the introduction contains a mixture of all these subjects, the choice of appropriate title may be more difficult. In that case perhaps the author needs to reconsider the contents and structure of this section. In this chapter, the introductory Sec. 10.1 is entitled 'General comments'; perhaps either 'The need for effective communications' or 'The nature of communications' would have been better headings!

(B) *Literature Review*

If the published literature related to a project is extensive, the inclusion of a separate Literature Review section is recommended. This section should contain a brief, critical assessment of the relevant publications in the field. The importance of each reference should be established and the main points, findings, conclusions, etc., should be emphasized. This section should provide an overview of the work that had previously been carried out, and it should be easy to read and use. The literature review may need to be presented in several sub-sections, each dealing with a particular aspect of the present work. Publications are normally discussed in chronological order, from earliest to most recent, and an asterisk used to denote particularly significant information.

If the published literature is not extensive, a separate literature review may be omitted. Publications related to the development of the project are then discussed in the introductory section; those relating specifically to the associated theory are discussed in the theory section.

(C) *Theory*

The theory relevant to the project is presented and/or developed in this section. The theory may include a personal derivation (usually included in full) or a summary of previously published work. In the latter case, only important equations are included if complete details are readily available elsewhere (include a full reference). Do *not* present experimental results in this section. The theory associated with engineering reports usually includes mathematical equations. The following recommendations are made for their presentation.

(i) All variables should be defined; symbols and units to be used should be explained. Symbols may be identified either following the equation in which they are first used, or in a collective Nomenclature included near the end of the report (or sometimes in both places).

(ii) In general, it is preferable to separate equations from the text on separate lines.

(iii) Short equations may sometimes be included within sentences.

(iv) Equations are numbered *only* if they are to be referred to within the report. The equation number is included after the equation and on the same line, e.g. . . . (5.4).

(v) Standard SI units should be used. If appropriate, other units may be included and enclosed in brackets, e.g. 5275 J (5 Btu). If non-SI units are used, e.g. from nomographs, then SI units *must* also be given.

(D) *Equipment*

This section (sometimes referred to as Apparatus) may be considered separately or combined with the Experimental Procedure. It is intended to describe fully the equipment and its layout, placing emphasis upon any important aspects or features. Usually a diagram and associated written text is included. In some situations an accurate scaled engineering drawing is required (see details in Sec. 9.5.2), however a diagram including appropriate marked dimensions is often sufficient. A schematic illustration of the layout is used to provide a quick appreciation of the experimental system. Standard items of equipment e.g. analytical equipment, etc., are not usually drawn although their use should be discussed.

The diagram should be discussed and expanded upon in the written text. Essential information such as materials of construction, dimensions, safety features, etc., should be included. The detail included in this section should be comprehensive enough to make it possible for the equipment to be reproduced (and hence your data obtained!).

(E) *Experimental Procedure*
This section is sometimes called the Operating Procedure or Experimental Method. It should begin with a brief overview of the entire operation. Subsequently, individual aspects should be described clearly and in sufficient detail for a competent reader to be able to reproduce the experiments, and obtain the same results.

The reasons for adopting specific procedures are explained, and particular aspects, e.g. safety features, are discussed. The procedure should be reported in quantitative terms, *not* what should have been done qualitatively.

(F) *Results*
This section contains the data obtained from the experiment, usually transformed and presented in tables, graphs, etc. The results are gathered together according to a particular variable being studied. Theoretical and/or empirical correlations are often included on a graph for comparison. A table, figure or graph should be included in the report as soon as practicable after it is first referred to in the text. Every table or graph included in the body of the report (excluding appendices) *must* be referred to in the text at least once. Each table, figure or graph is usually included on a separate page. Sample calculations are included in the Appendices (see part (K) below), *not* in the Results section.

The visual presentation of information has been discussed in detail in Chapter 9. The following specific points are reiterated briefly:

 (i) Each table, figure or graph *must* be self-explanatory; this requires inclusion of an *informative title* and sufficient sub-titles or legends. This is *very* important!
 (ii) For complicated drawings, parts may be numbered and listed in a legend.
 (iii) Tables requiring a few lines may be included in the text.
 (iv) Scales should be chosen so that curves fill the graph paper, but do *not* extrapolate data beyond the measured range! Axes do not have to start at zero, but they must be clearly marked.
 (v) Axes must be labelled with the correct quantities and units.
 (vi) Curves do not need to pass through all data points, but should be 'representative' of the data.

(G) *Discussion*
The Discussion refers to discussion of the results presented in the previous section. This part of the report *discusses* the results; it does *not* simply state observations, obvious points or isolated known facts. It is a discussion of *why* something occurred rather than merely what was observed!

The Results and Discussion sections are often presented together under a combined heading. In certain situations it may be preferable to present them separately, e.g. when the reported results are extensive. When discussing results it is important to state any assumptions, sources of errors and the extent of agreement with theory or other published work. Any exceptions should be explained and supported by experimental data and/or references. Explain why the graphs look the way they do. This section should also include an analysis of errors and make use of appropriate statistical methods to explain the significance of the data.

(H) *Acknowledgements*
This section is used to express appreciation for any assistance provided in the project. People are more likely to help in the future if you acknowledge and thank them for help already given.

(I) *Nomenclature*
A table of nomenclature (if included separately) provides an alphabetical list of all symbols used in the report, stating (in English) the name and meaning, and the units. Other symbols and Greek letters are included after the English symbols. The use of any subscripts or superscripts is also defined. The nomenclature should be consistent with common usage as recommended by the appropriate standards association.

(J) *References and Bibliography*
The Reference list (sometimes called Literature Cited) should only contain references that are cited in the text. A Bibliography is a list of support literature, background or further reading; this literature is not cited in the text.

Any reference in a report (cited or support literature) must be so specific that the reader could easily locate the information. References to books must include: all authors and all initials; the full title and edition; the volume (if more than one); publisher, place and date of publication; any pages cited (either given in the text or included in the full reference). The International Standard Book Number (ISBN) is optional but sometimes useful. References to journal articles must include: all authors and all initials; the full title of the article; the journal name (or accepted standard abbreviation); volume number (underlined or in italics); part number (in brackets); inclusive page numbers for the article; year of publication.

References to lecture notes should include: the course and unit titles and designation numbers; the lecturer; the particular topic; the date; and any other relevant information. References to anonymous authors, personal communications, or where incomplete information is available should be clearly stated.

References in the reference list should be presented as follows:

Ray, M. S., *Elements of Engineering Design: An Integrated Approach*, Prentice-Hall International, Hemel Hempstead, UK, 1985.

Note: Pages are cited in the text; authors' initials may be written first, i.e. M. S. Ray; this is the first edition.

Konstantinov, L., Joosten, J. and Neboyan, V., 'Nuclear Power and the Electronics Revolution', *IAEA Bulletin*, **27** (3), pp. 3–6, 1985.

Note: Authors' initials may be written before the surname; the abbreviation '*et al.*' should not be used in the reference list—all names and initials should be included.

References within the report may be cited in one of two ways. The method chosen should be adhered to throughout the report. In the first method, references are numbered in the order in which they are cited throughout the report. In the text, use the following reference notation as preferred:

   ... as demonstrated by Ray[8].
   ... as demonstrated by Ray (8).

... as previously demonstrated (ref. 8).
... as previously demonstrated [8].
... as previously demonstrated[8].

For this system, the Reference list is presented in numerical order with the appropriate number before the first author's name. The disadvantage of this referencing system is that any reference included in the report at a late stage requires a renumbering of all subsequent references and the reference list.

In the second method, the reference may be cited by using the first author's name; any second author is also included, and *et al.* for three or more authors; the date and suitable designation is included if the author has published several items. The notation in the text may be:

... as supported by Ray, ...

or

... as supported by Ray and Jones (1986) ...

or

... as supported by Ray *et al.* (1985, ref. B) ...

An alphabetical reference list is prepared and included. For the examples given above, the list would contain references by Ray and Jones in 1986 and other years, or by Ray *et al.* with two references (at least) in 1985, designated ref. A and ref. B in the list. The disadvantage of alphabetical referencing is the extra 'bulk' required in the text compared with use of a number system. However, inclusion of an additional references does not disrupt the numbering system (although the additional reference must be inserted in the list). The choice of referencing method may be a personal decision, or it may be specified by the company funding the work.

(K) *Appendices*
The Appendices contain all the supportive information that is associated with a project, but which is not essential or appropriate for inclusion in the main body of the report. The following information may be included in Appendices:

  (i) Useful data found in the literature but *not* readily available.
 (ii) Sample calculations; a once-through detailed demonstration of all calculations relevant to the report.
(iii) Manufacturers' literature, data sheets or instrument specifications.
 (iv) Copies of computer programs and print-outs.
  (v) All experimental data and some calculated data, usually in tabular form.
 (vi) The project directive.
(vii) Copies of any relevant personal communications.
(viii) Relevant standards, codes of practice, detailed drawings, etc.

In some cases it may be preferable to include the appendices as a separate volume of the main report, especially if they contain a large amount of information.

10.3.7 The short-form report

Whereas the formal report provides the full documentation for a project, a short-form report aims to provide specific but less detailed information. This type of report is particularly useful for providing a progress report for a project, and for communication with a reader who has some knowledge of the work. The Short Report is a technical report that conforms to the requirements previously discussed

(Secs 10.3.2 to 10.3.6). It should not be referred to as an Informal Report, as this title best describes the Memorandum-type of communication (see Sec. 10.3.8).

The typical contents and presentation of a Short-form Report are as follows:

Title Page and Abstract (or Summary)—combined on one page
Equipment and Experimental Procedure—include only if necessary, i.e. detailing new and original developments
Results and Discussion—Tabulate data and the calculated results; use graphical presentations; concise and relevant discussion
Conclusions and Recommendations—all reports have at least one conclusion!

Other sections may be included if they are relevant and necessary. Individual sections of a Short Report should conform to the same requirements as described in Sec. 10.3.6 for the Formal Report.

The length of different types of reports depends upon the subject under consideration. However, formal engineering reports often contain more than 25 pages and short-form reports less than this number. In certain situations, a short-form report may be adequately presented in 5 to 10 pages. The short report should only include information that is immediately relevant, and this depends upon the level of knowledge of the reader.

10.3.8 The memorandum

The memorandum is an example of an informal report; other examples include progress notes, letters, estimated project cost statements, survey results sheets, etc. An informal report is used to present results without including detailed information. Companies often produce standardized forms to be used for recording information related to particular activities, e.g. time cards for recording the number of hours spent on particular projects during a week.

A memorandum (often abbreviated to 'memo') provides specific information; it is addressed to someone familiar with the work so that detailed explanations can be omitted. A memorandum may be required each week outlining the progress made in a particular project; a collection of such reports may form the basis of an interim report. These memoranda can also provide the basis for the formal report to be written at the conclusion of a project. A memorandum is assigned a title (or a subject heading) and a file number for information storage. Sub-headings are rarely used within a memorandum and the information is presented as a carefully structured document. Paragraphs are used to identify the different aspects of be communicated; each paragraph usually presents one aspect of the report.

A memorandum is rarely longer than two typewritten pages, and is often limited to one page. The importance of the 'mini-memo' was discussed in the introduction to Sec. 10.3. The memorandum is probably the type of report most frequently used by engineers, both for requesting and for communicating information. Busy executives sometimes make important decisions based upon the information presented in an informal report. It is worth spending some time in the careful planning and presentation of such reports, since their importance and usefulness often far exceeds that indicated by their length.

EXERCISES

10.1 Read through some of the written work you have prepared over the last 2 years. Try to identify the words that you often misspell, and common grammatical errors.

Assess this written work critically in view of the suggestions made in this chapter.

Determine whether the written work was appropriate for the communication required and the nature of the information.

10.2 Ask a colleague to assess critically the written work in Exercise 10.1. Compare and discuss your findings.

10.3 Evaluate examples of the following means of communication based upon the recommendations in this chapter. If possible compare your own work with an independent example.
   (a) A formal report.
   (b) A short report.
   (c) A memorandum.
   (d) An oral presentation.
   (e) A laboratory report.
   (f) A tutorial problem.
   (g) A newspaper article.
   (h) An article in an engineering journal.
   (i) A television documentary.
   (j) A radio documentary.
   (k) Questions in Parliament.
   (l) A government application form, e.g. for a passport or driving licence.

10.4 MARKETING, ADVERTISING, SALES AND SERVICING

There are two other categories of written communication that concern the engineer. These are aspects of communications that are intended to induce sales of a product and those that are related to the after-sales services. Although the presentation of these types of communications is usually undertaken by people with specialist knowledge, the engineer who has been responsible for the development of a project often becomes involved in their preparation. The engineer may prepare the draft for a servicing manual or offer advice regarding the technical accuracy of a marketing or advertising brochure; however the written material is usually prepared by a specialist in those types of communications. The engineer may then be required to check the technical aspects of the final presentation.

In order to achieve sales of a product, a company will require expertise in marketing, advertising and sales. These may be provided by departments within the company or by specialist organizations. All three aspects involve communication, in this case between the company and prospective customers.

The marketing department should identify who will buy the product, at what price, and what features of the final article are considered positive or negative. Market research and the marketing of a product are specialized functions, and some consideration of these aspects of a project should be included as early as possible, and later as a continuous activity.

Marketing involves communication with the customers and with the project team, thus ensuring that the customer will want what is produced, and that it is possible to produce the required item. It will be necessary to print introductory leaflets, sales brochures, technical sales literature, press releases and technical articles. As with all forms of communication it is essential to produce written information that is appropriate to the particular reader, and to ensure that this person receives the information.

Advertising involves selecting the best method of making the customer aware that the product exists, and also inducing the customer to buy this particular product rather than an alternative, or none at all! Society is now bombarded by advertising from many sources, including television, radio, newspapers, journals, handouts, hoardings, stickers, word of mouth, packaging, etc. The job of the advertising department is to select the most effective type of advertising medium, to decide the nature of the message to be conveyed, and the format and promotional campaign.

Finally, the sales department should be kept adequately informed of all relevant details of the product, so that potential customers can be persuaded to buy it as soon as it becomes availabe. It is important to brief the sales department thoroughly so that enquiries relating to non-standardized uses of the product can be answered efficiently and accurately. There should always be liaison (feedback and feed-forward) between sales, marketing and engineering personnel, so that the advantages of the product can be exploited and any disadvantages are minimized or quickly eliminated.

The after-sales services for a product are concerned with packaging, installation, mainten-ance and servicing. Communications in these areas are mainly written instructions and, as previously emphasized, they should be prepared for a particular reader. The instructions should be clear, simple and unambiguous, but also as brief as possible and yet adequate for most situations. They should clearly state what is allowable and forbidden, and any emergency procedures.

The preparation of maintenance and servicing information will depend upon whether these tasks are to be performed by the supplier (specialist personnel) or the customer (general expertise). Some instructions may be oral, either from a supervisor to a workman or by telephone between the supplier and the customer.

10.5 MODELS AND PROTOTYPES

The construction of models and prototypes is part of the design and development stage of a project. Model-making is really an extension of the work of the drawing office, and building a prototype is a part of the project engineering function. However, both of these representations of the final product provide a useful form of visual communication which should be exploited to its full advantage.

Models can vary from simple glued cardboard components to accurately machined scale models that include many of the features of the final (envisaged) product. Models may provide only size and layout representation, e.g. a housing estate model or a chemical plant model to assist in piping layout. Alternatively, models may have moving parts, e.g. opening doors on a model car, or they may be illustrative working models, e.g. water flowing through a model of a dam.

A prototype is usually a full-scale, fully operational unit built to conform to the detailed design specifications. It is used to test the design data and any assumptions that were made, to obtain experimental results, to provide an accurate product specification and to test the operation of the product to determine any unsatisfactory features. It can be used to provide an accurate visual representation of the final product which will be useful for the designer, the production engineer and the potential customer. It will also enable the marketing and advertising departments to obtain realistic photographs of the product for use in their promotional literature.

10.6 ORAL AND VISUAL COMMUNICATION

There are many visual aids that can be used to supplement an oral presentation (see Chapter 9). In some cases sound may be an integral part of the visual material, e.g. a video film, or it may be provided separately by the speaker.

There are also many audio-visual aids available, therefore it will be important to ascertain the costs of alternative systems. The following list is not exhaustive but will provide the reader with an appreciation of what is available:

(a) overhead projector transparencies (acetate sheets) and felt-point pens;
(b) slides (35 mm);
(c) slides (35 mm) and tape commentary;
(d) boards for magnetic and stick-on materials;
(e) cutaway model sections;
(f) visual-cast reflection projector (epidiascope);
(g) film strip projector, with or without sound;
(h) video films;
(i) closed-circuit television.

It will be necessary, before deciding upon a particular form of audio-visual presentation, to determine the preparation time and the facilities required for its use.

10.7 SUMMARY

Having studied this chapter, there are two important points that the reader should appreciate. First, prepare material for a specific audience and decide upon the most efficient and effective presentation. Second, prepare audio and visual material *thoroughly*.

Finally some rules for effective communication. Hopefully the reader will already have thought of these:

(a) Decide the purpose of the communication.
(b) Make an appropriate presentation.
(c) Make the message short, sharp and to the point.
(d) Continually re-evaluate your communications.
(e) Practise listening skills.

Evaluate the effectiveness of the communications that are used both in your work and non-work situations.

EXERCISES

The information presented in this chapter is intended to provide some guidelines for the effective presentation of oral and written reports. The most useful exercises that can be performed are:

(a) prepare an oral report;
(b) prepare different types of written reports;
(c) present these reports (a and b) to an audience;
(d) obtain critical assessment from a colleague;
(e) re-evaluate the reports (a and b) and learn from previous mistakes.

It will be more beneficial to prepare reports for your own work, e.g. experimental studies, rather than for textbook-derived situations.

Note: When preparing technical reports, aim for clarity and honesty (reveal rather than conceal)!

REFERENCES

1. Peters, T. J. and R. H. Waterman Jr., *In Search of Excellence* (*Lessons from America's Best Run Companies*), Harper and Row, New York, 1981.
   (One of these lesssons was the use of short concise memos.)
2. Rosenstein, A. B., R. R. Rathbone and W. F. Schneerer, *Engineering Communications*, Prentice-Hall, Englewood Cliffs, New Jersey, 1964.
3. Woodson, T. T., *Introduction to Engineering Design*, McGraw-Hill, New York, 1966.
   (See Chapter 17, page 321, 'Communication and the Engineer', and also Chapter 18, page 347, 'Oral Reporting'.)
4. Vidosic, J. P., *Elements of Design Engineering*, Ronald Press Co., New York, 1960.
5. *Roget's Thesaurus* is available in several versions, for example:
   Browning, D. C. (ed.), *Everyman's Thesaurus of English Words and Phrases*, Pan, London, 1972.
   Chapman, R. L. (ed.), *Roget's International Thesaurus*, 4th edn, Harper and Row, New York, 1984.
   Dutch, R. A. (ed.), *Roget's Thesaurus of English Words and Phrases*, Longman, London, 1962.
   Roget, P. M., *Roget's Thesaurus of Synonyms and Antonyms*, Borden Publishing Co., Alhambra, CA, 1972.
   Roget, P. M., *Roget's Thesaurus of English Words and Phrases*, St Martin's Press, New York, 1978.
   Roget, S. M. (ed.), *Roget's Thesaurus of Synonyms and Antonyms*, Maxi Books, Sydney, 1980.
6. Bragg, G. M., *Principles of Experimentation and Measurement*, Prentice-Hall, Englewood Cliffs, New Jersey, 1974.
   (As title, including reporting results and report writing (with examples of reports).)
7. Ulrich, G. D., *A Guide to Chemical Engineering Process Design and Economics*, Wiley, New York, 1984.

BIBLIOGRAPHY

Anderson, J., B. H. Durston and M. Poole, *Thesis and Assignment Writing*, Wiley, New York, 1970.
Anderson, W. S. and D. R. Cox, *The Technical Reader*, Holt, Rinehart and Winston, New York, 1980.
Bailey, R. F., *A Survival Kit for Writing English*, Longman, Hawthorn, Australia, 1976.
Barnett, M. T., *Elements of Technical Writing*, Delmar Publishing Co., New York, 1974.
Barrass, R., *Scientists Must Write, A Guide to Better Writing for Scientists, Engineers and Students*, Chapman and Hall, London, 1978.
   (The title says it all; easy reading, highly recommended.)
Beard, R. M., *Research into Teaching Methods in Higher Education Mainly in British Universities*, 2nd edn, Society for Research into Higher Education, London, 1968.
Beard, R. M., *Course Design, Teaching Methods and Departmental Decisions*, University of London, Institute of Education, 1974.
Beard, R. M., *Teaching and Learning in Higher Education*, 3rd edn, Penguin, London, 1979.
   (All three publications by Beard are concerned with 'effective' teaching methods and contain much practical advice concerning all forms of communication.)
Brichta, A. M. and P. E. M. Sharp, *From Project to Production*, Pergamon, Oxford, 1970.
   (See Chapter 10, pages 250–261, 'Models and Prototypes'; also Chapter 11, pages 274–282, concerning literature associated with a product.)
Brinkworth, B. J., *An Introduction to Experimentation*, 2nd edn, The English Universities Press, London, 1973.
   (As title, including presentation and analysis of results (Chapter 6) and reporting the work (Chapter 7).)
Campbell, W. G., S. V. Ballou and C. Slade, *Form and Style: Theses, Reports, Term Papers*, 6th edn, Houghton Mifflin, Massachusetts, 1982.
Chessell, P. and H. Birnstihl, *Essay Writing: A Guide*, Sorrett Publishing, Malvern, Australia, 1976.
Coleman, P. and K. Brambleby, *The Technologist as Writer: An Introduction to Technical Writing*, McGraw-Hill, New York, 1969.
Damhurst, W. A., *Clear Technical Reports*, Harcourt Brace Jovanovich, New York, 1972.
Garvey, W. D., *Communication: The Essence of Science*, Pergamon, Oxford, 1979.
   (Sub-title: 'Facilitating information exchange among librarians, scientists, engineers and students'—good as background reading.)
Geiger, G. H., *Supplementary Reading in Engineering Design*, McGraw-Hill, New York, 1975.
Gilbert, M. B., *Clear Writing: A Business Guide*, Wiley, New York, 1983.
Glorfeld, L. E., D. A. Lauerman and N. C. Stageberg, *A Concise Guide for Writers*, 4th edn, Holt, Rinehart and Winston, New York, 1977.

Hall, W. C. and R. Cannon, *University Teaching*, Advisory Centre for University Education, University of Adelaide, Adelaide, South Australia, 1975.
   (Good advice and analysis of communication using lectures, tutorials and audio-visual aids.)
Herbert, A. J., *The Structure of Technical English*, Longman, London, 1977.
   (Intended to teach language not engineering, but uses engineering examples. A workbook or programmed text on how to write.)
Holtz, H. and T. Schmidt, *The Winning Proposal: How to Write It*, McGraw-Hill, New York, 1981.
Houp, K. W. and T. E. Pearsall, *Reporting Technical Information*, 5th edn, Glencoe Pub. Co., New York, 1984.
Johnson, K., *Communicate in Writing*, Longman, London, 1981.
King, L. S., *Why not Say it Clearly? A Guide to Scientific Writing*, Little, Brown and Co., Boston, MA, 1978.
Leggett, G., C. D. Mead, W. Charvat and R. S. Beal, *Handbook for Writers*, 8th edn, Prentice-Hall, Englewood Cliffs, New Jersey, 1982.
Lester, J. D., *Writing Research Papers*, 2nd edn, Scott, Foresman and Co., Glenview, IL, 1976.
Lyerly, R. H., *Essential Requirements for the College Research Paper*, Harper and Row, New York, 1966.
Markel, M. H., *Technical Writing, Situations and Strategies*, St Martin's Press, New York, 1984.
Mathes, J. C. and D. W. Stephenson, *Designing Technical Reports: Writing for Audiences in Organisations*, Bobbs–Merrill, New York, 1976.
Michaelson, H. B., *How to Write and Publish Engineering Papers and Reports*, 2nd edn, ISI Press, Philadelphia, 1986.
Mills, H., *Commanding Paragraphs*, Scott, Foresman and Co., Glenview, IL, 1977.
Mitchell, J., *How to Write Reports*, Collins, London, 1974.
Mitchell, J., *How to Write Reports*, Collins, 1974.
Scharf, B., *Engineering and its Language*, Frederick Muller, London, 1971.
   (A reference book describing terms used in engineering. Headings such as 'plastics' and 'mechanical testing'.)
Sherman, T. A. and S. S. Johnson, *Modern Technical Writing*, 3rd edn, Prentice-Hall, Englewood Cliffs, New Jersey, 1975.
Sides, C. H., *How to Write Papers and Reports about Computer Technology*, ISI Press, Philadelphia, 1984.
Sutton, B., *Clear Thinking: Exercises in English Expression*, Angus and Robertson, Australia, 1970.
Ulman, J. N. Jr and J. R. Gould, *Technical Reporting*, 3rd edn, Holt, Rinehart and Winston, New York, 1972.
   (A comprehensive text dealing with written communications.)
Weisman, H. M., *Basic Technical Writing*, 5th edn, Charles E. Merrill Publishing Co., Columbus, Ohio, 1985.
Weiss, E. H., *The Writing System for Engineers and Scientists*, Prentice-Hall, Englewood Cliffs, New Jersey, 1982.
Woelfle, R. M. (ed.), *A Guide for Better Technical Presentations*, Institute of Electrical and Electronics Engineers, New York, 1975.
Woodford, F. P. (ed.), *Scientific Writing for Graduate Students. A Manual on the Teaching of Scientific Writing*, Rockefeller University Press, New York, 1968.
   (Recommended, each to read, good notes on presentation of tables and figures, and also oral presentations.)
Woodson, T. T., *Effective Communication for Engineers*, McGraw-Hill, New York, 1975.
Yaggy, E., *How to Write Your Term Paper*, 4th edn, Harper and Row, New York, 1980.
   (Recommended—a student's manual.)

JOURNAL ARTICLES

Most of the following articles were published in *Chemical Engineering* journal. They are general articles concerning report writing and communication, and they are not specific to the chemical engineering profession. Other disciplines such as mechanical and electrical engineering have their own particular journals, and similar articles are often published.

Effective Communication for Engineers, volume of selected papers from *Chemical Engineering* journal, McGraw-Hill, New York, 1974.
Fair, J. R., *Diction and the Engineer*, Chem. Eng., vol. 76, no. 14, 1969, pages 114–117.
Hine, E. A., *Write in Style: Be Clear and Concise*, Chem. Eng., vol. 82, no. 27, 1975, pages 41–45.
Hissong, D. W., *Write and Present Persuasive Reports*, Chem. Eng., vol. 84, no. 14, 1977, pages 131–134.
Hughson, R. V., *Writing for Publication: Main Road to Recognition*, Chem. Eng., vol. 82, no. 27, 1975, pages 49–52.
Johnson, T. P., *Fast Functional Writing*, Chem. Eng., vol. 76, no. 14, 1969, pages 105–110.
Marbach, M. G., *Better Visuals Make Speeches Better*, Chem. Eng., vol. 84, no. 6, 1977, pages 141–144.
Michaelson, H. B., *Strategic Choices for the Engineer Author*, Chem. Eng., vol. 93, no. 15, 1986, pages 50–60.

Mintz, H. K., *How to Write Better Memos, Chem. Eng.*, vol. 77, no. 2, 1970, pages 136–139.

Popper, J., *Six Guidelines for Fast, Functional Writing, Chem. Eng.*, vol. 76, no. 14, 1969, pages 118–123.

Power, R. M., *Engineers Can't Write? Wrong! Hydrocarbon Processing*, vol. 52, no. 3, 1973, pages 105–110.

Quackenbos, H. M., *Creative Report Writing, Chem. Eng.*, vol. 79, no. 15, 1972, pages 94–98; vol. 79, no. 16, 1972, pages 146–150.

Vinci, V., *Watch Your Words, Chem. Eng.*, vol. 78, no. 18, 1971, pages 112, 114.

Vinci, V., *Ten Report Writing Pitfalls: How to Avoid Them, Chem. Eng.*, vol. 82, no. 27, 1975, pages 45–48.

IN THE LABORATORY: DESIGNING AND PERFORMING EXPERIMENTS

SCOPE

Part Five considers aspects of engineering experimentation beyond the actual measurements, experimental methods and reporting of results. Chapter 11 discusses the practical considerations when undertaking a laboratory-type project, and the aspects of planning, organization and safety that should become apparent from such a study.

The philosophy behind a laboratory course is considered in Chapter 12, including some of the alternatives available and the changes that could be made. Some of the comments are more applicable to the lecturer, but it is useful for students also to consider these aspects of laboratory work. Students should learn from everything they do and everything that occurs around them.

PLANNING AND PERFORMING EXPERIMENTS

CHAPTER OBJECTIVES

To encourage the reader to consider:
1. the types of laboratory work that can be performed, the aims of an experimental study, and the constraints in terms of time and available information;
2. how the success of a practical study is influenced by particular aspects such as planning, organization and safety.

QUESTIONS

- What planning stages should precede the experimental work of a laboratory project?
- Which factors may inhibit the progress of a project, e.g. time, information, etc.?
- Why is experimental work performed?
- Which short-term and long-term plans are required for experimental studies?
- What aspects of laboratory organization influence the success of a project?
- What safety considerations are important for laboratory work?

11.1 GENERAL COMMENTS

Every engineering and science-based course includes a significant amount of time to be spent on laboratory and project studies. The student should progress from straightforward demonstration-type experiments in the early part of the course to 'open-ended' project work. The laboratory programme should be designed so that the experiments progressively require a greater contribution from the student in terms of deciding what is to be achieved and how this is to be done. This type of plan will enable the student to gain not only a better understanding of the taught material in a course, but also to acquire the decision-making skills and practical abilities that an engineer or scientist must possess. It is only when the student undertakes a laboratory project or design project that the 'knowledge' associated with engineering experimentation is transformed into valuable experience and expertise. Unfortunately the transition from 'standard' laboratory experiments to project work often occurs abruptly at the start of the final year of a course. This chapter is intended to provide some useful information and ideas for the student who has not been offered any detailed advice regarding the development of experimental work and the particular aspects to be considered.

Chapters 11 and 12 would also provide useful reading for the lecturer who is responsible not only for developing laboratory experiments but also for the supervision of laboratory project work. Sometimes it is erroneously assumed that the student has acquired the skills necessary for competent organization and direction of laboratory studies. Some instruction is usually desirable and these notes are intended to fulfil that function at an introductory level.

All experimental work requires varying degrees of planning and organization, and appropriate practical experience. Each experimental study has different requirements and different problems; one set of rules or instructions is not applicable in every situation. These notes present an overall view of laboratory work, the factors that will influence the successful completion of a project, and the decisions that may need to be made.

11.2 TYPES OF LABORATORY WORK

Before engaging in a laboratory investigation it is worth while reviewing the different types of experimental work, and the skills that may be acquired by performing each type.

(a) *Demonstrations* are generally performed using equipment designed for simple operation and easy measurements. Any modifications should be easily carried out and new readings obtained quickly. An appropriate time will need to be allowed for equipment having a complicated start-up procedure, or which takes a long time to establish steady-state conditions. However, this latter type of equipment can be used to investigate the possibility of using automatic control features, and for preparing improved start-up and operational procedures.

(b) *Non-quantitative investigations* involve the examination of equipment in order to gain an appreciation of the method of operation. For example, dismantling and reassembling valves and pumps, or observation of injected particles in a flow system. Quantitative results are not obtained.

(c) *Standard items of equipment* are used to introduce a particular subject area of a course, or to help reinforce material already covered in lectures.

(d) *Extended laboratory studies* require the development of standard or traditional laboratory experiments, often involving equipment modifications, in order to obtain more extensive results. The interpretation of results is presented in a more detailed manner and may necessitate further analysis. This is a step towards more open-ended investigations.

(e) *Laboratory projects* are open-ended investigations (within certain practical constraints) that allow the investigator to determine the direction of the work and the approach to be followed. No investigation is completely open; the major constraints are usually time and money.

These types of laboratory investigations ((a) to (e)) should be seen as a natural progression for the student to follow. Laboratory projects are usually carried out in the final year of a degree course and their success depends upon the experience gained in the other types of experimental work ((a) to (d) above). Sometimes the student is not adequately prepared during previous years of the course to accept the full responsibility and direction required for the successful completion of a project. These notes are intended to provide some useful guidance for these students.

11.3 THE AIMS OF LABORATORY INVESTIGATIONS

In industry, experimental work is performed in order to obtain information that is useful to the operation of a company. In teaching institutions, laboratory studies are performed so that the student:

(a) obtains a better understanding of the lecture material;
(b) acquires laboratory skills;
(c) develops investigative abilities;
(d) becomes aware of the planning and organizational aspects associated with practical investigations.

Upon conclusion of a project, the student should have gained an appreciation of the problems and pitfalls associated with practical work, and obtained some knowledge of the technical considerations and experimental techniques that determine the success of a project. It is important that a student learns from mistakes and problems. In industrial situations, reliable results are needed quickly and at the minimum cost (i.e. no mistakes!).

The student should be prepared (and encouraged) to question and assess all aspects of an investigation that is proposed by a supervisor, and also to make suggestions and modifications that seem appropriate. Even the specialist sometimes makes mistakes, or fails to see the best way of performing a task. The success of a project ultimately rests with the investigator (the student), who should therefore be interested and involved in the work and have some control over the decisions that are made.

The experimental work performed by a student should provide an opportunity to:

(a) obtain, evaluate and summarize necessary information;
(b) view the subject in a general or qualitative manner;
(c) gain technical and practical experience;
(d) test the performance of equipment;
(e) obtain experimental results;
(f) identify general trends;
(g) compare data with theoretical predictions, and with other available data;
(h) propose reasons for any disagreements with item (g), and if possible test the validity of these proposals.

11.4 TIME CONSTRAINTS

The successful completion of any laboratory investigation requires careful planning in relation to the time required to be spent, and allocation of the time that is available. Time is required not only to perform the experiments but also to carry out a wide range of associated activities, including background reading, performing calculations, report writing, etc. Often the actual experimental work takes less time than all the other tasks to be performed.

The time required to complete a project should be calculated as accurately as possible. If only a limited time is available, then the time to be spent on different aspects of the work must be determined. Appropriate time allocation depends very much upon having suitable experience in similar studies. The time required is generally underestimated by inexperienced investigators usually because they optimistically assume that 'what could be done, will be done'. The late delivery of equipment, lack of spare parts, unavailability of technical assistance, etc., all lead to delays. Personal illness, national disputes such as postal strikes, equipment breakdowns, etc., are setbacks that plague every investigator. It is prudent to make conservative estimates of the time required, i.e. allow extra time to account for unforeseen circumstances. The person awaiting the results of an investigation may prefer to receive this information sooner, but he will be less impressed by delays and excuses than by delivery of the results by an agreed date.

If the equipment required for a project already exists, problems may occur if it has not been used recently. Problems invariably occur whenever equipment is modified. The modifications themselves are a source of problems and time delays, however the operation of modified equipment rarely proceeds as intended—at least not without further testing and modification. If new equipment is to be assembled, delays in obtaining the equipment and unforeseen operating problems should be expected. Often the time that is spent getting equipment functioning correctly is longer than the subsequent time required to obtain the desired results.

The often quoted Murphy's law should be remembered: 'Whatever can go wrong, will go wrong', and the corollary that 'Murphy was an optimist'! However, experimental studies should not be viewed too pessimistically, otherwise the novice may feel disinclined even to begin. What is required is a realistic approach involving careful estimates, adequate preparation and an innovative approach. Problems can sometimes be overcome, or the effects of possible actions predicted, by using inventive and temporary arrangements.

The following suggestions may prove useful to the student considering a project-type experimental study for the first time.

(a) Spend some time deciding what needs to be done and what can be done; evaluate the available information; the decisions that are made should be reviewed several times. Obtain advice from colleagues who are knowledgeable in that subject area. Obtain the opinions of people who have *no* knowledge of the subject; they can sometimes make important observations or suggestions. All this preparation may take a considerable amount of time and effort; the time spent in these activities will depend upon how much time is available. However, time spent in preparation is rarely wasted and often proves essential for the success of the project.

(b) Make estimates of the time required to carry out each stage of the project, try to adhere to these estimates and review the progress made and the time taken at regular intervals. Do not consider that the preparation/research/evaluation stage (part (a) above) can be conducted 'as the project proceeds'; allocate a separate and definite time for this activity. Sometimes the student becomes overwhelmed by the information acquired and the decisions to be made, and it becomes difficult to proceed to the next stage. Decide the date when the development stage will begin—and adhere to this decision.

(c) Obtain time estimates for any work to be performed by technicians, contractors, etc., and for any equipment to be delivered and/or installed. Check at regular intervals that these agreements can be fulfilled.

(d) Keep a planner, i.e. diary or wallchart, to compare the estimates that are made for completion of different tasks with what is actually achieved.

(e) Determine from the supervisor the date when the project *must* be completed, and when it *should* be completed. Ask to be informed immediately regarding any changes in these dates, or any additional factors that may affect the work schedule.

(f) Allow additional time to get the equipment operating satisfactorily after it is assembled. Students often assume that equipment will operate as required as soon as it is built; this is rarely the case and 'de-bugging' can be a time-consuming activity.

(g) Allow sufficient time to repeat experimental results (if it is possible to reproduce the conditions) so that operating errors and random errors can be identified.

(h) When equipment is operating satisfactorily, it is sensible to complete a set of experimental results even if this means working late at night or over a weekend. Shutting down a successful operation is tempting fate; do as much work as possible while things are going well!

11.5 INFORMATION

Questions

What information is needed?
Where is the necessary information?

These questions summarize the two main tasks/problems facing the student responsible for an experimental study. Information will be needed at many stages of the investigation and there are numerous sources that could be used. The problems for the student are the specification of the information and determining its location.

11.5.1 Specification

The first task is to specify the information required. The information necessary to perform an investigation (successfully), and whether it can be obtained, depends mainly upon the nature of the experiment. However, it also depends upon which stage of the experiment is being performed, the student's awareness and knowledge, and the amount of forward planning that has been carried out.

To help the student decide exactly what information is needed, it is useful to consider the various stages that occur in an investigation. These stages usually include:

(a) background;
(b) theory and data;
(c) equipment;
(d) experimental/operation;
(e) results, i.e. recording data/calculation/presentation;
(f) evaluation of results;
(g) discussion of results;
(h) implications/evaluation of the investigation.

Some of these stages require specific information available only from particular sources, e.g. equipment details are usually available only from the supplier or manufacturer. The student begins by reading and collecting background information which is generally oriented towards the goal of designing and building the equipment necessary to perform the practical work. The conclusion of the project may also require background reading in order to assess the implications and advantages of the work from a social, political and legal viewpoint. The student should try to evaluate the work performed in terms of its application, modification and possible future uses.

The student must plan ahead and obtain and assess relevant information in advance. The theory to be used and tested must be studied before the results are obtained, in order to ascertain exactly which values need to be recorded and how the data can be analysed. Similarly, the equipment literature and experimental plan are required before the practical work is begun.

11.5.2 Location

Having identified the information required to perform the investigation, it is necessary to discover where that information is available. Certain aspects of the work require reference to specific sources of information, other aspects are documented in many different sources. Table 11.1 provides details of the more readily available sources of information, and also the particular types of information provided by each source.

The sources in Table 11.1 are not listed in order of importance, and item 9 is probably the most important for the student. This source is readily available and provides specific relevant information. The university library is an easily accessible source of information but even if the student is familiar with the classification system, the catalogues, and the material and services available, specific information is not always instantly obtainable.

The information available immediately is the project description, specification or instruction sheet prepared by the supervisor. This specification may require modification as the project

Table 11.1 Sources and types of information

(From Ray, *Elements of Engineering Design*, Prentice-Hall International (UK) Ltd, 1985)

Information source	Type of information
1. Libraries: public, company, university, government, private	Books, pamphlets, journals, newspapers, magazines, standards (BS, ASTM, etc.); reference sources, e.g. *Engineering Index, Citation Index, Engineering Abstracts*
2. British Lending Library, Boston Spa, W Yorkshire, UK or Library of Congress, 10 First Street, SE, Washington DC 20402, USA	*All* information in 1
3. The Patent Office, 25 Southampton Buildings, London, UK or US Patent and Trade Mark Office, Box 9, Washington DC 20231, USA	Source of patents and related information, and may have a technical library
4. Government	Publications and reports of government departments, e.g. Trade, Industry, Agriculture, and of government agencies, e.g. (in the UK) the Statistical Office, National Physical Laboratory, The Design Council; (in the USA) the National Science Foundation, National Research Council, National Academy of Engineering, NASA (all in Washington DC). Government research centres. Foreign governments' embassies or consultants
5. Universities and various professional institutions	Various. Can approach departments directly
6. Private consultants	Private/confidential information
7. Private companies	Advertising and product literature, technical data sheets, technical advice and quotations
8. International organizations	Publications (reports) of UNESCO, WHO, IAEA, etc.
9. Personal experimental study, previous reports, specification from supervisor, personal discussions	Specific

proceeds, or as various practical problems become apparent. However, the specification should not be altered without consultation between the investigator and the supervisor, and all changes must be agreed between both parties. The specification may consist of a few sentences describing the *possible* objectives of the work, with the intention that a detailed specification is approved after initial investigation of the proposal. Alternatively, there may be a full and detailed instruction sheet describing what is to be achieved. How this specification is to be achieved (within certain time and cost constraints) may be left to the investigator.

Other useful information includes reports describing similar or associated experimental studies, technical information regarding equipment and materials available from manufacturers, and specific company information that may be provided in response to personal requests (and sometimes considered confidential and not for publication). The obvious and most frequently used sources of information are textbooks, journals, conference proceedings, patents, standards, government reports, etc. (i.e. published information that is generally available).

The library may provide either the information itself or details of the location of the information. Whenever books or papers are requested from another library or organization, two factors need to be considered. These are the cost of the material, e.g. a loan charge or photocopying cost, and the length of time the information will take to arrive. If a large number of references are required, photocopying and postage charges (especially from overseas) can consume a significant portion of the budget for a project. The provisions of the Copyright Act need to be observed when obtaining photocopies of information. Although it is usually possible to identify where a particular article is available, it may take considerable time before the material is received.

It is now possible to perform literature searches using an on-line computer search. The advantage of using such a system in terms of the time that can be saved is immense, and also a wider and more detailed search can be performed than would be possible manually. However, there are also several possible disadvantages. First, a computer search will only produce useful references if adequate preparation is carried out to determine the 'key words' that describe the subject area under investigation. Identifying the relevant key words or phrases requires an understanding of the subject and a knowledge of the associated terminology. The assistance of a librarian experienced in performing computer literature searches is invaluable.

The second disadvantage is the number of references the computer may identify that are associated with a particular key word. In some cases there may be thousands! It would impractical to obtain, file, cross-reference and store such a large number of references, without even considering reading them and obtaining useful information. The task is therefore to reduce the number of possible references to a manageable number while only including those that are directly relevant to the project, and also to avoid losing valuable information. This can be achieved by using a general key word which generates a large number of references, and then reducing this number by taking sub-sets of references associated with more specific key words. This 'reduction' process requires knowledge of the subject and the advice of an experienced librarian. It is unlikely that more than 50 references would be manageable (and therefore useful), although more extensive lists may be useful if the scope of the project is likely to be extended and several investigators become involved.

The third disadvantage is access to the references that the computer identifies. Usually only 10 references are obtained immediately from an on-line printer because of the cost involved; a full list is then obtained from the database location, preferably by airmail. If you perform a manual literature search in your nearest library, the material is available immediately. However, when the print-out from the computer search arrives, it may still be necessary to order and wait for the articles that are published in specialist publications. The print-out from a computer search usually contains the abstract from a journal article and this should provide some assistance in determining which references are likely to be of particular importance.

Finally, titles and abstracts are often misleading and references may appear more important than they are in practice. For example, an article may only contain the development of a mathematical model whereas what may be wanted for a project is details of other experimental studies. The title or abstract may not identify this problem. It is unwise to delay work on a project until particular references are obtained; their importance is often overestimated and their substance is often a disappointment. This comment is particularly true of patents, and articles published in foreign languages. Although patents contain a large amount of important information (and they are often underutilized by academic researchers), they rarely contain information that is essential for an investigation to proceed. The main problem for an article published in a foreign language is the extremely high cost of obtaining a translation, assuming that a competent technical translator can be found. The time and cost involved usually mean that a translation of the abstract, the conclusions and relevant tables of data are the best that can be achieved (i.e. afforded). This limited information is often sufficient to determine whether a full translation is necessary.

In summary, the experimenter must determine a course of action based upon decisions related to the information that is required and where the information can be found. Information that is not easily accessible should be requested as soon as possible but the project should not be delayed except for the acquisition of *essential* data. There is a wide range of available information including company literature, patents, manufacturers' operating manuals, foreign publications, etc. The usefulness of each type of information can vary considerably.

11.6 EXPERIMENTATION

Questions

Why is it necessary to perform experiments?
What specific aims are the experiments intended to achieve?
What approaches and methods will help to achieve the desired aims?

11.6.1 Why experiment?

A simple concise answer is in order to learn and to progress, and more specifically to obtain experimental data to be used as the 'raw material' of engineering designs. Unless ideas are tried and tested, their true value and usefulness will not be known. There are several situations in which it is deemed necessary to carry out experiments. First, in order to test:

(a) the validity of a theoretical analysis;
(b) a set of proposals or ideas that are combined to form a design concept;
(c) an idea that is the result of deduction or intuition.

It could be argued that all knowledge is based upon experimental study. Whether this is absolutely true is immaterial; what is obvious is that rapid advances in science and technology depend upon practical investigations. In industrial situations it is not feasible to produce and use (or sell) an item based upon, a 'build–use–modify' strategy. An item that is produced based only upon calculations, good ideas and optimism will not be a success; reliable experimental data and rigorous practical testing are also required.

Second, experiments are performed to obtain basic knowledge. These experiments may determine the physical, chemical and mechanical properties, e.g. thermal conductivity, reaction with acids and the hardness of a material respectively, and other data for particular materials. This type of data provides a basis on which calculations and decisions can be made.

Finally, experiments help to find the answers to problems that cannot be solved (easily) using a theoretical or mathematical approach.

11.6.2 Aims

The first stage in any investigation based upon experimentation is to identify the specific aims to be achieved by the practical work. These should include not only the immediate aims but also any long-term objectives of the overall project. Some speculation/estimation is involved regarding future events; if this is performed at the start of a project it may prevent wasted time, effort and money later.

An experiment may be conducted to satisfy a single objective or it may incorporate many different aspects. However, an experiment is usually conducted for one of two reasons: either to test or to investigate. Testing is performed in order to verify or disprove an idea. An investigative experiment will be initiated with some preconceived ideas about the possible outcome. These ideas may be a disadvantage, both in terms of the approach taken to the work and in the evaluation and interpretation of the findings. Experiments that produce inconclusive results are usually considered to be failures, although it can be useful to discover that an experiment did not yield particular (expected) results.

The experimenter should always remember that experimental results never prove or disprove an idea absolutely. The results indicate that under a certain set of conditions various phenomena will probably be observed. There are many variables that affect an experiment, some obvious and some hidden. Our interpretations and conclusions are based upon what is seen and recorded, both of which are subject to extraneous influences, alternative interpretations and errors.

11.6.3 Assessment of results

Whatever type of experiment is performed and whatever its aims, the results must satisfy three basic and fundamental conditions. The results must be:

(a) accurate;
(b) reliable;
(c) reproducible.

The accuracy of the final results depends upon any errors in the calculations, and also upon the measurements recorded during the experiment. The qualities of measurements are discussed in Chapter 2. A wide variety of measuring instruments is available and the correct choice is often not immediately obvious. Fortunately, specialized help is readily available from would-be salesmen. However, it will be a good idea to ascertain exactly what is required and obtain some knowledge regarding the performance of the options available, *before* consulting particular manufacturers (see Sec. 2.2).

To summarize the ideas presented in Sec. 2.2, the experimenter must decide the level of accuracy and the precision required from a measuring instrument. The accuracy is the closeness of the readings to the actual input. The precision of an instrument is its ability to produce the same reading each time it receives the same input value. Other features are important when selecting an instrument, such as sensitivity, easy calibration, easy scale reading, whether portable or fixed, etc. It is important to decide the type of readings that need to be recorded, e.g. steady operation (set values) or batch-type operations (where measurements are a function of time); also whether these requirements are likely to change during the course of an investigation.

The results must be reliable, and this depends upon the accuracy of the recorded values and is influenced by the planning that has preceded the actual experimental work. Reliable results are dependable and trustworthy, at least according to the dictionary! Sometimes results are obtained which the experimenter knows, from intuition and experience, are 'just not right', often 'too good to be true'! If the results are accurate and reproducible, then any errors are due to particular features of the equipment or its operation. It is necessary to identify the cause of an error (if it actually exists), otherwise difficulties may arise when the results are used in a different situation.

Readings should always be repeated (if possible), to ensure that they are reproducible. Results should also be obtained over a range of different operating conditions to check that the values which would be expected, i.e. predicted, are actually obtained.

Results and readings should always be checked, repeated and analysed to ensure that they conform to the requirements of accuracy and reliability, and to show that they are also reproducible.

11.6.4 Planning the experimental programme

When preparing a plan for an experimental study, the question to be asked and answered is:

How can it be ensured that what is obtained is what is required and needed?

The way to do this is to spend sufficient time planning the entire investigation, and in particular the experimental work. Making alterations to a scheme once the work has begun may be expensive and time-consuming, and will not always be easy or possible. In order to plan and carry out the experimental work efficiently, it is necessary to specify clearly the intended scope of the investigation, and the questions to be answered. It is necessary to understand and analyse the problems in terms of the verification to be made, the properties to be measured and the physical principles involved. From this analysis it is possible to decide upon the experimental approach, the variables to be measured and/or controlled, and the instrumentation required.

The experimenter must ascertain the amount of data and the particular readings required. Associated with this task is the necessity to decide the range of values for each variable that is to be investigated and recorded. This decision determines the quantity of data recorded. The class intervals chosen for incremental changes in the recorded values are important as they also affect the volume of data produced and its usefulness. If the class intervals are too small, then large quantities of (unnecessary) data are produced. However, if the intervals are too large then certain trends and effects may not become apparent.

As discussed in Sec. 11.6.3, the recording instruments must possess the required accuracy and provide the readings in an appropriate form. It may be possible to obtain a permanent automatic readout, or the results may be fed directly to a computer for storage and processing, etc. If these facilities are not available then some time should be spent planning the data recording stage of the experiment, i.e. the data record sheets, time required for data recording, etc.

The activities outlined above that make up the planning stage of an experiment do not necessarily occur in a pre-determined sequential order, rather they form an interdependent process. For example, one objective at the outset of a project may be to record a particular range of property values. However, if the necessary instrumentation is too expensive this objective may have to be modified. Alternatively, if the data processing time is found to be too long, then one solution would be to purchase the necessary automatic data processing and recording equipment.

This may require the use of funds previously allocated to a different aspect of the project, and further modification and compromise is required. The whole process of planning and decision-making requires some flexibility, and all the alternatives available should be determined. As in most engineering situations, the final choice will involve 'trade-offs' between the alternatives, and ultimately a compromise will have to be accepted.

11.6.5 Constraints

There are two significant factors acting as constraints on the progress of a project and the results that are obtained: these are time and economics.

Time constraints occur at three stages of a project:

(a) planning, design and implementation stage;
(b) actual experimental work;
(c) recording, computation and analysis of results.

As early as possible the experimenter should prepare detailed time schedules for each of these stages. Any subsequent revisions should be recorded, and their implications assessed. Whenever possible, accurate estimates of the time required for an activity should be obtained, e.g. equipment delivery, instrument calibration and testing, availability of shared computer facilities, etc.

Economic constraints are always present and influence all aspects of the work. The two main considerations are fixed costs such as equipment purchase and long-term computer leasing, and variable costs which are time- or quantity-dependent. Examples of variable costs are wages, electricity, materials used, maintenance, etc. There are more uncertainties related to variable costs, and the ranges (high and low) of these cost estimates should be determined, and any changes monitored at regular intervals.

11.6.6 Checking

Checking should be performed at all stages of a project but it is especially important for the experimental stage. Reliable results are not necessarily obtained just because the experimental procedures are well planned or the instruments are capable of providing accurate readings, although both these factors will help! There are many instances where mistakes or errors can be made, but errors in the theoretical development or the design calculations are not always easy to find. However, reliable decisions have to be made and checking is an important task to be performed.

Checking must be performed for all mathematical aspects of the work; this includes the arithmetic, algebraic and analytical features. Arithmetic or computational checks have become less time-consuming now that personal computers and microprocessors are more widely used in experimental work. However, it is necessary to ensure that the computer has been told to do what is *actually* required. The types of checking that are carried out are:

(a) repeating checks, i.e. following the same procedure;
(b) reversed checks, i.e. in the opposite order of operation;
(c) verification using an alternative method of solution.

Any assumptions made in the development of the experimental programme should be verified, whether for the theory to be applied or a scheme of work. The experimenter should identify at the outset the assumptions that are made, the range of values over which they apply and also whether they are necessary. When results are obtained, the validity and effects of the assumptions have to be determined.

Finally, verification should be made by applying common sense, or what could probably be referred to more appropriately as engineering sense. This means that the solution to the problem (or the results obtained) are practical, i.e. they could be implemented, and they are consistent with experience. There are several ways of performing this type of check, including dimensional consistency, observing trends and limit checks (see Sec. 6.1.2).

11.6.7 Estimation

Students often have only a vague idea of the difference between estimation and guessing. The jargon phrase of 'guestimation' has even been introduced. However, ignoring this latter irrelevant popularization (by definition only a guess or an estimate can be made), the engineer should only ever practise one of these activities, and that is estimation.

Guessing is an emotionally directed speculation, it has no basis in experience. However, estimation is based upon associated knowledge, past experience and careful consideration/speculation. The engineer makes estimates every working day, and remembers that they are estimates and not certainties. The engineer should *never* use guessing in professional work, if success is to be attained!

Students sometimes think that because experimental work is investigative, and therefore the end result is never certain, estimating and guessing are acceptable means of decision-making. For the reasons stated above only estimation should be employed, and only if it is based on sound experience and knowledge. It is only the end result in an experimental study that is uncertain; the detailed planning and the procedures to be used are based on established ideas and data.

11.7 LABORATORY PLANNING, ORGANIZATION AND SAFETY

Questions

Do short-term and long-term plans exist for the laboratory equipment and the work to be performed?

Are the needs of the investigator considered in any plans?

Is the laboratory organized efficiently and effectively?

Does the laboratory organization consider both the facilities and the persons carrying out the experimental work?

What aspects of safety apply to the use of the laboratory?

Is safety continually monitored and assessed?

11.7.1 Comments

The undergraduate student is unlikely to become involved in aspects of laboratory development. However, responsibility often comes quickly after graduation and the young engineer in industry may soon be in charge of a laboratory or the development of new facilities. Project-type laboratory studies can provide some useful experience for the student in terms of laboratory organization and development.

Having designed the apparatus and planned the experimental study, the location and operation of the equipment should be determined *before* the equipment is assembled, and in relation to existing facilities. The student will probably have little influence on the actual siting of equipment, however considering the relevant factors can be a useful exercise. Subsequent operation usually provides some valuable ideas regarding laboratory organization, and the opportunity to learn some important lessons from relatively inexpensive mistakes. The planning and organizational considerations required for the establishment and operation of a laboratory or a major item of equipment must be continually reassessed and appropriate changes implemented.

Planning and organization go hand-in-hand, aspects of one will affect the other. There are particular features of each activity that can be identified and discussed. Planning involves looking at what is available and what may become available, and then deciding what could be done and finally what will be done. Planning is decision-making concerned with the future—deciding strategies to be implemented today and tomorrow. Organization is decision-making concerned with today—identifying the resources that are available now and deciding how they can best be organized and used.

11.7.2 Planning

There are many aspects of planning that must be considered by the experimenter and the laboratory supervisor. A task of primary importance is to prepare short-term and long-term plans for the laboratory. A short-term plan would probably be for two years at most, whereas long-term planning may cover say 5 years.

Proposals should be made for all aspects of the laboratory work, even if the recommendation is only to continue as at present. The plans should include the utilization of laboratory equipment, both present and future, and facilities such as computer terminals, analytical instruments, testing equipment, etc., provided by other units or departments. Consideration must also be given to the purchase of new equipment and provision of extra facilities. This should be carried out as part of, and as support for, departmental policy and decisions. New equipment should not be purchased merely because it is new, improved, or readily available, or simply because funds are available. It should fit into and become part of the aims of the laboratory and the plan that has been adopted.

People are also one of the resources of a department or a laboratory facility, and their needs should also be considered. The personnel involved include the laboratory supervisor, the safety officer, technicians, workshop staff, cleaners, maintenance staff, etc. An important resource that is often forgotten is the investigator (i.e. the student). All these resources should be included in the planning operation with respect to their present functions, future requirements, effective utilization, etc.

Long-term planning has a different emphasis—it is concerned mainly with identifying new developments and trends in which it is both useful and necessary to gain experience; for example, the increased use of microprocessors, sophisticated instrumentation, automatic control and alarm systems, etc.

Whatever plans are developed and implemented, the most important consideration will be that they are well prepared and thought-out, and that they do not conflict with other existing policies or aims, e.g. those of departments, professional institutions, etc.

11.7.3 Organization

Organization is in some ways an easier task than planning as it deals with the arrangement and/or use of existing items. The following represent some examples of the organizational aspects of the laboratory.

The layout or siting of particular equipment must be decided after considering access, safety, maintenance, etc. New equipment should not be purchased if it cannot be properly accommodated within a laboratory. A central bench area is often useful/necessary for the location of smaller ancillary measuring instruments, etc., and to incorporate cupboard space for storage. The number of items of equipment often increases without sufficient thought being given as to how they can be accommodated satisfactorily within the space available.

It may be desirable to locate all sensitive measuring instruments, e.g. electric balances, microscopes, etc., in a separate room; however, this may require additional expenditure to cover any incurred costs such as a controlled environment. This is an example of the connection between an organizational decision and the planning aspects. The laboratory should be organized so that there is easy access to all necessary tools, chemicals and minor items, e.g. stopwatch, glassware, etc., however they should also be secure from theft. An area of the laboratory may need to be designated as a discussion/writing/calculating area, so that these tasks can be performed in reasonable comfort. Gangways and areas to be used for other functions must be maintained for easy access.

The laboratory should be organized so that the workers are comfortable and enjoy their work and surroundings. This means adequate ventilation and heating to provide an acceptable level of personal comfort (also consider the effects upon the experiments!), and appropriate noise control which is also a safety requirement. The laboratory should be both interesting and informative; this can be achieved by using notice boards, wall charts, wall-writing boards (water-soluble felt pens are especially easy to use). All old information should be periodically removed otherwise a poor impression will be obtained. The appearance and the 'quality' of the laboratory will be improved if it is kept clean, and all equipment and associated items are left in an acceptable condition at the end of an experimental session.

All these aspects require organization to achieve the required results and impressions. They will enhance the work performed, and the workers' attitude towards laboratory work.

11.7.4 Safety

Safety can be presented in terms of rigid, defined rules, or it can be implemented by the development of common sense and responsibility. The choice depends mainly upon the personal preference of the laboratory manager, provided that legal responsibilities are also observed. Definite rules and regulations can define exactly what is, and is not, acceptable. However, every rule is open to interpretation and misinterpretation, and once rules are introduced they tend to proliferate. The often quoted advantage is that the enforcers have powerful arguments and authority at their disposal, and may be less at risk in terms of legal liability. Training and experimental work are meant to develop the student's awareness and a sense of personal responsibility. These aims are not likely to be achieved if there is never any opportunity to exercise discretion and decision-making.

I would suggest the following as a minimum set of laboratory rules, to be given to the students and to be implemented—rules that are not enforced are better not made into rules!

No eating or drinking inside the laboratory, even if not performing experimental work.
No smoking.
Safety eye glasses to be worn at all times, for all experiments.
All persons must register in and out of the laboratory, even for short periods.
Helmets and other protective clothing to be worn when necessary.
Sensible clothing and footwear to be worn at all times.

The reasons for these rules should be self-evident; the aim should be for acceptance of these rules (by education and example) rather than enforcement and punishment.

Other aspects of laboratory safety can be made optional and at the workers' discretion. For example, wearing laboratory coats, hard hats, or safety shoes can all be covered by the phrase: 'suitable clothing should be worn at all times, and particular safety garments as deemed necessary'. Similarly: 'persons entering the laboratory are expected to act in a reasonable, controlled and safe manner at all times'. Any suggestions or rules should be issued in writing. It is important that laboratory workers develop their own sense of responsibility, and yet still have some guidelines for reference.

Two final points regarding safety. First, there should always be an experienced and competent person present when laboratory work is being performed. This person will be expected to take appropriate action in an emergency and should be familiar with basic first aid, and the procedures for dealing with fires, etc. Second, the actual laboratory safety procedures should be implemented by a departmental safety officer or the laboratory supervisor, and agreed with the permanent full-time safety officer of the institution. Safety should be a process and activity that is continually monitored and improved.

11.8 SUMMARY

Different types of experimental investigations are performed; these include demonstrations, non-quantitative studies, the use of standard laboratory apparatus to obtain data, and open-ended experimental projects.

Laboratory investigations are performed for various reasons; in industry the main reason is the need to obtain data. In teaching institutions the laboratory work helps to develop investigative abilities and practical skills, and provides the opportunity to become aware of the planning and organizational aspects of experimental work.

Time is the major constraint for any experimental work; it is important to determine the time available and the time required to complete a project. One of the first tasks is to make estimates of the time required for each stage of a project, and to review (and revise if necessary) this time schedule at regular interals.

The success of a project will depend upon the information that provides the basis for the study. Before any information is obtained, it is necessary to decide what information is required and where it is located. Different types of information are required for the various stages/activities associated with an investigation. The investigator needs to identify all possible sources of information, especially those that are easily accessible.

Experiments are performed in order to test ideas, theories and designs, to obtain basic knowledge, and to solve problems that cannot be solved (easily) by mathematical analysis. The aim of an experiment is either to test or to investigate; the results obtained should be accurate, reliable and reproducible.

Planning is an essential aspect of any experimental study. It is necessary to define the scope of the investigation, the variables to be measured and/or controlled, the instruments required and the quality and quantity of the data obtained. The main constraints are time and cost.

Checking should be performed at all stages of the project, and not only for mathematical and computational errors. An engineer can make estimates based upon knowledge and experience, but should *never* resort to guessing.

A laboratory course should teach the student not only experimental methods and techniques, but also an awareness of laboratory planning and organization. Short-term and long-term plans should be devised and implemented for the laboratory equipment and the work to be carried out. Organization of a laboratory involves determination of the physical location of equipment, and the conditions and environment within the laboratory.

Safety may be presented as a complete set of rules to be observed, or it may consist of a brief set of rules supplemented by recommendations. The latter approach helps to develop an awareness and appreciation of the importance of adequate safety provisions.

EXERCISES

Perform the following exercises (in retrospect) for a laboratory project that has been completed, or is nearing completion. Also perform these exercises as the next laboratory project begins, and as it continues. Upon completion of this new project, assess the full range of exercises that have been considered and decide whether an awareness of these activities has helped the performance of the project.

11.1 Define the type/nature of the project.

11.2 Identify the aims of the project, i.e. how *you* will benefit by performing this work.

11.3 Prepare a time schedule for *all* aspects of the work to be performed, including the early planning stages. Identify and plan for any obvious time delays.

11.4 Identify the information that is required, both essential and desirable, for completion of the project. Locate the sources of this information and estimate the time required before it becomes available. Assess the importance of the information that is obtained for the completion of the project.

11.5 State the reasons for performing the experimental work associated with the project, and list the aims of the work.

11.6 What approaches and methods are adopted in order to achieve the desired results?

11.7 Describe briefly the short-term and long-term plans relevant to the project.

11.8 Identify the aspects of laboratory organization that help and hinder the progress of the work to be performed.

11.9 How do (or should) safety considerations affect the successful completion of the project?

11.10 Compare the time schedule prepared at the start of the project with the actual time spent on various activities.

BIBLIOGRAPHY—CHAPTERS 11 AND 12

Anderson, V. L. and R. A. McLean, *Design of Experiments: A Realistic Approach*, Marcel Dekker, New York, 1974.

Daniel, C., *Applications of Statistics to Industrial Experimentation*, Wiley, New York, 1976.

Davies, O. L. (ed.), *The Design and Analysis of Industrial Experiments*, 2nd edn, Longman, London, 1978.

Dayton, C. M., *Design of Educational Experiments*, McGraw-Hill, New York, 1970.

Diamond, W. J., *Practical Experiment Designs for Engineers and Scientists*, Van Nostrand Reinhold, New York, 1981.

Finney, D. J., *Introduction to the Theory of Experimental Design*, University of Chicago Press, Chicago, 1975.

Kempthorne. O., *The Design and Analysis of Experiments*, Robert E. Krieger Publishing Co., Melbourne, Fla, 1975.

Kirk, R. E., *Experimental Design*, 2nd edn, Brooks/Cole Publishing Co., Monterey, CA, 1982.

Montgomery, D. C., *Design and Analysis of Experiments*, 2nd edn, Wiley, New York, 1984.

Myers, J. L., *Fundamentals of Experimental Design*, 3rd edn, Allyn and Bacon, Newton, MA, 1979.

Ogawa, J., *Statistical Theory of the Analysis of Experimental Design*, Marcel Dekker, New York, 1974.

Peterson, R. G., *Design and Analysis of Experiments*, Marcel Dekker, New York, 1985.

Schiller, R. W. (ed.), *Instrumentation Curriculum Guide for the Two Year Post Secondary Institution*, Instrument Society of America, Research Triangle Park, NC, 1977.

Subramanyam, K., *Scientific and Technical Information Resources*, Marcel Dekker, New York, 1981.
Turley, R. V., *Understanding the Structure of Scientific and Technical Literature*, Library Association Publishers, London, 1983.
Tuve, G. L. and L. C. Dumholdt, *Engineering Experimentation (A Laboratory Manual)*, McGraw-Hill, New York, 1966 (revised edition).
Winter, B. J., *Statistical Principles in Experimental Design*, 2nd edn, McGraw-Hill, New York, 1979.
Yates, F., *Experimental Design: Selected Papers*, Hafner Press, New York, 1970.

TWELVE

THE LABORATORY COURSE

CHAPTER OBJECTIVES

To encourage the reader to consider:
1. the objectives associated with a laboratory session or course;
2. the training that should be obtained by performing laboratory studies, and the attributes and skills that can be developed;
3. the factors acting as constraints on the development of a laboratory course;
4. possible alternative approaches and changes in laboratory classes.

QUESTIONS

- What are the objectives of a laboratory session, or an entire course?
- What training is provided by a laboratory course? What skills are developed?
- What factors act as constraints on the development of a laboratory course?
- What changes and new approaches could be implemented in your laboratory course?

12.1 INTRODUCTION

The subject of this book is Engineering Experimentation and although this material can be taught purely as a set of lectures, such a course only becomes meaningful when the theory is used in laboratory classes. Separate laboratory sessions do not have to be devised in order to emphasize and reinforce the material presented in this book; the laboratory classes associated with any of the subjects taught in an engineering or science course should suffice as suitable examples.

The aim should be to apply the principles put forward in this book to the interpretation of experimental results. It will not be possible to use and test all the ideas and suggestions in every experiment, however it should be possible to apply the principles gradually for a variety of different experiments.

This chapter is intended to draw attention to aspects of the laboratory course related to the teaching of engineering experimentation. Some traditional and established practices are challenged and some new ideas are proposed for consideration. Some of these ideas may be considered as 'heresy' by some lecturers, however my intention is not to inflame a revolution or to be pilloried for my ideas but rather to plant the seeds of a few possibilities. Students are continually being asked to accept new ideas, develop new approaches to problem-solving, and generally 'be creative and innovative'. I am only suggesting that perhaps both students and lecturers should adopt these traits, and laboratory sessions provide an ideal environment in which to implement some different ways and means.

12.2 OBJECTIVES

What should be the objectives of a particular laboratory session, or a complete laboratory course?
It should be:

(a) useful;
(b) interesting;
(c) demanding;
(d) satisfying;
(e) varied;
(f) capable of extension.

This list is not exhaustive or written in order of importance. All of these points (a) to (f) should be self-explanatory. They should provide both the criteria for the selection of experiments to be performed, and a basis for evaluation of 'success' at the end of a laboratory course.

One point worth mentioning is that if a laboratory exercise is not interesting then the students will get little benefit from it. To have a chance of being interesting, an experiment must be at an appropriate level for the ability and knowledge of the student; it must be neither too simple nor too difficult. There are two factors that often seriously reduce the interest and enthusiasm of the student. First, the use of experiments that are time consuming owing entirely to the need to record large quantities of data manually. Second, the requirement to submit lengthy formal laboratory reports for a large number of experiments (often simply because of convention). Both of these factors should be evaluated in a laboratory course and suitable experiments and appropriate reporting requirements should be adopted.

An experiment may not appear to be particularly interesting, but what is required is self-motivation by the student. This may be stimulated by the challenge of the work and the way in which the 'problem' is presented, and/or by encouragement from the lecturer. Stimulating or provoking questions may provide the spark for enthusiasm, perhaps even mention of the need to demonstrate the ability for self-motivation, i.e. the self-starter so frequently mentioned in job descriptions for young graduates. In certain situations it must sometimes be accepted that not all students will demonstrate the desired amount of interest, but hopefully this is a rare rather than a regular occurrence.

12.3 TRAINING

What training should a laboratory course provide and what attributes/skills should the student develop?
It is worth remembering that most science and engineering students are employed in industry

after completion of their course of study. Very few students proceed to higher study or obtain employment as academics. Therefore, the laboratory course should aim to provide practical, useful and relevant training for the student. The course should provide suitable training in the application of experimental methods for the analysis of results and appropriate practical experience in experimental procedures; it should also provide the opportunity to emphasize and explain the lecture material and to test the validity of certain ideas and theories. However, the course should go beyond these essential aims and should include experience and awareness of new developments and techniques that are becoming important in industry.

At the end of a laboratory course, the students should be able to work (more) independently and have confidence in their ability. If the laboratory programme is carefully planned and if appropriate assistance and advice are available, then the advantage of performing laboratory studies is that problem-solving and investigative skills are developed. The initial reaction to a laboratory exercise should be 'how much time is available?' and 'when can I begin?', rather than 'what should I do?' or 'where are the information and instructions?'. It is hoped the necessary confidence and enthusiasm can be developed.

12.4 CONSTRAINTS

What factors act as constraints on the development of a laboratory course?
The following factors may act as constraints on the development of a laboratory course:

(a) Space available and required in the laboratory; this may be alleviated by appropriate planning and the reorganization of existing facilities.
(b) Finance usually restricts what can be bought or experimental programmes that can be undertaken. A financial plan and a list of priorities are required.
(c) Personnel includes supervisors, students, technicians, support staff, etc. In some situations the investigator will need to learn new skills in order to complete a project.
(d) Student enrolments can act as constraints. Small groups of students may mean that a course is uneconomic, large numbers of students may present problems because of the restricted space available and safety considerations. This problem can usually be overcome by suitable timetabling.
(e) Courses are approved and reviewed by a course regulating committee, e.g. the Council for National Academic Awards (CNAA). A course may also be accredited by an appropriate professional institution. Both of these bodies consider the content of a course and other relevant details such as facilities, staffing, etc. The laboratory work associated with a course must conform to the requirements of both parties involved. Although professional recognition is desirable, it is not usually a condition for approval to run a course.
(f) Staff attitudes and ideas may also act as constraints on new developments.

12.5 NEW IDEAS

Where will the impetus for change originate?
New ideas and implementation of a policy for change should come from the staff responsible for organizing and supervising the laboratory course. However, the students are an important (and often forgotten) source of new ideas as well as providing critical appraisal of the existing programme. A laboratory course is organized for the students and performed by them, so it seems

sensible to elicit their opinions. It may take some time to find out exactly what were the problems (and strengths) of a course. Also what may be seen as simple problems can sometimes be very difficult to change.

Many problems or grievances in relation to laboratory work are due to poor communications between the supervisor and the student. The supervisor should recognize that the main expectation of the student is for security, i.e. a desire to be told what to do. This is due mainly to the way in which students are taught throughout school, however any system that changes to student-directed teaching too quickly will probably be criticized.

12.6 SOME PROPOSALS

The following are just a few of the proposals that could be made for changing the requirements of a laboratory course. I am not saying these ideas should be implemented, or that they will necessarily be successful or popular, however they may be useful for getting more out of a laboratory course. These ideas are directed towards the laboratory supervisor, however it will be useful for students to consider their advantages and disadvantages and to voice their opinions.

(a) Abandon 'standard' laboratory experiments and include practical demonstrations as part of the lecture courses. Set laboratory project-type investigations with the students having more control and responsibility for the direction of the work.
(b) Require only one or two formal reports to be submitted; otherwise stipulate just 'Results and Discussion' sections for the majority of experimental reports, with a maximum limit of (say) eight pages.
(c) Do not award marks for reports but give a standard mark for attendance during the full laboratory session.
(d) Provide only a one sentence description of the experiment to be performed. Allow the student to gain experience in determining the information required, the direction of the work, the report to be submitted, etc.

Society advances mainly by technological innovation; this occurs by questioning the validity of established ideas and adapting and changing situations as the need arises. New ideas and alternatives need to be discussed and tested, otherwise we can hardly expect future students to be adaptable and innovative.

12.7 FINAL OBSERVATIONS

With planning and foresight, the laboratory provides an excellent working situation in which the student can gain experience in conducting practical investigations and develop the skills associated with problem-solving. Students should develop an independent working attitude and confidence in their abilities. Laboratory work also provides an opportunity to develop 'management potential' by learning to work with people having different skills and ideas.

However, a laboratory course differs in one fundamental way from a lecture course. If students become disinterested in a lecture course, they can stop attending but still continue to study the subject matter. Experimental work must be conducted in the laboratory and encouragement and positive advice should be available when students become discouraged. Students need perseverance to make experiments work correctly; this is also part of their training.

The close contact that the supervisor has with students in the laboratory gives him or her the dual role of teacher and tutor.

This chapter may seem to be directed more to the lecturer than the student. However, successful laboratory work requires the participation and cooperation of both parties, and student feedback can make a valuable contribution. The student should consider (and question) the way in which a laboratory course is organized, and make use of this knowledge when conducting project-type experimental studies.

EXERCISES

12.1 List the objectives of a particular laboratory session, or an entire course.

12.2 Identify the training provided by a laboratory course and the attributes/skills that are (or should be) developed.

12.3 Discuss the factors that act as constraints on the development of a laboratory course.

12.4 Assess your laboratory course in terms of the implementation of new ideas, the introduction of new developments and new technology, and the ability and encouragement to provide feedback and to influence the organization of the course.

Note: Refer to the Bibliography at the end of Chapter 11.

FINAL COMMENTS

This book is about engineering experimentation. Within the scope of the book I have included:

(a) an introduction to measurements and instrument systems;
(b) details of some common measuring instruments and methods for pressure, flow and temperature measurement;
(c) basic methods and techniques for the interpretation of data;
(d) guidelines for the presentation of data;
(e) ideas and strategies for laboratory project work.

I have not included details of a wide variety of measurement methods and instruments, or information relating to particular laboratory experiments; these are specialist topics for particular engineering disciplines. I have also omitted any coverage of computer methods for data analysis; this topic should be covered in detail in particular courses and tailored to the needs of individual departments.

I hope that this book will serve as a set text for courses dealing with engineering experimentation, basic instrumentation, data analysis and information presentation. It should also serve as a student handbook throughout an engineering or science course, providing a handy reference source, advice and ideas when performing laboratory-based studies.

Engineering experimentation is similar to design work in that experience and hindsight are great advantages. However, the novice who has not yet acquired these attributes can still succeed by combining careful planning, common sense and a realistic approach, with an acceptance of the need to obtain (reliable) advice and to learn from past mistakes.

Remember that the work can always be improved upon—but hopefully what has been done is at least acceptable (and correct). Finally, learn from everything that you do and see; do not dismiss anything but rather be selective in what you accept and use.